Lecture Notes in Computer Science 9517

Commenced Publication in 1973
Founding and Former Series Editors:
Gerhard Goos, Juris Hartmanis, and Jan van Leeuwen

More information about this series at http://www.springer.com/series/7409

Qi Tian · Nicu Sebe
Guo-Jun Qi · Benoit Huet
Richang Hong · Xueliang Liu (Eds.)

MultiMedia Modeling

22nd International Conference, MMM 2016
Miami, FL, USA, January 4–6, 2016
Proceedings, Part II

 Springer

Editors

Qi Tian
University of Texas at San Antonio
San Antonio, TX
USA

Nicu Sebe
Department of Information Engineering
University of Trento
Povo, Trento
Italy

Guo-Jun Qi
EECS
University of Central Florida
Orlando, FL
USA

Benoit Huet
EURECOM
Sophia-Antipolis
France

Richang Hong
Hefei University of Technology
Hefei, Anhui
China

Xueliang Liu
School of Computing and Information
Hefei University of Technology
Hefei, Anhui
China

ISSN 0302-9743   ISSN 1611-3349   (electronic)
Lecture Notes in Computer Science
ISBN 978-3-319-27673-1   ISBN 978-3-319-27674-8   (eBook)
DOI 10.1007/978-3-319-27674-8

Library of Congress Control Number: 2015957238

LNCS Sublibrary: SL3 – Information Systems and Applications, incl. Internet/Web, and HCI

Printed on acid-free paper

This Springer imprint is published by SpringerNature
The registered company is Springer International Publishing AG Switzerland

Preface

The 22nd International Conference on Multimedia Modeling (MMM 2016) was held in Miami, USA, January 4–6, 2016, and was hosted by the University of Central Florida at Orlando, USA. MMM is a leading international conference for researchers and industry practitioners to share their new ideas, original research results, and practical development experiences from all multimedia-related areas. University of Central Florida is a Space-Grant university and has made noted research contributions to digital media, engineering, and computer science.

MMM 2016 featured a comprehensive program including three keynote talks, eight oral presentation sessions, two poster sessions, one demo session, five special sessions, and the Video Browser Showdown (VBS). The 168 submissions from authors of 20 countries included a large number of high-quality papers in multimedia content analysis, multimedia signal processing and communications, and multimedia applications and services. We thank our 130 Technical Program Committee members who spent many hours reviewing papers and providing valuable feedback to the authors. From the total of 117 submissions to the main conference and based on at least three reviews per submission, the program chairs decided to accept 32 regular papers (27.8 %) and 30 poster papers (25.6 %). In total, 38 papers were received for 5 special sessions, with 20 being selected, and 11 submissions were received for a demo session, with 7 being selected. Video browsing systems of nine teams were selected for participation in the VBS. The authors of accepted papers come from 17 countries. This volume of the conference proceedings contains the abstracts of three invited talks and all the regular, poster, special session, and demo papers, as well as special demo papers of the VBS. MMM 2016 included the following awards: the Best Paper Award, the Best Student Paper Award, and the winner of the VBS competition.

The technical program is an important aspect but only provides its full impact if complemented by challenging keynotes. We were extremely pleased and grateful to have three exceptional keynote speakers, Wen Gao (ACM/IEEE Fellow), Chang Wen Chen (IEEE Fellow), and Changsheng Xu (IEEE Fellow), accept our invitation and present interesting ideas and insights at MMM 2016.

We are heavily indebted to many individuals for their significant contributions. We thank the MMM Steering Committee for their invaluable input and guidance on crucial decisions. We wish to acknowledge and express our deepest appreciation to the organizing chairs, Xueliang Liu and Luming Zhang, the special session chairs, Wen-Huang Chen, Haojie Li and Rongrong Ji, the panel chair, Tat-Seng Chua, the demo chairs, Cathal Gurrin and Björn Þór Jónsson, the VBS chairs, Klaus Schöffmann and Werner Bailer, the publicity chairs, Yu-Gang Jiang, Shuicheng Yan, Hengtao Shen, Zhengjun Zha, and Sheng Wu, the publication chairs, Na Zhao and Zechao Li, and last but not least the Webmaster, Jun He. Without their efforts and enthusiasm, MMM 2016 would not have become a reality. Moreover, we want to thank our sponsor the University of Central Florida. Finally, we wish to thank all committee members, reviewers, session chairs, student volunteers, and supporters. Their contributions are much appreciated.

January 2016

Guo-Jun Qi
Benoit Huet
Richang Hong
Nicu Sebe
Qi Tian

Organization

MMM 2016 was organized by the University of Central Florida, USA.

MMM 2016 Steering Committee

Phoebe Chen	La Trobe University, Australia
Tat-Seng Chua	National University of Singapore
Shiqiang Yang	Tsinghua University, China
Kiyoharu Aizawa	University of Tokyo, Japan
Noel E. O'Connor	Dublin City University, Ireland
Cess G.M. Snoek	University of Amsterdam, The Netherlands
Meng Wang	Hefei University of Technology, China
R. Manmatha	University of Massachusetts, USA
Cathal Gurrin	Dublin City University, Ireland
Klaus Schoeffmann	Klagenfurt University, Austria
Benoit Huet	Eurecom, France

MMM 2016 Organizing Committee

General Co-chairs

Qi Tian	University of Texas at San Antonio, USA
Nicu Sebe	University of Trento, Italy

Program Co-chairs

Guojun Qi	University of Central Florida, USA
Benoit Huet	Eurecom, France
Richang Hong	Hefei University of Technology, China

Organizing Co-chairs

Xueliang Liu	Hefei University of Technology, China
Luming, Zhang	National University of Singapore, Singapore

Special Session Co-chairs

Wen-Huang Cheng	Academia Sinica, Taiwan
Haojie Li	Dalian University of Technology, China
Rongrong Ji	Xiamen University, China

Demo Session Co-chairs

Cathal Gurrin	Dublin City University, Ireland
Björn Þór Jónsson	Reykjavík University, Iceland

Publication Co-chairs

Na Zhao   National University of Singapore, Singapore
Zechao Li   Nanjing University of Science and Technology, China

Publicity Co-chairs

Yu-Gang Jiang   Fudan University, China
Shuicheng Yan   National University of Singapore, Singapore
Hengtao Shen   University of Queensland, Australia
Zhengjun Zha   Chinese Academy of Sciences, China
Sheng Wu   Google, USA

Panel Chair

Tat-Seng Chua   National University of Singapore, Singapore

Video Search Showcase Co-chairs

Werner Bailer   Joanneum Research, Graz, Austria
Klaus Schoeffmann   Klagenfurt University, Austria

Web Master

Jun He   Hefei University of Technology, China

Technical Program Committee

Selim Balcisoy   Sabanci University, Turkey
Yingbo Li   Ecole Normale Superieure, France
Lifeng Sun   Tsinghua University, China
Cathal Gurrin   Dublin City University, Ireland
Haojie Li   Dalian University of Technology, China
Rainer Lienhart   University of Augsburg, Germany
Rossana Damiano   University of Turin, Italy
Zheng-Jun Zha   Institute of Intelligent Machines, CAS, China
Vincent Charvillat   University of Toulouse, France
Liqiang Nie   National University of Singapore, Singapore
Wolfgang Hurst   Utrecht University, The Netherlands
Wei-Guang Teng   National Cheng Kung University, Taiwan
Bo Yan   Fudan University, China
Werner Bailer   Joanneum Research, Austria
Mei-Ling Shyu   University of Miami, USA
Luiz Fernando Gomes   Catholic University of Rio de Janeiro, Brazil
  Soares
Joemon Jose   University of Glasgow, UK
Mylene Farias   University of Brasilia, Brazil
Wolfgang Huerst   Utrecht University, The Netherlands

Sponsors

University of Central Florida

Google

FX Palo Alto Laboratory

Springer Publishing

Springer

Contents – Part II

Video Browser Showdown

Contents – Part I

Poster Papers

Special Session Poster Papers

Special Session Poster Papers
(continued)

Transfer Nonnegative Matrix Factorization for Image Representation

Tianchun Wang[1](✉), TengQi Ye[2], and Cathal Gurrin[2]

[1] School of Software, TNList, Tsinghua University, Beijing, China
wtc13@mails.tsinghua.edu.cn
[2] Insight Centre for Data Analytics, Dublin City University, Dublin, Ireland
yetengqi@gmail.com, cgurrin@computing.dcu.ie

Abstract. Nonnegative Matrix Factorization (NMF) has received considerable attention due to its psychological and physiological interpretation of naturally occurring data whose representation may be parts-based in the human brain. However, when labeled and unlabeled images are sampled from different distributions, they may be quantized into different basis vector space and represented in different coding vector space, which may lead to low representation fidelity. In this paper, we investigate how to extend NMF to cross-domain scenario. We accomplish this goal through TNMF - a novel semi-supervised transfer learning approach. Specifically, we aim to minimize the distribution divergence between labeled and unlabeled images, and incorporate this criterion into the objective function of NMF to construct new robust representations. Experiments show that TNMF outperforms state-of-the-art methods on real datasets.

Keywords: Nonnegative matrix factorization · Transfer learning · Image representation

1 Introduction

The development of online images and videos has created a compelling demand for advanced technologies for organizing and analyzing the multimedia content. As a powerful technique for finding succinct representations of stimuli and capturing high-level semantics of visual data, *Nonnegative Matrix Factorization* (NMF) can represent images using the combination of nonnegative low-dimensional "basis" vectors. This makes NMF an ideal dimensionality reduction algorithm for image processing [1], face recognition [2,3], and image clustering [4], where it is natural to consider the object as a combination of the parts.

One major problem of NMF is how to improve the quality of the part-based representation while maximally preserving the signal fidelity. To achieve this goal, many works have been proposed to leverage more knowledge. Kuang et al. [4] provide a symmetric NMF method for clustering based on graph. Ding et al. [15] proposed a semi-nonnegative NMF while relaxing the constraint on

© Springer International Publishing Switzerland 2016
Q. Tian et al. (Eds.): MMM 2016, Part II, LNCS 9517, pp. 3–14, 2016.
DOI: 10.1007/978-3-319-27674-8_1

the basis vectors. Ding et al. [16] also show that when the Frobenius norm is used as divergence, NMF is equivalent to a relaxed form of K-means clustering, which justifies the use of NMF for data clustering. However, when unlabeled and labeled data are sampled from different distributions, they may be quantized into different coding vectors corresponding to different low-dimensional feature representations. In this case, the basis vectors learned from the feature space of labeled images cannot effectively indicate the features of unlabeled images with high fidelity, and also the unlabeled images may reside far away from the labeled images under the new representation. This distribution difference will greatly challenge the robustness of existing NMF algorithms for cross-distribution image classification problems [23].

Transfer learning [5] has recently become a hot research topic that the labeled training data and unlabeled test data are sampled from different probability distributions. This is a very common scenario in web applications, since training and test data are usually collected in different time periods, or under different conditions. In this case, standard classifiers trained on the labeled data may fail to make reasonable predictions on the unlabeled data. To boost the generalization performance of supervised classifiers across different distributions, Pan et al. [6,7] proposed to extract a low-dimensional feature representation through which the probability distributions of labeled and unlabeled data are drawn close and achieves much better classification performance by explicitly reducing distribution divergence.

Inspired by recent efforts in both matrix factorization and transfer learning, this paper presents a novel Transfer Nonnegative Matrix Factorization (TNMF) algorithm to construct robust representations for classifying cross-domain images accurately. We aim to minimize the distribution divergence between labeled and unlabeled images with a nonparametric distance measure. Specifically, we incorporate this criterion into the objective function of TNMF to make the new representations of the labeled and unlabeled images close to each other. In this way, the induced representations are made robust for cross-distribution image classification problems. Moreover, to enrich the new representations with the intrinsic nonlinear structure of data space, we also incorporate the Hessian regularization term of coefficients [9,11,12] in our objective function. Different from Laplacian term [8,14], Hessian has a richer null space and drives the solution varying smoothly along the manifold. Experimental results verify the effectiveness of the TNMF approach.

2   Related Work

In this section, we review related efforts including NMF and transfer learning.

Recently, NMF has been a hot research topic of computer vision. Ding et al. [15] proposed a semi-nonnegative algorithm where only one matrix factor is restricted to contain nonnegative entries, while it relaxes the constraint on the basis vectors. Ding et al. [16] also show that when the Frobenius norm is used as a divergence, NMF is equivalent to a relaxed form of K-means clustering, which justifies the use of NMF for data clustering. Heiler et al. [17] derive optimization schemes for NMF based on sequential quadratic and second order cone programming. Our work aims to discover a shared basic matrix which can encode both labeled and unlabeled data sampled from different probability distributions. To improve the quality of low-rank representation, researchers have modified the nonnegative constraint [18], adding graph regularization [8], etc. Our approach aims to construct robust low-dimensional representations for image classification problems across domains, which is different from the previous works.

Aiming at transfer knowledge between the labeled and unlabeled data sampled from different distributions, transfer learning [5] has received extensive research focus. To achieve this goal, Pan et al. [7] proposed a Transfer Component Analysis (TCA) method to minimize the reconstruction error of the input data by reducing the discrepancy between the labeled and unlabeled data. Quanz et al. [19] have explored knowledge transfer of sparse feature, which is a more restricted procedure and prone to overfitting. In addition, Wang et al. [20] extends NMF to cross-domain senario. They assume that a small set of the target domain data are labeled and an SVM classifier is learned across domains. Different from them, our approach is based on the setting that there is no labeled data in the target domain, which is more restricted. Moreover, our work additionally incorporates the Hessian term of coefficients [9] in the objective function, which can discover more discriminating representations for classification tasks.

3 Preliminaries

In this section, we briefly introduce a basic knowledge of NMF and Hessian Regularization.

3.1 Nonnegative Matrix Factorization

Non-negative Matrix Factorization (NMF) [2] is a matrix factorization algorithm that focuses on the analysis of data matrices whose elements are nonnegative. Given data matrix $X = [x_1, \ldots, x_N] \in R^{M \times N}$, each column of X is a sample vector. The goal of NMF is to find a basic matrix $U = [u_{ik}] \in R^{M \times N}$ and a coding matrix $V = [v_{jk}] \in R^{N \times K}$ and the production of them can well approximate the original matrix X:

$$X \approx UV^T.$$

The commonly used cost function is the square of the Euclidean distance between two matrices give by:

$$\min_{U,V} \| X - UV^T \|^2 = \sum_{i,j} \left(x_{ij} - \sum_{k=1}^{K} u_{ik}v_{ik} \right)^2. \tag{1}$$

The objective function in Eq. (1) is convex in either U or V. Therefore, it can be solved by alternatingly optimizing one variable while fixing the other one.

3.2   Hessian Regularization

To make the basis vectors respect the intrinsic geometric structure underlying the input data, Cai et al. [8] proposed a Laplacian regularized NMF (we call it LapNMF) method which is based on the assumption that if two data points x_i and x_j are close in the intrinsic geometry of data distribution, then their coding vectors v_i and v_j are also close.

Table 1. Notations and descriptions used in this paper

Notation	Description
X	Input data matrix
U	Basic matrix
V	Coding matrix
M	MMD matrix
H	Hessian regularization matrix
D_l, D_u	Labeled/unlabeled data
n_l, n_u	Labeled/unlabeled example index
m	Dimensionality of shared feature space
λ	Trade-off parameter of Hessian regularization
μ	Trade-off parameter of MMD regularization

However, Laplacian regularization is short of extrapolating power. The null space of the graph Laplacian is a constant function along the compact support of the marginal distribution and thus the solution of the Laplacian regularization is biased toward a constant function. In contrast to Laplacian, Hessian can properly exploit the intrinsic local geometry of the data manifold and has a richer nullspace to make the learned function vary linearly along the underlying manifold [9, 10]. The discretization of the Hessian regularization for encoding the local geometry of unlabeled samples is achieved as follows [9, 11, 12]:

(1) Finding the k-nearest neighbours N_p of the j-th unlabeled sample x_j and centralizing the neighbourhood by taking x_j off from the k-nearest neighbours. This centralization makes x_j to be the origin of the tangent space $T_{x_j}(M)$.

(2) Estimating the orthonormal coordinate system of the tangent space $T_{x_j}(M)$ by the eigenspace U of the neighborhood N_p of p associated with the largest d eigenvalues. This can be implemented by conducting singular value decomposition on $X^j = [x_i - x_j]_{i=1}^{k}$, where in x_i is the i-th sample in the k-nearest neighbours N_p.

(3) Using Gram-Schmidt orthonormalization, we take the $(d+1)$-dimensional nullspace off from the matrix $H^j = [1, u_1, \ldots, u_1 u_1, \ldots, u_d u_d]$ and get $\hat{H}^j$. Therefore, the Frobenius norm of $\hat{H}^j$ is given by $(\hat{H}^j)^T \hat{H}^j$.

(4) Accumulating $(\hat{H}^j)^T \hat{H}^j$ over all images and then resulting the Hessian regularization matrix H.

Inspired by the recent progress of Hessian regularization method [11–13], in this paper, we integrate the minimizing criterion $tr(VHV^T)$ for preserving the geometric structure into Eq. (1) and get the Hessian Regularized NMF (HeNMF) given by:

$$\min_{U,V} ||X - UV^T||^2 + \lambda tr(VHV^T) \tag{2}$$

where λ is the regularization parameter to trade off the weight between NMF and geometric preservation.

4 Transfer Nonnegative Matrix Factorization

In this section, we will present the Transfer Nonnegative Matrix Factorization (TNMF) algorithm for image representation, which extends HeNMF by taking into account the minimization of distribution divergence between labeled and unlabeled data.

4.1 Problem Definition

Given labeled data $\mathcal{D}_l = \{(\mathbf{x}_1, y_1), \ldots, (\mathbf{x}_{n_l}, y_{n_l})\}$ with n_l examples, unlabeled data $\mathcal{D}_u = \{\mathbf{x}_{n_l+1}, \ldots, \mathbf{x}_{n_l+n_u}\}$ with n_u examples, denote $X = [\mathbf{x}_1, \ldots, \mathbf{x}_n] \in \mathbb{R}^{m \times n}$, $n = n_l + n_u$ as the input data matrix. Assume that the labeled and unlabeled data are sampled from different probability distributions in an m-dimensional feature space. Therefore, we can define the problem our problem as follows:

Problem 1 (Transfer Nonnegative Matrix Factorization): Given labeled data $\mathcal{D}_l$ and unlabeled data $\mathcal{D}_u$ under different distributions, we aim to learn a coding matrix V, and basic matrix U, to construct a robust representation for original images sampled from $\mathcal{D}_l$ and $\mathcal{D}_u$.

Notations and descriptions used frequently in this paper are summarized in Table 1.

4.2 Proposed Approach

In order to make NMF capture the representation across different distributions, we expect that the basis vectors can represent the shared knowledge underlying both labeled and unlabeled domains. However, the difference of extracted coding vector spaces between labeled and unlabeled data will still be significantly large. Thus a major computational problem is to reduce the distribution difference by explicitly minimizing some predefined distance measures. To achieve this goal, a common strategy is to make the probability distributions of labeled and unlabeled data close to each other in the low-rank representation. Therefore, by representing all data points X with the learned coding matrix V, the probability distributions of the coding vectors for the labeled and unlabeled data should be close enough. In this paper, we adopt the joint distribution adaptation *Maximum Mean Discrepancy* [14,23] as the nonparametric distance measure to compare different distributions, which minimize both the distance of marginal and conditional distribution divergence between labeled and unlabeled data in the coding vector spaces:

$$
\left\| \frac{1}{n_l^{(c)}} \sum_{x_i \in \mathcal{D}_l^{(c)}} v_i - \frac{1}{n_u^{(c)}} \sum_{x_j \in \mathcal{D}_u^{(c)}} v_j \right\|^2 = \sum_{i,j=1}^{n} v_i^T v_j M_{ij}^{(c)} = tr\left(V M^{(c)} V^T \right)
$$

where $M^{(c)}$, $c \in \{1, 2, \ldots, C\}$ are MMD matrices given by:

$$
M_{ij}^{(c)} = \begin{cases} \frac{1}{n_l^{(c)} n_l^{(c)}}, & x_i, x_j \in D_l^{(c)} \\ \frac{1}{n_u^{(c)} n_u^{(c)}}, & x_i, x_j \in D_u^{(c)} \\ -\frac{1}{n_l^{(c)} n_u^{(c)}}, & \begin{cases} x_i \in D_l^{(c)}, x_j \in D_u^{(c)} \\ x_j \in D_l^{(c)}, x_i \in D_u^{(c)} \end{cases} \\ 0, & otherwise \end{cases} \tag{3}
$$

Therefore, the joint distribution adaptation MMD matrix can be computed as $M = \sum_{c=0}^{C} M^{(c)}$, where $n_l^{(0)} = n_l$, $n_u^{(0)} = n_u$, $D_u^{(0)} = D_u$, $D_l^{(0)} = D_l$. Integrating the MMD regularization into Eq. (2), we obtain the objective function of TNMF:

$$
\mathcal{O} = \min_{U,V} \left\| X - UV^T \right\|^2 + tr\left(V(\lambda H + \mu M)V^T \right) \tag{4}
$$

where μ is the trade-off adaptation regularization parameter weighting between HeNMF and distribution matching. Setting μ to 0, the TNMF degenerates to HeNMF. Hence, the geometric preservation and distribution matching is unified into the NMF objective.

Moreover, the adaptation regularization in Eq. (4) is significant to make TNMF robust across different probability distributions. According to [25], MMD will asymptotically approach zero if and only if the two distributions are the same. By minimizing this adaptation regularization, TNMF can match the distributions between labeled and unlabeled data based on the coding vector space.

4.3   Optimization

We use gradient descent as optimization algorithm for minimizing the objective function in Eq. (4). For this problem, gradient descent leads to the following update rules:

$$
u_{ik} \leftarrow u_{ik} + \eta_{ik} \frac{\partial \mathcal{O}_1}{\partial u_{ik}} \tag{5}
$$

$$
v_{jk} \leftarrow v_{jk} + \delta_{jk} \frac{\partial \mathcal{O}_1}{\partial v_{jk}} \tag{6}
$$

where η_{ik} and δ_{jk} referred as step size parameters. As long as η_{ik} and δ_{jk} are sufficiently small, the objective would reduce gradually to the desired minimum. We can use some tricks to set size parameters automatically. Following [8], we let $\eta_{ik} = -u_{ik}/2\left(UV^TV\right)_{ik}$, hence, we can write Eq. (5) as

$$
\begin{aligned}
&u_{ik} + \eta_{ik}\frac{\partial \mathcal{O}_1}{\partial u_{ik}}\\
&= u_{ik} - \frac{u_{ik}}{2\left(UV^TV\right)_{ik}}\frac{\partial \mathcal{O}_1}{\partial u_{ik}}\\
&= u_{ik} - \frac{u_{ik}}{2\left(UV^TV\right)_{ik}}\left(-2\left(XV\right)_{ik} + 2\left(UV^TV\right)_{ik}\right)\\
&= u_{ik}\frac{\left(XV\right)_{ik}}{\left(UV^TV\right)_{ik}}
\end{aligned}
\tag{7}
$$

Similarly, the update formulation of Eq. (6) can also be get easily. It is clear that the updating rules in this problem are special cases of gradient descent with an automatic step parameter selection. The multiplicative updating rules guarantees that U and V are non-negative and Eqs. (5) and (6) would ultimately converge to a local minimum.

5   Experiments

In this section, we perform image classification experiments on benchmark datasets widely used for visual domain adaptation to evaluate TNMF.

5.1   Dataset Description

USPS[1] dataset consists 9,298 images of size 16×16. **MNIST**[2] dataset has 70,000 examples of size 28×28. From Fig. 1, we see that USPS and MNIST follow very different distributions. They share 10 semantic classes, each corresponding to one digit. Following [14], we randomly sample 1,800 images in USPS to form the training data and 2,000 images in MNIST as test data to construct a dataset *USPS vs MNIST*. We rescale all images to size 16×16 and represent each image by a 256-dimensional feature vector.

   MSRC[3] dataset contains of 4,323 images and 18 classes, provided by Microsoft Research Cambridge. **VOC2007**[4] dataset contains 5,011 images annotated with 20 concepts, collecting from digital photos on Flickr[5]. The MSRC and VOC2007 datasets used in our experiments share the following 6 semantic classes: aeroplane, bicycle, bird, car, cow, sheep. Following [21], all 1,269

---

[1] http://www-i6.informatik.rwth-aachen.de/~keysers/usps.html.
[2] http://yann.lecun.com/exdb/mnist.
[3] http://research.microsoft.com/en-us/projects/objectclassrecognition.
[4] http://pascallin.ecs.soton.ac.uk/challenges/VOC/voc2007.
[5] https://www.flickr.com.

Fig. 1. First line shows examples of USPS, MNIST, MSRC, VOC2007 Second line shows examples of Caltech256, Amazon, Dslr and Webcam

images in MSRC and all 1,530 images in VOC2007 are selected to construct the dataset *MSRC vs VOC*. We uniformly rescale all images to be 256 pixels in size and extract 128-dimensional dense SIFT features with VLFeat [22]. A 240-dimensional codebook is created, where K-means clustering is used to obtain the codewords.

Caltech256[6] is a standard database for object recognition consists of 30,607 images and 256 categories. **Office** [24] dataset consists of three domains: Amazon (images from online merchants), Webcam (low-resolution images by a web camera), and DSLR (high-resolution images by a digital SLR camera). It has 4,652 images with 31 classes. In these experiments, we adopt the 800-dimensional SURF features released by Gong et al. [24]. By randomly selecting two different domains as source domain and target domain, we construct 12 cross-domain datasets, *e.g., Caltech vs Amazon, Caltech vs Webcam, . . . , Dslr vs Webcam*.

5.2   Performance on Cross-Domain Datasets

We compare our TNMF approach with three NMF related competing rivals and basic NN classifier for image recognition problems:

(1) 1-Nearest Neighbor Classifier (NN)
(2) Nonnegative Matrix Factorization (NMF) + NN
(3) Laplacian regularized NMF (LapNMF) [8] + NN
(4) Hessian regularized NMF (HeNMF) + NN

In our experiments, we choose NN as the base classifier since it does not require tuning parameters for cross-validation. NN classifier is trained on the labeled source data for classifying the unlabeled target data. Since labeled and unlabeled data are sampled from different distributions, it is impossible to tune

---

[6] https://www.eecs.berkeley.edu/~jhoffman/domainadapt.

Table 2. Accuracy (%) on cross-domain datasets

	NN	NMF	LapNMF	HeNMF	TNMF
USPS vs MNIST	36.50	29.45	31.01	37.45	**39.20**
MSRC vs VOC	25.95	26.86	26.67	27.06	**28.82**
Caltech vs Amazon	21.82	29.85	29.96	30.27	**31.00**
Caltech vs Webcam	13.90	16.27	17.63	18.98	**22.03**
Caltech vs Dslr	8.92	19.75	20.38	21.66	**23.57**
Amazon vs Caltech	24.31	26.98	27.43	28.41	**29.03**
Amazon vs Webcam	**24.07**	21.69	17.29	21.02	21.69
Amazon vs Dslr	18.47	19.11	21.66	23.57	**28.03**
Webcam vs Caltech	**24.31**	20.48	20.93	22.80	22.44
Webcam vs Amazon	27.35	27.66	27.56	28.39	**29.12**
Webcam vs Dslr	56.05	54.14	52.87	52.87	**57.32**
Dslr vs Caltech	22.26	25.73	25.20	25.91	**26.36**
Dslr vs Amazon	21.71	24.74	24.63	25.89	**26.30**
Dslr vs Webcam	44.07	51.19	54.24	49.15	**55.59**

the model parameters using cross-validation. Therefore, we evaluate all methods by grid search for the optimal parameter settings which gives the highest classification accuracy on all datasets, and report the best results of each method. We set the trade-off parameter of Hessian regularization λ by searching $\lambda \in \{10^{-2}, 10^{-1}, 1, 10, 10^2\}$ and adaptation regularization parameter μ by searching $\mu \in \{10^{-4}, 10^{-3}, 10^{-2}, 10^{-1}, 1, 10, 10^2, 10^3, 10^4\}$. Our proposed TNMF is run 5 times independently and we report their average. The classification accuracy on test data is used as the evaluation metric, which reads

$$Accuracy = \frac{|x : x \in \mathcal{D}_t \wedge f(x) = y(x)|}{|x : x \in \mathcal{D}_t|} \qquad (8)$$

where $\mathcal{D}_t$ is the set of test data, $y(x)$ is the ground truth of x, $f(x)$ is the label predicted by the classification algorithm.

The classification performance of TNMF as well as baselines on all datasets are listed in Table 2. When comparing with the baseline methods, we set the Hessian regularization trade-off parameter λ as 0.1, and the MMD regularization trade-off parameter μ as 1. The number of iterations is set to 300. From the results, we can find that on all datasets TNMF gaining an improvement of 6.55 % compared to HeNMF and significantly improves 19.36 % than NN, which illustrates the significant role of adaptation regularization. Moreover, the classification accuracy of HeNMF improves 4.02 % than LapNMF which reflects that Hessian can exploit more intrinsic geometry than Laplacian.

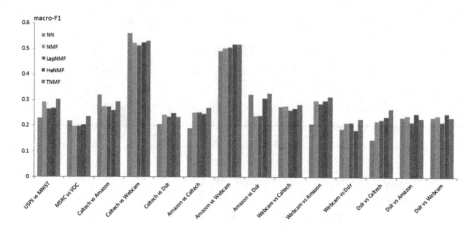

Fig. 2. Macro-F1 measure on cross-domain datasets

Moreover, we choose F1 measure to evaluate the performance of our proposed method as well as baselines. Due to the fact that the datasets have more than 2 classes, macro-averaged evaluation is considered, which is given as follows:

$$Macro - F1 = \frac{2 * Macro - Precision * Macro - Recall}{Macro - Precision + Macro - Recall} \tag{9}$$

where Macro-Precision and Macro-Recall are obtained respectively as:

$$Macro - Recall = \frac{\sum_{i=1}^{c} R_i}{c} \tag{10}$$

and

$$Macro - Precision = \frac{\sum_{i=1}^{c} P_i}{c} \tag{11}$$

where c is the number of classes, R_i and P_i are recall and precision for the i-th class. The results of Macro-F1 score of each methods on all the cross-domain datasets are shown in Fig. 2.

From Fig. 2, we can observe that our method perform better than all the baseline methods as a whole, while a little weak on several datasets. On average, the Macro-F1 measure of TNMF is 1.49 % higher than HeNMF and 3.17 % than the basic NN classifier.

6   Conclusion

In this paper, we extend NMF to cross-domain scenario and propose a novel Transfer Nonnegative Matrix Factorization (TNMF) approach for low-dimensional image representation. An important advantage of TNMF is that it can explore as many necessary learning objectives as possible in cross-domain image representation task, yet still remain simple to implement practically.

Experimental results on several benchmark datasets show that TNMF outperforms state-of-the-art NMF methods.

Acknowledgements. The research was supported by the Irish Research Council (IRCSET) under Grant Number GOIPG/2013/330.

References

1. Luong, T.X., kyeong Kim, B., Lee, S.-Y.: Color image processing based on nonnegative matrix factorization with convolutional neural network. In: 2014 International Joint Conference on Neural Networks (IJCNN), pp. 2130–2135 (2014)
2. Lee, D.D., Seung, H.S.: Learning the parts of objects by non-negative matrix factorization. Nature **401**, 788 (1999)
3. S. Z., Li, X., Hou, Zhang, H., Cheng, Q.: Learning spatially localized, parts-based representation. CVPR. IEEE Comput. Soc. **1**, 207–212 (2001)
4. Kuang, D., Park, H., Ding, C.H.Q.: Symmetric nonnegative matrix factorization for graph clustering. SDM **12**, 106–117 (2012)
5. Pan, S.J., Yang, Q.: A survey on transfer learning. IEEE Trans. Knowl. Data Eng. **22**(10), 1345–1359 (2010)
6. Fox, D., Gomes, C.P. (eds.): In: Proceedings of the Twenty-Third AAAI Conference on Artificial Intelligence, AAAI 2008. AAAI Press, Chicago, 13–17 July 2008
7. Pan, S.J., Tsang, I.W., Kwok, J.T., Yang, Q.: Domain adaptation via transfer component analysis. IEEE Trans. Neural Netw. **22**(2), 199–210 (2011)
8. Cai, D., He, X., Han, J., Huang, T.S.: Graph regularized non-negative matrix factorization for data representation. IEEE Trans. Pattern Anal. Mach. Intell. **33**(8), 1548–1560 (2011)
9. David, L.D., Carrie, G.: Hessian eigenmaps: new locally linear embedding techniques for high-dimensional data (2003)
10. Bengio, Y., Schuurmans, D., Lafferty, J.D., Williams, C.K.I., Culotta, A. (eds.): Advances in Neural Information Processing Systems 22. In: 23rd Annual Conference on Neural Information Processing Systems 2009. Proceedings of a meeting held 7–10 Dec 2009. Vancouver, British. Curran Associates Inc (2009)
11. Liu, W., Tao, D.: Multiview hessian regularization for image annotation. IEEE Trans. Image Process. **22**(7), 2676–2687 (2013)
12. Tao, D., Jin, L., Liu, W., Li, X.: Hessian regularized support vector machines for mobile image annotation on the cloud. IEEE Trans. Multimedia **15**(4), 833–844 (2013)
13. Liu, W., Liu, H., Tao, D., Wang, Y., Lu, K.: Multiview Hessian regularized logistic regression for action recognition. Signal Process. **110**, 101–107 (2015)
14. Long, M., Wang, J., Ding, G., Pan, S.J., Yu, P.S.: Adaptation regularization: a general framework for transfer learning. IEEE Trans. Knowl. Data Eng. **26**(5), 1076–1089 (2014)
15. Ding, C.H.Q., Li, T., Jordan, M.I.: Convex and semi-nonnegative matrix factorizations. IEEE Trans. Pattern Anal. Mach. Intell. **32**(1), 45–55 (2010)
16. Ding, C., Li, T., Peng, W.: On the equivalence between non-negative matrix factorization and probabilistic latent semantic indexing. Comput. Stat. Data Anal. **52**(8), 3913–3927 (2008)

17. Heiler, M., Schnörr, C.: Learning sparse representations by non-negative matrix factorization and sequential cone programming. J. Mach. Learn. Res. **7**, 1385–1407 (2006)
18. Liu, H., Wu, Z., Cai, D., Huang, T.S.: Constrained non-negative matrix factorization for image representation. IEEE Trans. Pattern Anal. Mach. Intell. (2012). To appear
19. Quanz, B., Huan, J., Mishra, M.: Knowledge transfer with low-quality data: a feature extraction issue. IEEE Trans. Knowl. Data Eng. **24**(10), 1789–1802 (2012)
20. Wang, J.J., Sun, Y., Bensmail, H.: Domain transfer nonnegative matrix factorization. In: 2014 International Joint Conference on Neural Networks, IJCNN, pp. 3605–3612. Beijing, 6–11 July 2014
21. Wang, C., Blei, D., Li, F.-F.: Simultaneous image classification and annotation (2009)
22. Vedaldi, A., Fulkerson, B.: VLFeat: an open and portable library of computer vision algorithms. In: Proceedings of the International Conference on Multimedia, MM 2010, pp. 1469–1472. ACM, New York (2010)
23. Long, M., Wang, J., Ding, G., Sun, J., Yu, P.S.: Transfer feature learning with joint distribution adaptation. In: 2013 IEEE International Conference on Computer Vision (ICCV), pp. 2200–2207. IEEE (2013)
24. Gong, B., Shi, Y., Sha, F., Grauman, K.: Geodesic flow kernel for unsupervised domain adaptation. In: CVPR, pp. 2066–2073. IEEE (2012)
25. Gretton, A., Borgwardt, K.M., Rasch, M.J., Schölkopf, B., Smola, A.: A kernel two-sample test. J. Mach. Learn. Res. **13**, 723–773 (2012)

Sentiment Analysis on Multi-View Social Data

Teng Niu[1], Shiai Zhu[1](✉), Lei Pang[2], and Abdulmotaleb El Saddik[1]

[1] MCRLab, University of Ottawa, Ottawa, Canada
{tniu085,elsaddik}@uottawa.ca, zshiai@gmail.com
[2] Department of Computer Science, City University of Hong Kong,
Kowloon Tong, Hong Kong
leipang3-c@my.cityu.edu.hk

Abstract. There is an increasing interest in understanding users' attitude or sentiment towards a specific topic (e.g., a brand) from the large repository of opinion-rich data on the Web. While great efforts have been devoted on the single media, either text or image, little attempts are paid for the joint analysis of multi-view data which is becoming a prevalent form in the social media. For example, paired with a short textual message on Twitter, an image is attached. To prompt the research on this interesting and important problem, we introduce a multi-view sentiment analysis dataset (MVSA) including a set of image-text pairs with manual annotations collected from Twitter. The dataset can be utilized as a valuable benchmark for both single-view and multi-view sentiment analysis. With this dataset, many state-of-the-art approaches are evaluated. More importantly, the effectiveness of the correlation between different views is also studied using the widely used fusion strategies and an advanced multi-view feature extraction method. Results of these comprehensive experiments indicate that the performance can be boosted by jointly considering the textual and visual views.

Keywords: Sentiment analysis · Multi-View data · Social media

1 Introduction

Sentiment analysis aims at predicting the attitude or opinion towards an event, topic or product from the user-contributed data (e.g., text, image or video). Many applications can benefit from the automatic identification of sentiment. For example, opinions contained in consumers' comments on the Web can be utilized by companies to improve their products or adjust marketing strategies. Politicians can also understand voters' reactions about their campaigns.

Previous works on sentiment analysis mainly focus on the data collected from review-aggregation resources, such as Internet-based retailers and forum Websites, where feedbacks from Web users are usually stereotyped texts or qualitative scores. Both the lexicon-based approaches which aggregate the sentiments of words and the statistical learning approaches are extensively investigated. While the structured and accurate data can better reflect the users' sentiment,

© Springer International Publishing Switzerland 2016
Q. Tian et al. (Eds.): MMM 2016, Part II, LNCS 9517, pp. 15–27, 2016.
DOI: 10.1007/978-3-319-27674-8_2

the collection is relatively small and only covers limited number of entities due to the fact that it is hard to involve users in the data collection. For example, customers may not comment their purchased products on the Web.

The popularity of social media, which is a convenient platform for sharing messages, introduces explosively increasing opinion-rich data freely available on the Web. Thus sentiment analysis on social data has received significant research attention recently. However, understanding social data contributed from uncontrolled Web users is challenging. Firstly, Twitter messages are usually short, and thus lack of strong evidences indicating the users' sentiment. Secondly, the posted messages are almost free-form texts having quite different styles, contents and vocabularies (e.g., abbreviations). Standard lexicon-based methods and statistical learning approaches are faced with the difficulty of sparse representations.

Besides the texts, portable devises integrated with camera have made the acquisition and dissemination of image/video much more convenient. The attractive visual data has been found to be more repostable than text messages [4]. Thus understanding affect in visual content is becoming an active research area. Inspired by the semantic concept detection, a widely used computational framework includes feature extraction and statistical modeling. Various visual features are investigated in the literature, ranging from handcraft low-level features (e.g., GIST and SIFT) [17] to middle-level features [7,15]. However, sentiment analysis of visual instance is challenging, since affective expression (e.g., "sentiment") is more abstract than the general concepts (e.g., "dog").

While significant progress has been made on sentiment analysis of social data, most of the efforts are devoted on single media type. However, social data usually has multiple views. For example, over 99 % of the images posted on Twitter are accompanied with texts [3]. In [17], a corpus analysis indicates that image and text in a message are positively correlated. The challenges in single view analysis are expected to be addressed by jointly considering the·multiple views in the social data. To the best of our knowledge, little attempts are paid to the sentiment analysis of multi-view data. One common obstacle is the lack of annotated data for model learning and performance evaluation. To prompt research on this interesting and important problem, we present a multi-view benchmark (MVSA) with rigorous manual annotations. A set of features extracted from single view and multiple views is also provided. In addition, we provide a good baseline using several state-of-the-art sentiment analysis methods and multi-view learning approaches. The main contribution of this work is in establishing a benchmark for sentiment analysis in multi-view social data. In addition, insightful discussions are provided for better understanding the limitations of existing solutions and the potentiality of exploring the correlations between different views through a comprehensive set of experiments.

2   Related Works

2.1   Sentiment Analysis Datasets

Coming up with the extensive research on text sentiment analysis, there is plenty of datasets available on the Web. These datasets are usually constructed for specific domains. For example, 50,000 movie reviews with annotations for positive and negative sentiment are provided in [8]. Recently, as many researchers turned their attentions to more timely and convenient social data, some corresponding datasets are proposed. A widely used one is STS [5], where training set consists of 1.6 million tweets automatically annotated as positive and negative based on emotioncons (e.g., ":)" or "=)"), and testing set includes 498 manually labeled tweets. While STS is relatively large, the labels are derived from unreliable emotioncons. In [11], a large dataset including manually labeled 20 K tweets is constructed for the annually organized competition in SemEval challenge. In these datasets, each message is attached with one label. However, each tweet may include mixed sentiments. In STS-Gold [12], both message-level and entity-level sentiments are assigned to 2,206 tweets and 58 entities. Besides the datasets for general sentiment analysis, there are other datasets constructed for specific domains or topics, such as Health Care Reform (HCR) dataset [13] including eight subjects and Sanders dataset[1] for four topics.

Comparing to textual data, very few datasets have been built for sentiment analysis on visual instances. In [18], a total of 1,269 Twitter images (ImgTweet) are labeled for testing their method. In [2], accompanying with the proposed emotional middle-level features (Sentibank), a small dataset including 603 multi-view Twitter messages with manual annotations is provided. Another related large-scale dataset is proposed in [6], which nevertheless is constructed for emotion detection on Web videos. To the best of our knowledge, there are no other datasets dedicatedly designed for multi-view sentiment analysis.

2.2   Sentiment Analysis Approaches

We can roughly categorize existing works on text sentiment analysis into two groups: lexicon-based approaches and statistic learning approaches. The former leverages a set of pre-defined opinion words or phrases, each of which is assigned with a score representing its sentiment. Sentiment polarity is the aggregation of opinion values of terms within a piece of text. Statistic learning approaches usually adopt a variety of supervised learning methods with some dedicated textual features. In addition, sophisticated nature language processing (NLP) techniques are developed to address the problems of syntax, negation and irony. These techniques are discussed in a comprehensive survey [9].

As a new and active research area, visual sentiment analysis adopts a similar framework with general concept detection, where sentiment classifiers are trained on visual features (e.g., "GIST"). In [18], a robust feature using deep

---

[1] http://www.sananalytics.com/lab.

neural networks is introduced into sentiment analysis. Similar to the semantic gap in concept detection, there is also an affective gap in sentiment analysis. To narrow down the gap, middle-level features defined on a set of affective atom concepts [20] or emotional Adjective Noun Pairs [2] are investigated.

While great progress has been made on sentiment analysis of textual or visual data, little effort is paid on the multi-view social data. A straightforward way [2,6] is to fuse features or prediction results generated from different views. However, it fails to represent the correlations shared by the multiple views, and thus losses important information for sentiment analysis. The most related works on learning cross-view or multi-view representations [16] may be helpful to handle this problem. For example, a joint representation of multi-view data is developed using Deep Boltzmann Machine (DBM) in [14]. However, it is still unclear whether these techniques are able to represent the complex sentiment-related context in the multi-view data.

3   The MVSA Dataset

3.1   Data Collection and Annotation

All the image-text pairs in MVSA were collected from the Twitter, which has over 300 million active users and includes 500 million new tweets per day[2]. We adopted a public streaming Twitter API (Twitter4J)[3]. In order to collect representative tweets, the stream was filtered by using a vocabulary of 406 emotional words[4]. In specific, only the tweets containing the keywords in the message or hashtags were downloaded. The vocabulary includes ten distinct categories (e.g., happiness and depression) covering almost all the felling of human beings. Some emotional words, such as happy and sad, frequently appear in the tweets. To balance the collected data among different categories, we used the keywords roundly and collected at most 100 tweets for one keyword at each round. In addition, the data collection was daily performed at several time slots within one day. We further extracted the image URLs within the messages to download the paired images. Only the text-image tweets with accessible images were kept for annotation.

Annotating sentiments on large set of image-text pairs is difficult, particularly when uncontrolled Web users may post messages without correlations between the image and text. To facilitate the annotation, we developed an interface. Each time, a image-text pair is shown to an annotator, who will assign one of three sentiments (positive, negative and neutral) to the text and image separately. Note that text and image in a message do not necessarily have same sentiment label. The annotations can be used for generating three subsets of data corresponding to text, image and multi-view respectively. Until now, we have received annotations for 4,869 messages. We only include the tweets that receive same

---

[2] https://about.twitter.com/company.
[3] http://twitter4j.org/en/.
[4] http://www.sba.pdx.edu/faculty/mblake/448/FeelingsList.pdf.

Table 1. Statistics of manually annotated datasets for tweet sentiment analysis.

Dataset	#Positive	#Negative	#Neutral	Data type
HCR	541	1,381	470	Text
STS	182	177	139	Text
SemEval	5,349	2,186	6,440	Text
STS-Gold	632	1,402	77	Text
Sanders	570	654	2,503	Text
ImgTweet	769	500	–	Image
Sentibank	470	133	–	Text + image
MVSA	1398	724	470	Text + image

labels on both text and image as the final benchmark dataset. All the data and annotations are released for public[5]. Table 1 lists the details of MVSA and several popular public datasets for tweet sentiment analysis. Comparing to other datasets, MVSA is already the largest dataset for multi-view data analysis. We will keep increasing the dataset by including more up-to-date messages, and the annotations will be regularly released.

3.2 Data Analysis

We have observed that there are inconsistent sentiments represented in user posted image and the corresponding text. This is because that the motivations of posting both text and image may be not always to enhance the sentiment or emotion of users. For example, text showed in Fig. 1(a) is the description of the event in the photo. The two views are visually related, rather than emotionally

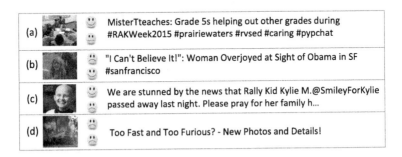

Fig. 1. Example tweets with both image and text. The top and bottom icons in the middle indicate sentiment (positive, negative or neutral) showed in image and text respectively.

---

[5] http://www.mcrlab.net/research/mvsa-sentiment-analysis-on-multi-view-social-data/.

related. Another reason is that sentiment expressed in the message is usually affected by the contexts of posting this message. For example, Fig. 1(b) shows a crying woman in the picture, while textual part indicates that the woman was overjoyed at seeing Obama. In contrast, Fig. 1(c) is a photo of smiling kid, however, the fact is that the kid passed away as described in the text. Besides these, image and text can enhance the users' sentiment. In Fig. 1(d), the negative sentiment of text is weak, which is strengthened by the attached image about a firing car in an accident. In Table 2, we list the percentage of agreements on the labels of text and image. The messages are grouped into ten categories based on the contained emotional words. We can see that only 53.2 % messages show same sentiments in their posted image and text. This poses significant challenges on multi-view sentiment analysis. In addition, users express their feeling about "happiness" (agreement of 64.6 %) more explicitly using happiness words and images.

Table 2. The percentage of messages with same labels in both textual and visual views. The messages are grouped into 10 categories based on the contained emotional keywords.

Category	Anger	Caring	Confusion	Depression	Fear	Happiness
Agreement (%)	47.5	64.6	47.5	48.7	50.9	64.6
Category	Hurt	Inadequateness	Loneliness	Remorse	**Overall**	
Agreement (%)	48.4	46.1	49.1	51.0	**53.2**	

We further analyze the accuracy of emotional words for indicating the overall sentiment of a Twitter message. We first manually label the 406 emotional keywords as positive, negative and neutral. The sentiment polarity of each tweet is same as that of the contained emotional keyword. Table 3 shows the results grouped by the 10 categories. We can see that the overall accuracy is only 30.6 %. The performance is especially poor for category "Confusion", where keywords are ambiguous for expressing human affect. Thus, more advanced technique is needed in sentiment analysis. Again, "happiness" performs much better than other categories.

Table 3. Performance of sentiment analysis using keyword matching method. The 406 emotional keywords are grouped into 10 categories.

Category	Anger	Caring	Confusion	Depression	Fear	Happiness
Accuracy (%)	19.4	50.7	12.3	20.7	18.6	51.4
Category	Hurt	Inadequateness	Loneliness	Remorse	**Overall**	
Accuracy (%)	25.5	15.2	20.6	19.9	**30.6**	

4   Predicting Sentiment in Multi-view Data

The state-of-the-art sentiment analysis systems follow the basic pipeline showed in Fig. 2, which is similar to many other recognition problems. The most important component is the feature extraction which converts different type of data into feature vectors. Then a statistical learning approach is conducted to train a classifier which is employed to predict the sentiment polarity of the input testing data. In the following, we will briefly introduce some representative features for different data types. In the experiment, the classifier is learned using the popular linear SVM due to its outstanding performance.

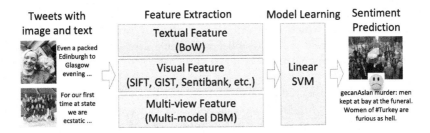

Fig. 2. The pipeline of sentiment analysis on social data.

4.1   Text-Based Approaches

Bag-of-Words (BoW) is a standard textual feature. With a pre-defined vocabulary including a set of terms (individual words or word n-grams), a document is represented as a feature vector, where each element can either be a binary value indicating the appearance of corresponding term or a counting value indicating the term frequency (TF). The vocabulary is usually constructed by selecting the most frequently appeared terms in the corpus. For sentiment analysis, emotion related terms can be also included. In addition, considering that the terms may have different importances, each term can be assigned with a weight such as the inverted document frequency (IDF). There are other variants designed for some specific applications. For text sentiment analysis, we will only test the TF and TF-IDF in the experiment.

Lexicon consists of emotional words which are manually assigned with sentiment scores. Besides the statistical learning approaches with textual features, lexicons can be directly utilized for determining the sentiment of tweet. Usually, a tweet is positive if the overall score is positive, otherwise the tweet is negative. In this paper, we consider two popular lexicons: SentiWordnet[6] and SentiStrength[7].

---

[6] http://sentiwordnet.isti.cnr.it/.
[7] http://sentistrength.wlv.ac.uk/.

4.2 Visual-Based Approaches

In this paper, we consider following visual features.

Color Histogram is the concatenation of three 256 dimensional histograms extracted from RGB color channels respectively.

GIST descriptor, which is helpful for scene classification, is a 320 dimensional feature. It includes the output energies of several filters (3 scales with 8, 8 and 4 orientations respectively) over 4×4 grids of an image.

Local Binary Pattern (LBP) descriptor is a popular texture feature. Each pixel is represented as binary codes (pattern) by comparing its value with that of neighbors. The feature vector is generated by counting the number of different patterns in the image.

Bag-of-Visual-Words (BoVW) borrows the idea of BoW except the words are SIFT descriptors which are computed on densely sampled image patches. Vocabulary includes centers of several groups generated by clustering a set of descriptors. In this paper, the vocabulary and spatial partition follow the same way used in [2].

Classemes [15] is a middle-level feature which consists of the outputs of 2,659 classifiers trained for detecting some semantic concepts (e.g., objects). Each dimension indicates the probability of the appearance of a category. Comparing to low-level features, Classemes represents an images at a higher semantic level.

Attribute is another middle-level feature, which represents abstract visual aspects (adjectives, e.g., "red" and "striped"), rather than the concrete objects used in Classemes. We adopt the 2,000 dimensional attribute proposed in [19], which is designed for representing category-ware attributes.

SentiBank [2] is an attribute representation dedicatedly designed for human affective computing. It includes 1,200 Adjective Noun Pairs (ANPs), e.g., "cloudy moon" and "beautiful rose", which are carefully selected from Web data and representative for expressing human affects. SentiBank is intuitively suitable for visual sentiment analysis.

Aesthetic feature is helpful for understanding the visual instance at more abstract level such as "beautiful". We adopt following aesthetic features used in [1]: dark channel feature, luminosity feature, S3 sharpness, symmetry, low depth of field, white balance, colorfulness, color harmony and eye sensitivity.

4.3 Multi-view Sentiment Analysis

Multi-view analysis makes use of the information extracted from both textual and visual aspects of a tweet. The most straightforward and standard way is either early fusion or late fusion. In early fusion, the features extracted from text and image are concatenated into a single feature vector. Late fusion combines the output scores of two models learned on textual and visual data respectively.

While both early and late fusion are able to boost the performance, the inherent correlations between two views is missing. Recently, multi-model learning method has showed strong performance in multi-view data analysis. In this paper, we adopt the approaches proposed in [14], where a multi-model Deep

Boltzmann Machine (DBM) is trained using textual and visual features as inputs. In [10], it has been showed to be helpful in detecting emotions from Web videos. We use a similar architecture including a visual pathway and a textual pathway. The input of the visual pathway is a 20,651 dimensional feature combining the Dense SIFT, HOG, SSIM, GIST, and LBP. On the other hand, BoW representation is utilized for generating input features for the textual pathway. The joint layer upon the two pathways contains 2,048 hidden units. The multi-model DBM is trained on 827,659 text-image pairs provided by the SentiBank dataset [2]. Details of the architecture can be referred to [10,14].

5 Experiments

We perform experiments on polarity classification of sentiment by using the data labeled as positive and negative in our constructed dataset. The dataset is randomly split into training and testing set by 50 %–50 %. We use accuracy and F-score for performance evaluation. F-score is computed on positive class (F-positive) and negative class (F-negative) respectively, and their average is denoted as F-average.

5.1 Results on Textual Messages

We first evaluate the performance of BoW feature using TF and TF-IDF strategies respectively. Figure 3 shows the accuracy with various vocabulary size. The overall trend of performance is that accuracy can be improved with larger vocabulary size. However, TF is less sensitive to the vocabulary size than TF-IDF. We observe that an appropriate size is 2,000 or larger. In the following, we fix the vocabulary size as 4,000. In addition, IDF weighting strategy hurts the performance for various vocabulary size, which is different from the conclusion in general document classification. The reason may be that the IDF assigns large weights to rare words, rather than the words reflecting human feelings. For example, "path" is assigned with larger weight than "good". This may pose negative impact on the sentiment analysis.

Table 4 further lists the results of lexicon-based approaches. For comparison, the results of TF and TF-IDF are also included. Generally, statistical learning approaches perform better than lexicon-based approaches. This indicates that sentiment analysis on tweets needs more advanced approaches and dedicated designs. However, there are many factors which may influence the performance of statistical learning approaches, such as the imbalance between positive and negative data. Our dataset includes more positive tweets. As a result, performance of negative class (F-negative) is relatively worse than that of positive class (F-positive) for both TF and TF-IDF. This is consistent to the observation in [12]. On the other hand, lexicon-based approaches are employed on each tweet independently. In some cases, it may be helpful to boost the performance of the rare class. Thus, F-negative of SentiStrength is better than that of TF or TF-IDF.

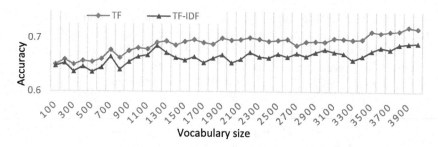

Fig. 3. Accuracy of BoW feature (TF and TF-IDF) on text sentiment analysis using different vocabulary sizes.

Table 4. Performance of different approaches for text sentiment analysis.

Method	Accuracy	F-positive	F-negative	F-average
SentiWordnet	0.603	0.640	0.557	0.598
SentiStrength	0.632	0.628	**0.636**	0.632
TF	**0.719**	**0.791**	0.569	**0.680**
TF-IDF	0.692	0.767	0.542	0.655

5.2 Results on Images

In addition to the visual features introduced in Sect. 4.2, we further test the early and late fusion of different features. LV-Late and LV-early fuse the four low-level visual features. V-Late and V-early combine all the eight features. The experiment results are showed in Table 5. For the low-level features, BoVW, which is

Table 5. Performance of different visual features for sentiment analysis.

Feature		Accuracy	F-positive	F-negative	F-average
Low-Level	Color histogram	0.652	0.772	0.263	0.518
	GIST	0.647	0.778	0.146	0.462
	LBP	0.653	0.787	0.066	0.426
	BoVW	0.667	0.775	0.354	0.565
Middle-Level	Classemes	0.650	0.747	0.432	0.589
	Attribute	0.680	0.782	0.400	0.591
	SentiBank	0.687	0.791	0.378	0.584
Aesthetic	Aesthetic	0.659	0.785	0.181	0.486
Fusion	LV-Early	0.673	0.780	0.366	0.573
	LV-Late	0.679	0.788	0.338	0.563
	V-Early	0.681	0.780	0.421	0.601
	V-Late	0.683	0.788	0.378	0.583

a preferred feature in many visual recognition tasks, only slightly outperforms others. This is because that visual appearances in a sentiment class is extremely diverse. Local features as SIFT in BoVW may be not representative for sentiment analysis. By combining the four features, which can capture different visual aspects of an image, LV-Late and LV-early further improve the results over each individual low-level feature.

For the middle-level features, both Attribute and SentiBank performs better than Classemes. This may indicate that the overall styles of the images or emotional concepts are more useful than concrete concept detectors defined in Classemes. In addition, Attribute and SentiBank, which model the semantic aspects of visual instances, achieve better performance than the low-level visual features. We can also see that aesthetic feature is an important aspect on visual sentiment analysis. However, fusion of all the features fails to boost the performance, as the result is dominated by some robust features (e.g., "SentiBank"). Furthermore, there is no winner between early fusion and late fusion with respect to the accuracy and F-average. Due to the fact that sentiment is much more abstract and extremely challenging to be represented from visual data, elaborative designs on feature selection and multi-feature fusion strategy are needed.

5.3 Results on Multi-View Data

In this section, we examine the effectiveness of different approaches on multi-view data. T-V-Early and T-V-Late respectively represent the early fusion and late fusion of TF feature and all the eight visual features. M-DBM is the performance of the feature using the outputs of the final joint layer in our learned multi-model DBM. Comparing Table 6 with Tables 4 and 5, we can see that jointly utilizing textual and visual information in the tweets can boost the performance even simply linearly fusing the features. In addition, the correlation between two views can be captured and represented by the multi-model DBM. Thus, M-DBM performs better than the results in Tables 4 and 5. We can also see that M-DBM performs better than T-V-Late and slightly worse than T-V-Early. This may be caused by the domain shift as multi-model DBM is trained on Flickr images. However, M-DBM feature (2,048 dimension) is much more compact than T-V-Early using a 12,655 dimensional feature. In general, Table 6 shows encouraging performances on multi-view sentiment analysis, which is worthy to be further investigated.

Table 6. Performance of different approaches for multi-view sentiment analysis.

Method	Accuracy	F-positive	F-negative	F-average
T-V-Early	0.752	0.825	0.572	0.699
T-V-Late	0.725	0.818	0.451	0.635
M-DBM	0.747	0.825	0.535	0.680

6   Conclusion and Future Work

We have introduced a new dataset called MVSA consisting of multi-view tweets for sentiment analysis. To the best of our knowledge, it is already the largest dataset dedicatedly constructed for multi-view sentiment analysis. While current dataset only includes several thousands of annotations, we will continuously update the data and regularly release the annotations on the most up-to-date tweets in the future. With this dataset, we have provided a pipeline for sentiment analysis on single-view and multi-view data. Extensive experiments are conducted to evaluate different features extracted from single view and multiple views, as well as their different combinations. Encouraging results suggest the joint analysis of both textual and visual information. Our result can be utilized as a baseline.

Besides the sentiment analysis, our dataset can also be used for investigating other interesting open issues, such as the subjective tweets detection. In addition, we have showed that there are many inconsistent labels between the text view and image view in the collected tweets. This suggests that future research should pay particular attention on the differentiation of emotional context with other contexts between two views, so that we can appropriately leverage the information from two views.

References

1. Bhattacharya, S., Nojavanasghari, B., Chen, T., Liu, D., Chang, S.F., Shah, M.: Towards a comprehensive computational model for aesthetic assessment of videos. In: ACM MM (2013)
2. Borth, D., Ji, R., Chen, T., Breuel, T., Chang, S.F.: Large-scale visual sentiment ontology and detectors using adjective noun pairs. In: ACM MM (2013)
3. Chen, T., Lu, D., Kan, M.Y., Cui, P.: Understanding and classifying image tweets. In: ACM MM (2013)
4. Chen, T., SalahEldeen, H.M., He, X., Kan, M.Y., Lu, D.: VELDA: Relating an image tweets text and images. In: AAAI (2015)
5. Go, A., Bhayani, R., Huang, L.: Twitter sentiment classification using distant supervision. Processing **150**(12), 1–6 (2009)
6. Jiang, Y., Xu, B., Xue, X.: Predicting emotions in user-generated videos. In: AAAI (2014)
7. Li, L.J., Su, H., Xing, E.P., Li, F.F.: Object bank: a high-level image representation for scene classification and semantic feature sparsification. In: NIPS (2010)
8. Maas, A.L., Daly, R.E., Pham, P.T., Huang, D., Ng, A.Y., Potts, C.: Learning word vectors for sentiment analysis. In: ACL (2011)
9. Pang, B., Lee, L.: Opinion mining and sentiment analysis. Found. Trends Inf. Retr. **2**(1–2), 1–135 (2007)
10. Pang, L., Ngo, C.W.: Multimodal learning with deep boltzmann machine for emotion prediction in user generated videos. In: ICMR (2015)
11. Rosenthal, S., Nakov, P., Kiritchenko, S., Mohammad, S.M., Ritter, A., Stoyanov, V.: SemEval-2015 Task 10: sentiment analysis in twitter. In: SemEval 2015 Workshop (2015)

12. Saif, H., Fernandez, M., He, Y., Alani, H.: Evaluation datasets for twitter sentiment analysis: a survey and a new dataset, the STS-Gold. In: ESSEM Workshop (2013)
13. Speriosu, M., Sudan, N., Upadhyay, S., Baldridge, J.: Twitter polarity classification with label propagation over lexical links and the follower graph. In: EMNLP Workshop (2011)
14. Srivastava, N., Salakhutdinov, R.: Multimodal learning with deep boltzmann machines. J. Mach. Learn. Res. **15**(1), 2949–2980 (2014)
15. Torresani, L., Szummer, M., Fitzgibbon, A.: Efficient object category recognition using classemes. In: Daniilidis, K., Maragos, P., Paragios, N. (eds.) ECCV 2010, Part I. LNCS, vol. 6311, pp. 776–789. Springer, Heidelberg (2010)
16. Xie, W., Peng, Y., Xiao, J.: Cross-view feature learning for scalable social image analysis. In: AAAI (2014)
17. You, Q., Luo, J.: Towards social imagematics: sentiment analysis in social multimedia. In: MDMKDD (2013)
18. You, Q., Luo, J., Jin, H., Yang, J.: Robust image sentiment analysis using progressively trained and domain transferred deep networks. In: AAAI (2015)
19. Yu, F., Cao, L., Feris, R., Smith, J., Chang, S.F.: Designing category-level attributes for discriminative visual recognition. In: CVPR (2013)
20. Yuan, J., Mcdonough, S., You, Q., Luo, J.: Sentribute: image sentiment analysis from a mid-level perspective. In: WISDOM (2013)

Single Image Super-Resolution via Convolutional Neural Network and Total Variation Regularization

Yanyun Qu[1($\boxtimes$)], Cuiting Shi[1], Junran Liu[1], Liying Peng[1],
and Xiaofeng Du[2]

[1] Department of Computer Science, Xiamen University, Xiamen, China
yyqu@xmu.edu.cn, shicuitingfei@foxmail.com,
ilevanaliu@gmail.com, 394964966@qq.com
[2] School of Computer and Information Engineering,
Xiamen University Techonology, Xiamen, China
xfdu@xmut.edu.cn

Abstract. We propose a super-resolution reconstruction model based on the fusion of convolutional neural networks and regularization constraints. Our model not only takes advantage of the convolutional neural network's prominent capability for nonlinear mapping between low-resolution and high-resolution images, but also takes the image inherent tendency to have bountiful repeated structural information into accounts. We derive our total variation regularization constraints based on the image local similarity and non-local similarity. Through coalescence of convolutional nerual network and delicately devised adaptive regularization constraints, our model yields a state-of-the-art restoration quality from a single image. Besides, our system can be expanded to tackle more low-level vision problems as well.

Keywords: Convolutional neural network · Nonlocal constraint · Image super resolution · Steering kernel constraint

1 Introduction

Rescent years have witnessed the remarkable development of the resolution of the digital camera. The prohibitively expensive price of constructing imaging chips and optical components to capture exceedingly high-resolution images has nonetheless made it impractical to real applications, e.g., widely used surveillance cameras and cell phone built-in cameras. In addition to the cost, the resolution can also be hampered by myraid unfavorable physical constraints. Therefore, it is desirable to address this problem by adopting Super Resolution (SR) algorithm, which can construct high-resolution (HR) images from a single or several observed low-resolution (LR) images with little cost, thereby increasing the high frequency components and minimizing visual artifacts.

The current SR methods which restore a single image are usually divided into two classes: interpolation based SR [1] and learning based SR. The inerpolation based SR methods are intuitive, simple and computationally effecient, e.g., the bicubic interpolation

Q. Tian et al. (Eds.): MMM 2016, Part II, LNCS 9517, pp. 28–38, 2016.
DOI: 10.1007/978-3-319-27674-8_3

is exceedingly prevelant in SR application. However, they cannot restore the texture well. Learning based SR methods, unlike the step-by-step forward interpolation-restoration method, approach the optimal reconstruction by utilizing the SR reconstruction steps stochastically, e.g. the simplest maximum likelihood (ML) [2, 3], maximum a posteriori (MAP) [4–6], joint MAP restoration [7, 8], Bayesian law [9]. These super-resolution approaches depended on assembling multiple frames that contain complementary spatial information. Consequently, without sufficient number of frames observed, the restroration resuts will be discounted dramatically.

In oder to alleviate the limitation, a nascent emerging methodology for single image SR reconstruction is to study abudant examples sampled from other irrelevent images. First came the research work proposed by Freeman et al. [10], predicting the high resolution image on the basis of the mapping relationship between the LR patches and the HR ones directly. Subsequently, many other example-based approaches [11–14] are proposed to represent the convoluted non-linear functions. No one could deny the fact that a neural network has gained popularity partially because of its potential capability to simulate the complicated mapping functions and perform parallel computations. With little human intervention, the network's weight factors can be tuned automatically by applying the back-propagation technique during the time consumming trainning process. In spite of the fact that neural network has been applied successfully in other fields, to our knowledge, it is still quite unsatisfactory for researchers to attempt to apply it to the SR restoration problems effeciently. In practice, the restoration efficacy is mainly restricted by the network's ineptitude of capturing sufficient informaion of adjacent pixels due to its structure. Besides, a dearth of futher exploiting the redundancies of an image ought to be responsible for the poor restructed results as well.

Unlike the architecture of the traditional neural network, a convolutional neural network (CNN), can gain more knowledge of a pixel's neighborhood and require less human operation on the image. In addition, in CNN, the shared weights and sparse connectivity make it a lightweight and efficient deep learning model. The unique structure fully explores the local similarity in a given image, which makes it have more advantages over the traditional ones. Its prediction accuracy, however, is limited by the fact that image prior distribution cannot be learned through the conventional layers. Efforts ought to be made to endow it with such ability through regularization constraints. In this paper, we propose a super-resolution reconstruction model through the fusion of the convolutional neural network and regularization constraints. The model takes full advantage of both CNN's prominent function mapping ability and the corresponding regularization constraints on the image prior distribution.

2   Overview of the SR Algorithm

The overall flowchart of our SR algorithm is illustrated in Fig. 1. To train a convolutional neural network for super-reoslution, we first feed the network on low-resolution (LR) and high-resolution (HR) patch pairs. During the feed forward process, the input maps are convoluted with learnable kernels to produce corresponding output maps. Meanwhile, the mean squared error (MSE) is to be computed at the output layer and propagated back to the previous layer. The network's parameters, namely the

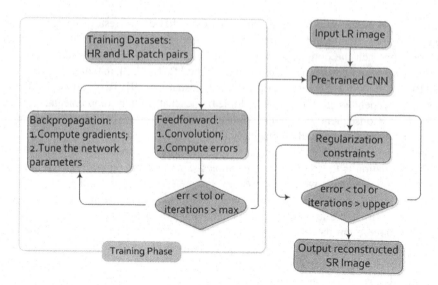

Fig. 1. The flowchart of Our SR

filters' weights and bias, are adjusted accordingly. The training phrase terminates when the accumulative error at the output layer is small enough or the maximum training iteration is reached.

Depending on the learned network, the high-resolution details of a LR image is firstly restored. Then, the output map's structural information can be procured by solving iteratively the optimal problem which is restricted by regularization constraints. Once the required terminating condition is satisfied, the algorithm will wind up yielding the final super-resolution reconstructed image.

3   Convolutional Neural Network for SR

3.1   Training Set Generation

100 natural high resolution images are downloaded from the Internet, which consist of different categories and background, and thus can provide a wide spectrum of texture details. Each HR image is firstly blurred with Gaussian filter and then downsampled with the desired scale to generate the counterpart, namely the LR image. Subsequently, via the bicubic interpolation, the LR image is interpolated into the blurred HR image with the desired scale. For each blurred HR image and its corresponding raw HR image, they are sampled randomly to form LR and HR patch pairs. Specifically, each training pair comprises a sampled blurred HR image patch and its corresponding patch in the original HR image with the same size. Image patches are extracted from the training image pairs to build the training dataset. We also preclude the image patches whose norm is close to zero since image patches are uninformative. Consequently, our training set is composed of roughly 21000 training pairs.

3.2   Convolutional Neural Network for SR

Figure 2 shows the architechture of the CNN used in our approach. The CNN comprises three convolutional layers. At each convolutional layer, the feature maps of the previous layer are convolved with learnable kernels and put through the rectified linear unit (ReLU) [15] activation function to form an output feature map. The output maps in each layer is the linear combination of the output feature maps. To illustrate, the first

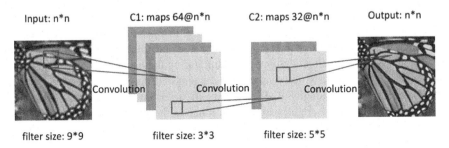

Fig. 2. Convolutional Neural Network Architecture for SR

convolution layer contains 64 learnable kernels to extract a richer representation of the input data. To put it another way, each subimage is convolved with the kernels to form 64 feature maps accordingly. These maps are further processed as the input data of the second convolutional layer. They are initially convoluted with a single filter and then the results are combined to form one specific output map. Since there are 32 filters at this layer, 32 distinct feature maps are formed. The network is intended to conclude with the desired reconstructed image, which, in this case, has the same size as the input one. To this end, the last convolutional layer only have one learnable kernel, combining all the output feature maps of the second layer. It's noteworthy that the learning process, via mini-batch stochastic gradient descent algorithm [16, 17], involves the computation of gradients of kernels and the alteration of filter weights. The trainning process stops when the terminating condition is satisified, as shown in Fig. 1. More theoretical details are to discussed in the following part.

For one thing, the convolution operation in the architecture of the aforementioned CNN can be formulated as follows,

$$m_j^l = f\left(\sum_{i \in M_j} m_i^{l-1} * k_{ij}^l + b_j^l\right) \tag{1}$$

where M_j represents the whole input maps. In our model, all the convolution operations concerned invole the padding around the input maps for keeping the size of the maps of each layer constant. Each output map is given an additive bias b. For a particular output map m_j^l at layer l, the input maps will be convolved with distinct kernels. In other words, athough the output maps m_j^l and m_k^l both sum over input map m_j^{l-1}, the kernels implemented on m_j^l are different from those implemented on m_k^l. The activation function f applied to each layer is the Rectified Linear Unit (ReLU) [15], which could achieve faster training speed and more favorable prediction results compared with the classical activation function, the sigmoid function.

For another thing, we treat the CNN for SR as an optimal problem. At the output layer, for a single training forward pass with a batch of N input subimages, the object function of the optimal problem can be formulated as

$$\mathrm{min}E = \mathrm{min}\frac{1}{N}\sum_{i=1}^{N}\|Y_i - X_i\|^2 \qquad (2)$$

where E is the Mean Squared Error (MSE), Y_i is the reconstructed image and X_i is the corresponding high-resolution image. The error will be minimized by adopting the mini-batch stochastic gradient descent algorithm during the backpropagation phase. To be specific, the error is used to compute the gradient and thus propagated back to the upper layer. Once the gradients of all the layers are determined, each layer's adaptable parameters will be tuned slightly to further enhance the overall performance of the network. Through repeated training epochs, the error is likely to be convergent to the plateau with desirable performance.

4   Regularization Constraints for SR

As we mentioned before, the CNN is versed in capturing the relationship between high-resolution patches and low-resolution patches. Nevertheless, it only focuses on the information contained in a single patch without taking into account the image prior distribution such as the relationship between a patch and its neighborhood. Therefore, further exploration of this problem need to be carried out.

In this paper, we design two regularization constraints to throw a light on this quandary, in an attempt to further ameliorate the image restoration quality. The first regularization term, the non-local similarity regularization constraint, is related to local image edge geometries. Another regularization term pertains to nonlocal image similarity. Explication on these terms is proffered in the following sections.

4.1   Non-Local Similarity Regularization Constraint

The core of the non-local similarity regularization constraint is that if two small patches are similar in a LR image, then the counterparts in the reconstructed image should be similar as well. As shown in Fig. 3, the textual structures given in the two red boxes in the LR image (the butterfly on the left side in Fig. 3) are similar. Hence, the corresponding shapes in the SR image (the butterfly on the right side in Fig. 3) should be alike too. To sum up, non-local similarity regularization constraint functions as an aid to minimize the discrepancy between the pairwise resembling patches in the output high-resolution image. In light of the fact that the nearer the distance between these patches is, the greater weight should be, we can form the non-local similarity regularization constraint formula as follows:

Non-local similarity regularization constraint

Fig. 3. Illustration of Non-local Similarity Constraint

$$R_{NL} = \|(I - W_{NL})x\|_1 \tag{3}$$

$$W_{NL}(i,j) = \begin{cases} \dfrac{w(i,j)}{\sum\limits_{j \in S_i} w(i,j)}, j \in S_i \\ 0, j \notin S_i \end{cases} \tag{4}$$

$$w(i,j) = \exp\{-\frac{\|G_\alpha \circ (x_i - x_j)\|_2^2}{2h^2}\} \tag{5}$$

where I is identity matrix, S_i symbolizes the set of similar pixels of x_i in a relative large area and $w(i, j)$ represents the weight. Supposing that patch A has center x_i while the center of patch B is x_j, as similarity between A and B is stronger and the distance is nearer, the weight will be greater. G_α denotes the Gaussian filter, and $\circ$ indicates the operation on each pixel. More details are explained in [23].

4.2 Local Similarity Regularization Constraint

The crux of local similarity regularization constraint is that the textual details in a single image should be smooth and successive. For instance, in Fig. 4, an edge of the butterfly circumscribed by the red circle ought to be successive and thus can be expressed as an approximate polynomial expression as well. Inspired by this idiosyncrasy of a specific image, we can apply this constraint to the low-resolution image to gain the desired high-resolution image, overcoming the unevenness or discontinuousness caused by the non-local similarity regularization constraint. These two constraints are complement to each other. For the local similarity regularization constraint, we use Steering Kernel Regression (SKR) [18] to express it:

Local similarity regularization constraint

Fig. 4. Illustration of Local Similarity Constraint

$$x_i = f(l_i) + \varepsilon_i, l_i = [l_{i1}, l_{i2}]^T \tag{6}$$

We use Taylor expansion to approximate the result of this regression. The final formula is shown as follows:

$$R_L = \|(I - W_L)x\|_1 \tag{7}$$

$$W_L(i,j) = \begin{cases} (e_1^T(\Psi^T K \Psi)^{-1} \Psi^T K)_{i,j}, j \in X_i \\ 0, j \notin X_i \end{cases} \tag{8}$$

$$K = diag[K_H(l_1 - l), K_H(l_2 - l), \cdots, K_H(l_p - l)] \tag{9}$$

$$\Psi = \begin{bmatrix} 1 & (l_1 - l)^T & vech\{(l_1 - l)(l_1 - l)^T\} & \cdots \\ 1 & (l_2 - l)^T & vech\{(l_2 - l)(l_2 - l)^T\} & \cdots \\ \vdots & \vdots & \vdots & \vdots \\ 1 & (l_p - l)^T & vech\{(l_p - l)(l_p - l)^T\} & \cdots \end{bmatrix} \tag{10}$$

$$K_H(l_i - l) = \frac{\sqrt{\det(C_i)}}{2\pi h_k^2} \exp\left(-\frac{(l_i - l)^T C_i (l_i - l)}{2h_k^2} \right) \tag{11}$$

where I represents identity matrix, X_i is the set of pixels surrounding x_i, which is used to solve Taylor expansion, and C_i stands the gradient covariance matrix. We only use the first item in Taylor expansion to approximate the result. More details are explicated in [23].

4.3 Fundamental Formula

A low-resolution (LR) image can be regarded as the result of degeneration of a high-resolution (HR) image which is blurred, down sampled and then degraded by superfluous noise. The relationship between HR and LR image could be expressed as the following formula:

$$y = DHx + n \tag{12}$$

where D is a down-sampling operator, H is a blurring operator, generally using Gaussian Blur, n symbolizes the noise that introduced in the LR image, x is the target HR image and y is the LR image.

According to formula (1), one of the classical super-resolution approaches is iterative back-projection (IBP) [19] which attempts to find an appropriate solution x to minimize the error, $\frac{1}{2}\|DHx - y\|_2^2$. However, it's hard for this method to recover a certain high-resolution image since the solution space of IBP is ill-posed. In our approach, the two regularization terms, as we mentioned in the former two sections, utilizing the prior knowledge and image redundancies, are incorporated in the optimal problem as follows:

$$\arg \min_{x} \left\{ \frac{1}{2} ||DHx - y||_2^2 + \lambda_{NL} R_{NL}(x) + \lambda_L R_L(x) \right\} \tag{13}$$

where $R_{NL}(x)$ is non-local regularization term specifying the prior knowledge of the HR image and $R_L(x)$ is the local regularization. λ_{NL} and λ_L are their balanced weights respectively.

5 Experimental Results

5.1 Datasets

For a fair comparison, we implement the comparable methods on the standard test sets. The SET5 [20] contains five images: woman, baby, butterfly, head and bird. These images include some details, such as the hair of the baby and the pattern of the butterfly, which can fairly judge the performance of our method. If our method can handle these images well, it will also perform well in most of the commonplace images. So, we use these five images to evaluate the performance of upscaling factors 2 and 4. The peak signal-to-noise ratio (PSNR) and the structure similarity (SSIM), as well as the visual effect of these images, constitute the criteria.

5.2 Results and Comparison

As was mentioned above, we use PSNR and SSIM as the criteria. We also consider the visual effect as one of the evaluation standards. We compare our method with the

Table 1. The results of PSNR (dB) and SSIM on the Set5 dataset

Set5 Images	Scale	Bicubic		SRCNN		OURS	
		PSNR	SSIM	PSNR	SSIM	PSNR	SSIM
Baby	2	37.066	0.952	38.301	0.964	**38.372**	0.964
Bird	2	36.808	0.972	40.642	0.985	**40.677**	0.985
Butterfly	2	27.434	0.916	32.203	0.961	**32.805**	0.966
Head	2	34.858	0.862	**35.638**	0.884	35.619	0.882
Woman	2	32.145	0.948	34.938	0.967	**35.313**	0.968
Average	2	33.662	0.930	36.344	0.952	**36.557**	0.953
Baby	4	31.777	0.856	**32.982**	0.878	32.972	0.872
Bird	4	30.180	0.873	31.976	0.902	**33.085**	0.924
Butterfly	4	22.099	0.738	25.070	0.843	**26.824**	0.895
Head	4	31.590	0.753	32.186	0.772	**32.723**	0.789
Woman	4	26.464	0.832	28.214	0.871	**29.838**	0.898
Average	4	28.422	0.811	30.086	0.853	**31.089**	0.876

state-of-the-art SR methods: Bicubic and SRCNN [21]. In terms of the details of the implementation, our model mainly benefits from the MatConvNet package [22], furnishing the model with the ability to work both on the GPU device and in the CPU environment. With approximate 21000 training patch pairs, the training process takes roughly 30 h, on a Telsa C2075 GPU.

As shown in Table 1, our model achieve the highest average PSNR, albeit the average SSIM is somewhat inferior to SRCNN. When it comes to the upscaling factor 4, our model achieve the best average performance in terms of both PSNR and SSIM. Note that our model could achieve better restoration results by excuting more training iterations. The results also further testify to our supposition that the local regularization constraint and non-local regularization constraint have an exceptional ability to capture

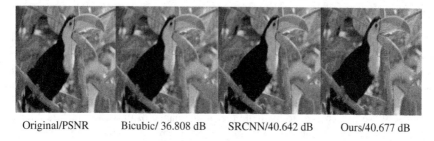

Original/PSNR   Bicubic/ 36.808 dB   SRCNN/40.642 dB   Ours/40.677 dB

Fig. 5. "Bird" image from Set5 with an upscaling factor 2

Original/PSNR   Bicubic/ 27.434 dB   SRCNN/32.203 dB   Ours/32.805 dB

Fig. 6. "Butterfly" image from Set5 with an upscaling factor 2

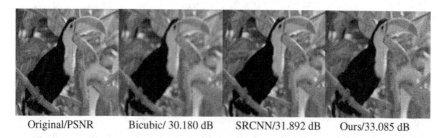

Original/PSNR   Bicubic/ 30.180 dB   SRCNN/31.892 dB   Ours/33.085 dB

Fig. 7. "Bird" image from Set5 with an upscaling factor 4

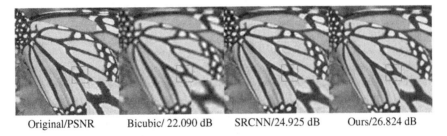

Original/PSNR   Bicubic/ 22.090 dB   SRCNN/24.925 dB   Ours/26.824 dB

Fig. 8. "Butterfly" image from Set5 with an upscaling factor 4

more potential structural information ingrained in the image itself than the convolutional neural network does. Figures 5, 6, 7 and 8 show the visual effects with the magnification factors 2 and 4 respectively. From our observation, our model also exhibits the tendency to recover sharper edge details and remove the vague or noise from the low-resolution images.

6 Conclusion

We propose a super-resolution reconstruction model for a single image based on the convolutional neural network and the regularization constraints. The convolutional neural network, unlike the traditional neural networks, has a unique and lightweight architecture partially resulted from the shared weights, which can capture more adjacent details around a pixel and have less neurons. It is the peculiarity that makes CNN a desirable deep learning architecture to address highly intricate SR image restoration problems more efficiently, compared with the state-of-the-art methods. It is true that the architecture of the CNN does have potential for recovering the lost details from the blurred and distorted low-resolution images, but further advancement of the results is resticted by its limited capability for exploiting other structural information embedded in the image. Taking structural prior distribution, the local similarity and the global similarity of an image, into accounts, we utilize the regularization constraints to further rectify the network's output. We extropolate that the model would be further improved if more work could concentrate on the architecture of the network and the design of more exquisite regularization constraints. The model, with further advacement, might be expanded to tackle real-time SR reconstruction problems in the near future.

References

1. Hou, H.S., Andrews, H.: Cubic splines for image interpolation and digital filtering. IEEE Trans. Acoust. Speech Signal Process. **26**(6), 508–517 (1978)
2. Irani, M., Peleg, S.: Super resolution from image sequences. In: Pattern Recognition, pp. 115–120 (1990)

3. Tom, B.C., Katsaggelos, A.K., Galatsanos, N.P.: Reconstruction of a high resolution image from registration and restoration of low resolution images. In: ICIP (1994)
4. Kaltenbacher, E., Hardie, R.C.: High resolution infrared image reconstruction using multiple, low resolution, aliased frames. In: NAECON (1996)
5. Hardie, R.C., Barnard, K.J., Armstrong, E.E.: Joint MAP registration and high-resolution image estimation using a sequence of undersampled images. ICIP 6(16), 1621–1633 (1997)
6. Schultz, R.R., Stevenson, R.L.: Extraction of high-resolution frames from video sequences. ICIP 5(6), 996–1011 (1996)
7. Chung, J., Haber, E., Nagy, J.: Numerical methods for coupled superresolution. Inverse Prob. 22(4), 1261–1272 (2006)
8. Tom, B.C., Katsaggelos, A.K.: Reconstruction of a high-resolution image by simultaneous registration, restoration, and interpolation of low-resolution images. In: ICIP (1995)
9. Michael, E.T., Christopher M.B.: Bayesian image superresolution. In: Proceedings of Advances in Neural Information Proceeding Systems, pp. 1279–1286 (2003)
10. Freeman, W.T., Jones, T.R., Pasztor, E.C.: Example-based super-resolution. Comput. Graphics Appl. 22(2), 56–65 (2002)
11. Sun, J., Zheng, N.N., Tao, H., Shum, H.Y.: Image hallucination with primal sketch priors. In: CVPR (2003)
12. Wang, Q., Tang, X., Shum, H.: Patch based blind image super resolution. In: ICCV (2005)
13. Chang, H., Yeung, D.Y., Xiong, Y.: Super-resolution through neighbor embedding. In: CVPR (2004)
14. Yang, J., Wright, J., Huang, T., Ma, Y.: Image super-resolution as sparse representation of raw image patches. In: CVPR (2008)
15. Nadir, V., Hinton G.E.: Rectified linear units improve restricted Boltzmann machines. In: ICML, pp. 807–814 (2010)
16. Lan, G.: An optimal method for stochastic convex optimization. Technical report, Georgia Institute of Technology (2009)
17. Lecun, Y., Bottou, L., Bengio Y., Haffner: Gradient-based learning applied to document recognition. In: Proceedings of the IEEE, vol. 86, no.11, pp. 2278-2324 (1998)
18. Takeda, H., Farsiu, S., Milanfar, P.: kernel regression for image processing and reconstruction. Image Process. 16(2), 349–366 (2007)
19. Goldstein, T., Osher, S.: The split Bregman method for L1-regularized problems. J. SIAM J. Imaging Sci. 2(2), 323–343 (2009)
20. Timofte, R., De V., Van Gool, L.: Anchored Neighborhood Regression for Fast Example-Based Super-Resolution. In: ICCV, pp. 1920–1927 (2013)
21. Dong, C., Loy, C., He K., Tang X.: Image super-resolution using deep convolutional networks. In: Pattern Analysis and Machine Intelligence (2014)
22. Vedaldi, A., Lenc, K.: MatConvNet - convolutional neural networks for MATLAB (2014). arXiv:1412.4564
23. Zhang, K., Gao, X., Tao, D., Li, X.: Single image super-resolution with non-local means and steering kernel regression. Image Process. 21(11), 4544–4556 (2012)

An Effective Face Verification Algorithm to Fuse Complete Features in Convolutional Neural Network

Yukun Ma, Jiaoyu He, Lifang Wu[✉], and Wei Qi

School of Electronic Information and Control Engineering,
Beijing University of Technology, Beijing 100124, China
{yukuner,hjy1993}@emails.bjut.edu.cn,
lfwu@bjut.edu.cn, qiwei7@sina.com

Abstract. Face verification for on line application is a difficult problem and many researchers have tried to solve it by convolutional neural network. Among of them, most works used the last-hidden layer as the feature of face, and abandoned the features in the lower layers which indicate local information. To remedy this, we extract features of all layers in the convolutional neural network, and fuse these features together after dimensionality reduction with PCA. Then these features are utilized for face verification with neural network classifier. Experiment results show that complete features can improve the verification rate effectively than using the last-hidden layer only.

Keywords: Deep learning · Convolutional neural network · Face verification · Principle component analysis

1 Introduction

With rapid development of network and communication technology, network security is becoming a challenging problem. It requires secure and reliable identity authentication. Traditional authentication methods such as password, ID card and so on, are easy to forget or lose. In comparison, biometrics including face, fingerprints, DNA, iris, face and so on are unforgettable and do not need to carry. Additionally, these biological features have high stability and uniqueness. In this paper, we focus on the problem of face verification.

In face authentication, feature selection is a crucial step. A set of good features could bring about good performance. In the early time, most of features in face recognition algorithm are fixed and handcrafted, such as LBP [1], SIFT [2] and Gabor [3], and the performance based on these features is limited. Recently, many researchers have tried to solve face recognition by convolutional neural network (CNN) [4–8], because deep neural networks can extract much better features by stacking multiple non-linear layers [9]. Among all these works, Sun Y learned features with the CNN and achieved precision of face verification around 97.45 % on LFW in [8], and CNN is proved to be more suitable for image data than other deep networks.

In CNN, the neurons of different layers indicate different information of inputting image. In [10], M.D. Zeiler used a multi-layered Deconvolutional Network to visualize

© Springer International Publishing Switzerland 2016
Q. Tian et al. (Eds.): MMM 2016, Part II, LNCS 9517, pp. 39–46, 2016.
DOI: 10.1007/978-3-319-27674-8_4

the input stimuli in the model. The visualization result shows that the lower layers neurons represent local features, and the higher ones correspond to more global features. For example, in image classification task, the second layer responds to corners and other edge/color conjunctions while the third layer captures similar textures, such as mesh patterns and text. And the last convolutional layer shows entire objects.

However, most work of CNN only use the last-hidden layer as features for face recognition. While the lower layers indicates pixel-wise and local information [10], which are also useful for later work. Rejection of these layers will unavoidably affect the performance of face verification unavoidably. In order to remedy this, we extract features in all layers of CNN network, which have been trained with supervision of identification signals, and fuse them together after dimension reduction with PCA. Then these features are utilized for face verification with Neuron Network (NN) classifier. The framework of the proposed method is shown in Fig. 1. Experiment results show that our method can improve the verification accuracy effectively than using the last-hidden layer only as feature.

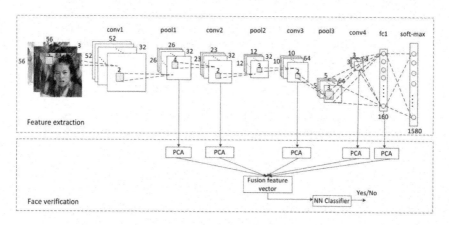

Fig. 1. Framework of the proposed face verification algorithm.

2   Related Work

Before 2006, most features in face recognition algorithm are fixed and handcrafted. Berg and Bellhumeur [11] trained classifiers to distinguish the faces from two different people. They used SVM classifiers which are shallow structures, and their learned features are still relatively low-level.

Hinton and Salakhutdinov put forward an effective way of initializing the weights that allow deep auto encoder networks to reduce the dimensionality of data [12]. From then on, a great deal of efforts have been made for face recognition with deep learning algorithm, and deep models such as CNN have been proved to be effective for extracting high-level visual features. To learn features in high-resolution images, Huang G.B. learned hierarchical representations for face verification with unsupervised convolutional deep belief networks [5]. Sun Y. et al. achieved accuracy around

91.75 % under the LFW unrestricted protocol with ConvNet-RBM which is supervised by the binary face verification signal [6]. In order to improve the accuracy, Sun Y. et al. found that deep models learn more effective feature with supervision of identification signals than verification signals, because identification signals contain more information. But this model needs a large number of training data and augmentation to learn effective feature. If training data is limited, feature learned will not be very good.

3 Methodology

3.1 Network Structure

CNN is a multi-layer neural network. Each layer is composed of a plurality of two-dimensional planes, and each plane is composed of multiple independent neurons. The network is designed to extract the parameters of each layer by experiment. And the training procedure is as follows: The first layer is trained without labels information and the parameters are learned. Then the following layers are trained one by one. After learning the $(n-1)^{th}$ layer, the output of the $(n-1)^{th}$ layer is as input of the nth layer. Then we train the CNN from top to bottom with the labeled data and then tune the network according to transmission errors. The advantages of CNN could be described that the network avoid the traditional recognition algorithm in the complex feature extraction and data reconstruction process. Thus, the image can be directly used as input without preprocessing.

We use a CNN network to extract features. As shown in Fig. 1, the CNN network is composed of four convolution layers, followed by the fully-connected layer and the softmax output layer. Every convolution layer is followed by a max-pooling layer, and the softmax output layer indicates identity classes.

The number of the rectangle denotes the number of feature maps, while the width and height denotes the dimension of the feature map. The input is $56 \times 56 \times k$, where $k = 3$ corresponds to RGB components of input image. If the input image is a gray level image, $k = 1$. We do convolution operation on input images with the convolutional filters. The number of feature maps is equal to the convolution filter number. Feature numbers continue to reduce from the feature extraction layers hierarchically to the full-connected layer(*fc1*). Referring to Sun's idea [8], the dimension of the *fc1* layer is fixed to 160. Because there are 1580 identities in the training set, the softmax output layer includes 1580 neuron.

The convolution operation is:

$$y^{j(r)} = \max\left(0, b^{j(r)} + \sum_{i} k^{ij(r)} * x^{i(r)}\right) \qquad (1)$$

x^i is the ith input map, and y^j is the jth output map. k^{ij} is the convolution kernel between x^i and y^j. Symbol '*' means convolution. b is the bias of the output map. $y = \max(0, x)$ means ReLU nonlinearity for hidden neurons. ReLU is proved to have better performance than the sigmoid function [13].

Max-pooling operation means that each neuron in y^i is the maximum value of an $s \times s$ region in x^i:

$$y^i_{j,k} = \max_{0 \le m,n < s} \left\{ x^i_{j \cdot s + m, k \cdot s + n} \right\}$$ (2)

The full-connected layer is represented as:

$$y_j = \max \left(0, b_j + \sum_i x_i w_{i,j} \right)$$ (3)

Where, x_i and y_j are i^{th} neuron in the input layer and the jth neuron in the output layer. $w_{i,j}$ is the weight between the neuron x_i and y_j.

The soft-max layer is represented as:

$$y_i = \frac{\exp(y'_i)}{\sum_{j=1}^{n} \exp(y'_j)}$$ (4)

Where y'_i is the full-connected result with the *fc1* layer as the input.

3.2   Feature Extraction

We extract several typical layers in the CNN network. Figure 2 shows the corresponding features in the layers of *pool1, conv2, local3, local4, fc1* for three of the input images. The lower layers are of pixel-wise and local features, and the higher layers correspond to global and abstract features. And for the same identity input, the neurons stimulation patterns are similar.

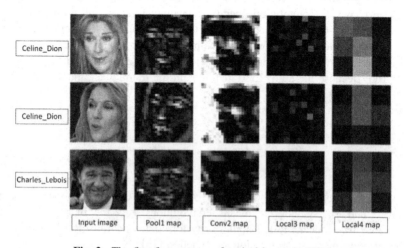

Fig. 2. The first feature map of typical layers in CNN.

The dimension of each layer of the input image is reduced to 150 by Principal Components Analysis (PCA) [14]. The PCA algorithm includes the following steps: First, the mean of the original dataset is computed and subtracted. Next, the covariance matrix R is calculated and the eigenvalues are obtained:

$$R = \begin{bmatrix} r_{11} & r_{12} & \cdots & r_{1n} \\ r_{21} & r_{22} & \cdots & r_{2n} \\ \vdots & \vdots & \cdots & \vdots \\ r_{n1} & r_{n2} & \cdots & r_{nn} \end{bmatrix} \tag{5}$$

$$r_{ij} = \frac{1}{n-1} \sum_{k}^{n} (x_{ki} - x_i)(x_{kj} - x_j) \tag{6}$$

Then, the eigenvectors are sorted according to the order of the eigenvalues and get the n largest eigenvectors. The dimension of original data can be reduced to n by the transfer matrix which is composed of the n largest eigenvectors.

By now, we could get 150 dimension features for each layer of whole CNN network. By fusing features of total five layers, we could get a 750-dimensional feature vector.

3.3   Verification

We choose a Neuron Network as verification classifier. The Neuron Network structure is shown in Fig. 3. It contains two full-connected layers and one softmax output layer. The neuron numbers of full-connected layers are separately 300, 80. Its soft-max output is 0/1(Yes/no).

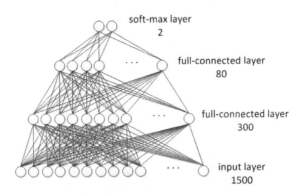

soft-max layer
2

full-connected layer
80

full-connected layer
300

input layer
1500

Fig. 3. The structure of the neural network used for face verification

4 Experiments

We train the model on MSRA-CFW(data set of celebrity faces on the web) [15] and test it on LFW(labeled faces in the wild) [16]. LFW is a database for studying face recognition in unconstrained environments. There are 13233 images of 5749, only 85 have more than 15 images, and 4069 people have only one image. It is not suitable for training. So we trained the CNN on MSRA-CFW. MSRA-CFW contains 202973 face images of 1580 celebrities.

We randomly choose 85 % cropped images per person in MSRA-CFW to train the designed CNN. And we use the remaining 15 % to test the CNN and learn the NN verification classifier. In order to augment the training data, we train the network over 10 patches of the original images, which include 4 corner patches, center patch and their horizontal reflections. This will produce better results than training over the original images alone. The testing identification error rate is 27.70 %.

Then, with the remaining 15 % images per person in MSRA-CFW, we match the 750-dimensional features into 1500-dimensional feature pairs and train the NN verification classifier. And we test the verification accuracy on LFW dataset. The result turns out to be 85.86 %, while it is 78.09 % when use the last-hidden layer as feature only.

We compare the proposed algorithm with the one using Euclidean distance. When the distance between the two features is larger than the threshold, we regard them as

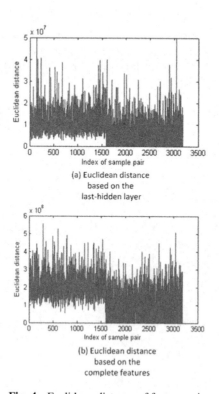

Fig. 4. Euclidean distance of feature pairs

different identities. We choose 1596 positive samples and 1596 negative samples, which means 1596 face pairs from a same identity and 1596 face pairs from different identities. We observe the Euclidean distance and find that there is obvious difference between positive samples and negative samples, as shown in Fig. 4. Figure 4(a) is the Euclidean distance based on the last-hidden layer features only. Figure 4(b) is the Euclidean distance based on the fusion features of all layers. The first half part samples are negative, and the latter part samples are positive. We can see that the difference of Euclidean distance between positive samples and negative samples is more obvious, which is more beneficial to verification.

Figure 5 shows the ROC curves of different algorithms. The four curves are separately based on Euclidean distance method using last-hidden layer, Euclidean distance method using complete feature, NN classifier using last-hidden layer and NN classifier using complete feature. The AUC (Area Under the ROC Curve) values are 0.7472, 0.7843, 0.8749, 0.9300, and the verification accuracy of these algorithms are 67.04 %, 72.34 %, 78.09 % and 85.86 %. The results show that, features in all layers are useful and the proposed approach fusing all these features could achieve better performance using both Euclidean distance and neuron network classifier.

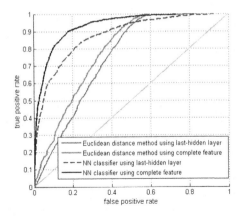

Fig. 5. Comparison of the ROC curves of verification

5   Conclusion

In this paper, we propose a new algorithm to extract the features based on CNN. Different from the typical algorithm using the last-hidden layer only previously, we choose several typical layers in CNN, which include convolution layers and the last-hidden layer. After dimensionality reduction by PCA, we fuse these layers into a whole feature of input image. The compared experimental results with last-hidden layer feature and complete feature show that the proposed algorithm is more effective.

References

1. Ojala, T., Pietikäinen, M.: Multiresolution gray-scale and rotation invariant texture classification with local binary patterns. PAMI **24**, 971–987 (2002)
2. Lowe, D.G.: Distinctive image features from scale-invariant keypoints. IJCV **60**, 91–110 (2004)
3. Wiskott, L., Fellous, J.-M., Krger, N., Malsburg, C.V.D.: Face recognition by elastic bunch graph matching. PAMI **19**, 775–779 (1997)
4. Chopra, S., Hadsell, R., LeCun, Y.: Learning a similarity metric discriminatively, with application to face verification. In: Proceeding CVPR (2005)
5. Huang, G.B., Lee, H., Learned-Miller, E.: Learning hierarchical representations for face verification with convolutional deep belief networks. In: Proceeding CVPR (2012)
6. Sun, Y., Wang, X., Tang, X.: Hybrid deep learning for face verification. In: Proceeding ICCV (2013)
7. Zhu, Z., Luo, P., Wang, X., Tang, X.: Deep learning identity-preserving face space. In: Proceeding ICCV (2013)
8. Sun, Y., Wang, X., Tang, X.: Deep learning face representation from predicting 10,000 classes. In: Proceeding CVPR (2014)
9. Xiong, C., Liu, L., Zhao, X., Yan, S., Kim, T.: Convolutional fusion network for face verification in the wild. IEEE Transactions on Circuits and Systems for Video Technology (2015)
10. Zeiler, M.D., Fergus, R.: Visualizing and understanding convolutional networks. In: Fleet, D., Pajdla, T., Schiele, B., Tuytelaars, T. (eds.) ECCV 2014, Part I. LNCS, vol. 8689, pp. 818–833. Springer, Heidelberg (2014)
11. Berg, T., Belhumeur, P.: Tom-vs-Pete classifiers and identity-preserving alignment for face verification. In: Proceeding BMVC, vol. 2, p. 7 (2012)
12. Hinton, G., Salakhutdinov, R.: Reducing the dimensionality of data with neural networks. Science **313**(5786), 504–507 (2006)
13. Krizhevsky, A., Sutskever, I., Hinton, G.: Imagenet classification with deep convolutional neural networks. In: Proceeding NIPS (2012)
14. Harrington, P.: Machine Learning in Action. Manning Publications, Greenwich (2012)
15. Zhang, X., Zhang, L., Wang, X.: Finding celebrities in billions of web images. IEEE Trans. Multimedia **14**(4), 995–1007 (2012)
16. Huang, G.B., Ramesh, M., Berg, T., Learned-Miller, E.: Labeled faces in the wild: a database for studying face recognition in unconstrained environments. Technical Report 07-49, University of Massachusetts, Amherst (2007)

Driver Fatigue Detection System Based on DSP Platform

Zibo Li[1], Fan Zhang[1], Guangmin Sun[1,2(✉)], Dequn Zhao[1],
and Kun Zheng[1]

[1] Department of Electronic Engineering, Beijing University of Technology,
Beijing 100124, China
gmsun@bjut.edu.cn
[2] Beijing Key Laboratory of Computational Intelligence and Intelligent System,
Beijing 100124, China

Abstract. Due to non-invasiveness, monitoring driver state in computer vision (CV) has become a major way to detect driver fatigue. In contrast to other researches, we brought the driver fatigue detection system on the basis of DSP platform, which can make contribution to application on the integrated system for vehicle. However, the conventional system cannot easily be transplanted into DSP due to its storage and computation capacity. Therefore, designing an algorithm that can detect the fatigue efficiently is goal in this study. As the most important part of system, the geometric relationship and shape information within near frontal face is employed in the eye detection part, which depicts the eyebrow, eye and nose. In experiment part, a self-made database is assembled to test the performance of system. As the results of experiment, the detection rate of eye is achieved at 92.71 % and driver fatigue state is obtained at 97.5 %.

Keywords: Driver fatigue · DSP platform · Eye detection

1   Introduction

With the increasing number of vehicles, driver fatigue has become a main cause of accidents. During past decades, numerous studies based on different signals are designed to alert driver to getting rid of fatigue state. Compared with the other methods, monitoring driver fatigue from facial image attracts lots of researchers' attention because image signal not only contains more information than other one dimensional signals but also can be acquired in a non-invasiveness way. [1–3] designed the driver fatigue system by CV methods. However, most of driver fatigue detection system that employs CV method is built by computer, which can be hardly transplanted into other programmable device, such as digital processing processor (DSP). In this study, a driver fatigue detection system in DSP platform is proposed because DSP shows high performance in embedding in other system (e.g. an auxiliary controller in vehicles) and outperforms other programmable devices in computation capacity. In terms of vehicle structure, electronic control unit (ECU) plays kernel role to control the operation of each parts. In contrast with computer, ECU can be treated as DSP. However, DSP shows a lower capacity of storage and computation but more portable than computer.

© Springer International Publishing Switzerland 2016
Q. Tian et al. (Eds.): MMM 2016, Part II, LNCS 9517, pp. 47–53, 2016.
DOI: 10.1007/978-3-319-27674-8_5

On the basis of this idea, Wang et al. [4] proposed the DSP-based fatigue detection system by making full use of hardware (i.e. infrared source) rather than CV method. Therefore, the CV method in DSP-based system should be designed carefully.

So far, conventional driver fatigue detection system is made up of three parts: face detection, eye detection and eye state classification. Among three parts, eye detection is the most critical part. As a topic for three decades, eye detection algorithm not only can be designed by different types of methods but also can be employed as an important parameter in the various applications, such as gaze estimation, face recognition, expression recognition, and so on [13–16, 17]. In terms of the statistic methods, the sparse classifier [1, 2], Adaboosted classifier [3] and Bayesian classifier [5] were adopted in separate or combined way. Although these algorithms can show a high performance in circumstance of accuracy and adaptation (i.e. occlusion or pose variation), the limitation of these methods is complex structure and large room costs for training phase. On the other hand, the structure-based method depicts the spatial association of facial objects or shape of eye. [6–8] utilized the high contrast distribution of eye region and built the eye detector by projection or contrast mapping. However, projection and contrast mapping are prone to decreasing the detection accuracy due to light variation. As far as driver fatigue detection, lots of studies are also discussed by proposing a novel eye detection method. Zhang and Zhang [9] tracked the eye coordinate by unscented Kalman filter. Eye position was firstly measured by image projection and then put into Kalman filter to track the accurate position. Hernan et al. [10] employed the active shape model (ASM) to track the eye. 113 points were assembled to build the coordinate model. Omer et al. [11] adopted the adaboost algorithm to recognize face and eye, and the circular Hough transform (CHT) method was proposed subsequently to detect the eye state. The above systems can estimate driver fatigue by accurate eye detection algorithm. But these algorithms share a high cost of computation and are hardly transplanted into DSP. To solve previous problems, a structure-based eye detection method is introduced in this paper.

The main contribution of this paper is proposing a driver fatigue detection system that can be transplanted into DSP platform. Similar to the conventional system, our DSP-based system is made up of three parts: face detection, eye detector and eye state detection. Among three parts of driver fatigue system, eye detection algorithm plays the most important role. Almost novel eye detection cannot be adopted in the DSP since the capability of DSP is lower than PC. In order to overcome this problem, a structure-based eye detection algorithm is used as the second contribution, which involves the geometric relationship within facial objects and shape information of eye as the basic elements. In addition, in order to keep driver's visual field clear, near frontal facial image is adopted as system's input. In terms of experiment, this system is transplanted into the DSP platform and tested in the normal inner-car environment.

The rest of paper is organized as follows: Sect. 2 is methodology part, which proposes details of each part of this driver fatigue detection system with some intermediate results. Section 3 illustrates information about experiments and shows the final statistics data. Section 4 ends up the article with conclusion.

2 Methodology

2.1 Face Detection

As the preprocessing part, the skin-color model is used as the face detector in our fatigue detection system. Although a better accuracy and robustness can be obtained by the famous Adaboost algorithm, the skin-color model relies on less parameter and shows an more efficient performance than Adaboost algorithm. In order to save the room of DSP, the skin-color algorithm is chosen. As a result, under adequately lighting levels, the accuracy of skin-color model can reach at 98 % during our experiments.

2.2 Eye Detection

After locating face, the near-frontal face region (shown by Fig. 1(a)) is analyzed on the basis of connective algorithm and eye shape feature. To decrease the wrong detection rate due to asymmetric facial region, searching field is diminished by extracting the center part of image to focus on single eye, as shown in Fig. 1(b). The gray-scale distribution of reduced region is simplified and can contribute to stressing eye region by binarization (Fig. 1(c)). Based on the binary image, eye can be further located by the connected analysis. Taking consideration of geometric relationship between eyebrow and eye, the rules for eliminating redundant region and selecting potential eye region are:

(a) face detection result;   (b) reduced facial region;   (c) binarization image

Fig. 1. Results of preparation step for eye detection

A. The algorithm is executed following the order from left-top to right-down during searching procedure.
B. Some components are eliminated under the following conditions:
  (a) regions connect to the border of image, aiming at the hair, nostril and glasses frame.
  (b) regions contain small dimension, aiming at noise region.
  (c) the length-width ratio of regions is strange, aiming at glasses frame regions are not removed by condition (a).
C. The geometric information of remaining components is recorded as potential eye regions.

After the connected analysis, the ratio of two ellipse axis based on moment [12] is employed to distinguish eye from other potential components because the shape of eye is similar to ellipse. Point set P is defined as ellipse. The three secondary moments of P are defined as (1–3), respectively:

$$\mu_{xx} = \frac{1}{S} \sum_{p \in P} (x_p - \bar{x})^2 \tag{1}$$

$$\mu_{yy} = \frac{1}{S} \sum_{p \in P} (y_p - \bar{y})^2 \tag{2}$$

$$\mu_{xy} = \frac{1}{S} \sum_{p \in P} (x_p - \bar{x})(y_p - \bar{y}) \tag{3}$$

Where S represents the dimension of P; $\bar{x}$, $\bar{y}$ are the center coordinate of center point of P.

Based on previous three moments, two axis a, b of ellipse P can be measured by (4–5):

$$a = 2\sqrt{2}\sqrt{\mu_{xx} + \mu_{yy} + \sqrt{(\mu_{xx} - \mu_{yy})^2 + 4\mu_{xy}^2}} \tag{4}$$

$$b = 2\sqrt{2}\sqrt{\mu_{xx} + \mu_{yy} - \sqrt{(\mu_{xx} - \mu_{yy})^2 + 4\mu_{xy}^2}} \tag{5}$$

To measure the shape difference between eyebrow and eye, a ratio of ellipse is defined as λ [11]:

$$\lambda = a/b \tag{6}$$

By making use of feature λ, a region can be treated as potential eye by the condition (7):

$$|\lambda_t - \lambda_E| < T \tag{7}$$

where λ_E is defined to model average shape of eye, T is the threshold.

However, in order to enhance the accuracy of eye detector, elliptical model cannot be the sole way and eye region can also be extracted through geometric relationship between eye and eyebrow. Based on the potential components resulting from connective analysis, eye region is detected in a bottom-to-top order and prone to being recognized firstly because eye is located under eyebrow. Components which meet the condition (7) are counted. When there is sole remaining region, the information of this region is directly treated as output of eye detector. When several components are recorded, the first component is detected as eye region. On the other hand, if all components cannot meet (7), algorithm will fail to detect the eye region.

2.3 Eye State Estimation

Eye state estimation consists of two parts: detecting state for single and integrated frames, separately. Similar to the famous standard PERCLOS, the height of eye is adopted to estimate eye state of single frame. As far as state detection is concerned in single frame, drowsiness is identified when height of current frame is lower than 70 % of trained threshold. Besides, assuming that the appearance of eye is kept, drowsiness is also identified when system fails to find the eye region. During testing, the first 30 frames states are used to train this threshold. The system's output is measured by a scheme (8) that fatigue is labeled when counted drowsiness frequency is greater than 3 among five frames.

$$state = \begin{cases} 0 & if \quad there\ are\ at\ least\ 3\ 'Drowsiness'\ frames\ in\ 5 \\ 1 & otherwise \end{cases} \tag{8}$$

3 Experiment and Results

In this section, two experiments are set to show the system's performance under the PC and DSP. The motivation of the testing in PC is for showing the accuracy of eye detection while the testing in DSP is for the performance of fatigue detection. In terms of measurement, detection rate is adopted in two experiments. In the following subsections, introduction of experiment setting and then some results will be introduced.

3.1 Experiment Setting

In terms of DSP, system makes use of the ADSP-BF609 EZ-KIT, which is a 500 MHz dual-core high performance DSP that make it ideal for vision analytics for automotive electronics. Both of the BF609's BlackFin cores are used, Core0 is responsible for system initialization and face detection. When a face is detected in Core0, it will send the coordinate information of detected face to a shared memory available for two cores. Due to the algorithm illustrated above, Core1 takes responsibility for locating eye region and identifying status of driver when eyes are closed and finally alert the driver. Because it is not convenient to acquire the sequences in DSP platform, recording detection rate of driver fatigue relies directly on the human-computer interaction interface of DSP. Based on this system, the experiment for driver fatigue detection is executed in the inner-car environment. 20 testing groups are implemented in this experiment. All subjects invited in this experiment are skilled drivers aged between 21 and 40 and physically fit and free from medication or alcohol.

On the other hand, a self-collected database is utilized in PC testing part. Each image in this database is a screenshot, which is randomly extracted during our experiment in DSP platform. This database contains 160 images, including 15 subjects. In this paper, a laptop with a typical CPU speed of 2.10 GHz is employed.

3.2 Experimental Results

On the basis of previous experiment setting, several experimental results are introduced in this subsection.

The result of eye detection is shown by Table 1 and Fig. 2. In order to discuss the performance of our algorithm, conventional and state-of-the-art methods are used as a comparison. According to the Table 1, accuracy of our method (92.71 %) is not the best one because the simple structure of our system can not solve some extreme case. Although decreased the detection accuracy, simple structure ensures the transferability of our system.

Table 1. Comparison of the eye detection rate

Methods	Detection rate (%)
Adaboost [3]	90.73
Projection [9]	62.30
2C SRC [1]	93.09
Proposed method	92.71

Fig. 2. Eye detection results

Based on accurate eye detection rate, experiment driver fatigue detection achieves detection rate at 97.5 %. Compared with the eye detection rate, the reason for enhancement of fatigue detection rate can be divided into two parts. On one hand, scheme (8) can make contribution to eliminating the influence of some miss-classified frames by considering integrated frames. On the other hand, some frames failing to detect eye region are linked to closed eye. Although contributing to wrong eye detection rate, these frames can be classified as a fatigue state as the final output.

4 Conclusion

In this paper, we propose a driver fatigue detection system and transplant it to the DSP platform. Similar to conventional setting, the structure of our system is made up of face detection, eye detection and driver fatigue detection. However, under the limitation of DSP in storage and computation capacity, many conventional algorithms cannot be adopted in DSP. Therefore, an effective system is introduced in this study. As the main part of the system, geometric analysis and shape moment are used to detect the eye in an efficient way. Because the DSP platform employed in experiments contains two cores, possibility of enhancing the speed of system is offered. Therefore, this advantage is employed in this paper. In the experiment part, DSP and PC are used for driver fatigue and eye detection test, respectively. As a result, our system achieves 92.71 % of eye detection rate and 97.5 % of driver fatigue detection rate. For the future work, regarding the unstable quality of binary image, we will find a outperform eye-map to replace the normal image segmentation algorithms.

References

1. Ren, Y., Wang, S., Hou, B., et al.: A novel eye localization method with rotation invariance. IEEE Trans. Image Process. **23**, 226–239 (2014)
2. Wright, J., Yang, A.Y., Ganesh, A., et al.: Robust face recognition via sparse representation. IEEE Trans. Pattern Anal. Mach. Intell. **31**, 210–227 (2009)
3. Li, S.Z., Chu, R., Liao, S., et al.: Illumination invariant face recognition using near-infrared images. IEEE Trans. Pattern Anal. Mach. Intell. **29**, 627–639 (2007)
4. Song, L., Yu, F.: Driver fatigue detection system based on DSP. Comput. Eng. Des. **33**, 519–522 (2012)
5. Everingham, M., Zisserman, A.: Regression and classification approaches to eye localization in face images. In: 7th International Conference of Automatic Face Gesture Recognition, pp. 2105–2112. IEEE Press, Southampton (2006)
6. Geng, X., Zhou, Z.H., et al.: Eye location based on hybrid projection function. J. Softw. **14**, 1394–1400 (2003)
7. Zhu, B.L., Feng, J.J., et al.: Multi-position eye location based on complex background. Appl. Res. Comput. **29**, 1977–1979 (2012)
8. Zheng, Y., Wang, Z.F.: Minimal neighborhood mean projection function and its application to eye location. J. Softw. **19**, 2322–2328 (2008)
9. Zhang, Z., Zhang, J.: A new real-time eye tracking based on nonlinear unscented Kalman filter for monitoring driver fatigue. J. Control Theor. Appl. **8**, 181–188 (2010)
10. Gercia, H., Salazar, A., Alvarez, D., Orozco, Á.: Driving fatigue detection using active shape models. In: Bebis, G., et al. (eds.) ISVC 2010, Part III. LNCS, vol. 6455, pp. 171–180. Springer, Heidelberg (2010)
11. Söylemez, Ö.F., Ergen, B.: Eye location and eye state detection in facial images using circular Hough transform. In: Saeed, K., Chaki, R., Cortesi, A., Wierzchoń, S. (eds.) CISIM 2013. LNCS, vol. 8104, pp. 141–147. Springer, Heidelberg (2013)
12. Zhang, W.Z., Wang, Z.C., et al.: Eye localization and state analysis for driver fatigue detection. J. Chongqing Univ. **36**, 22–28 (2013)
13. Zhao, S., Yao, H., Sun, X.: Video classification and recommendation based on affective analysis of viewers. Neurocomputing **119**, 101–110 (2013)
14. Shen, F., Shen, C., Shi Q., et al.: Inductive hashing on manifolds. In: 2013 IEEE Conference on Computer Vision and Pattern Recognition, pp. 1562–1569. IEEE Press, Portland (2013)
15. Yang, Y., Zha, Z., Gao, Y., et al.: Exploiting web images for semantic video indexing via robust sample-specific loss. IEEE Trans. Multimedia **16**, 1677–1689 (2014)
16. Yang, Y., Yang, Y., Huang, Z., et al.: Tag localization with spatial correlations and joint group sparsity. In: 2011 IEEE Conference on Computer Vision and Pattern Recognition, pp. 881–888. IEEE Press, Providence (2011)

Real-Time Grayscale-Thermal Tracking via Laplacian Sparse Representation

Chenglong Li[1], Shiyi Hu[2], Sihan Gao[1], and Jin Tang[1]([✉])

[1] School of Computer Science and Technology, Anhui University, Hefei, China
{chenglongli,tj}@ahu.edu.cn, sihangao@163.com
[2] School of Data and Computer Science, Sun Yat-sen University, Hefei, China
hushiyi@mail2.sysu.edu.cn

Abstract. Grayscale and thermal data can complement to each other to improve tracking performance in some challenging scenarios. In this paper, we propose a real-time online grayscale-thermal tracking method via Laplacian sparse representation in Bayesian filtering framework. Specifically, a generative multimodal feature model is induced by the Laplacian sparse representation, which makes the best use of similarities among local patches to refine their sparse codes, so that different source data can be seamlessly fused for object tracking. In particular, the multimodal feature model encodes both the spatial local information and occlusion handling to improve its robustness. With such feature representation, the confidence of each candidate is computed by the sparse feature similarity with the object template. Given the motion model, object tracking is then carried out in Bayesian filtering framework by maximizing the observation likelihood, i.e., finding the candidate with highest confidence. In addition, to achieve real-time demand in related visual information processing systems, we adopt the reverse representation and the parallel computation to improve tracking efficiency. Extensive experiments on both public and collected grayscale-thermal video sequences demonstrate accuracy and efficiency of the proposed method against other state-of-the-art sparse representation based trackers.

Keywords: Grayscale-thermal tracking · Sparse representation · Laplacian constraint · Real-time processing · Information fusion

1 Introduction

Visual tracking is an active research topic due to its wide range of applications, such as video surveillance, motion analysis, and object retrieval [18,27]. By introducing additional modal information, visual processing systems will be more robust against extremely challenging factors, such as low illumination, background clutter as well as bad weathers (snow, rain, and smog, etc.). Single-sensors usually are limited by their particular perception shortcomings, and less robustness than multimodal systems, which offers complementary sensing ranges and data frequencies, especially in varying environmental conditions [20].

© Springer International Publishing Switzerland 2016
Q. Tian et al. (Eds.): MMM 2016, Part II, LNCS 9517, pp. 54–65, 2016.
DOI: 10.1007/978-3-319-27674-8_6

Grayscale-thermal tracking has drawn a lot of attentions in the community with the popularity of thermal infrared sensors [8]. Thermal cameras, a kind of passive sensors, can capture the infrared radiation emitted by subjects with a temperature above absolute zero. Their imaging procedures are not sensitive to illumination condition, and the thermal information can provide strong complementary to visible spectrum information. On the other hand, when similar temperature objects or background occur (called thermal crossover), the grayscale information is more helpful to distinguish them. Therefore, effective fusion of grayscale and thermal data can achieve robust tracking performance in challenging scenarios.

Motivated by the advances of sparse representation for visual tracking [15, 16, 21, 28], we propose a real-time online grayscale-thermal tracking method via Laplacian sparse representation. Given the initial frame with ground truth, we first construct the dictionary from the object template, and fix it unchanged to avoid incorrectly update due to tracking failures or occlusions. Then, we generated the template feature model by Laplacian sparse representation, which makes the best use of local patch similarities to refine their codes to improve the robustness. In particular, the similarity between local patches is computed by both the grayscale and the thermal information, which provides a seamless integration of different modalities. We also take occlusion handling into account in generating the sparse feature model of object template.

Given the motion model of object, a candidate set is obtained, and object tracking is carried out in Bayesian filtering framework to maximize the likelihood function, i.e., find the candidate with highest confidence. More specifically, for each candidate, we employ the Laplacian sparse representation on the constructed dictionary to generate its sparse feature model, and its confidence can be computed by the sparse feature similarity with the object template. In this way, the MAP solution is achieved by finding the candidate with highest confidence. The flow chart of our tracking system is presented in Fig. 1.

To satisfy the real-time demands of most visual information processing systems, we utilize the reverse representation [30] and the parallel computation to improve the efficiency of the proposed method. It attributes to the fact that the number of local patches is usually much larger than the size of dictionary, and the sparse feature model of all candidates in one frame can be computed in a parallel way.

This paper makes the following contributions for grayscale-thermal tracking and related applications. First, it presents a general generative algorithm for grayscale-thermal tracking based on Laplacian sparse representation in Bayesian filtering framework. The proposed method can seamlessly integrate the grayscale and the thermal data for robust visual tracking. Second, it makes the best use of local patch similarities to refine their codes in the Laplacian sparse representation, which provides more robust solution against noises. Third, it leverages the reverse representation and the parallel computation to improve the efficiency of the proposed method, making it real-time to satisfy the demands of most visual information processing systems. Extensive experiments on both public and

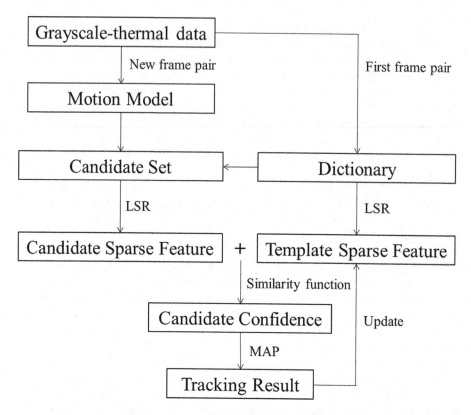

Fig. 1. The flow chart of our framework. LSR and MAP denote the Laplacian sparse representation and the maximum a posterior, respectively.

collected grayscale-thermal video sequences suggest that the proposed method outperforms other state-of-the-art sparse representation based methods in both accuracy and efficiency.

The rest of this paper is organized as follows. In Sect. 2, the relevant methods to our works are introduced. In Sect. 3, we review the Bayesian filtering framework for object tracking. The proposed algorithm is presented in Sect. 4, and the experimental results are shown in Sect. 5. The final Sect. 6 concludes this paper.

2   Related Work

Grayscale object tracking has been extensively studied in computer vision community. The most successful stories come from the application of the various machine learning techniques, including SVM [11,24], boosting [1,10], low-rank or sparse representation [14,23,26,30] and correlation filter [6,12,25]. Most of these methods, however, only work on grayscale data and thus suffer from the aforementioned challenges.

Grayscale-thermal tracking has drawn a lot of attentions in the community with the popularity of thermal infrared sensors [2–5,13,15,21]. Conaire et al. [4] proposed a framework that can efficiently combine the grayscale and the thermal features for automated surveillance applications. Cvejic et al. [5] first investigated the impact of pixel-level fusion of the grayscale and the thermal videos, and then accomplished a means of a particle filter which fuses a color cue and the structural similarity measure. A pedestrian tracker designed as a particle filter was introduced by Leykin et al. [13] based on the proposed background model, in which each pixel was represented as a multi-modal distribution with the changing number of modalities for both grayscale and thermal input. These methods adopt simple integration on grayscale-thermal data, and usually suffer from aforementioned challenges.

As increasing attentions of *sparse representation* in visual tracking, some methods have applied it into grayscale-thermal tracking to achieve the promising results. Wu et al. [21] proposed a data fusion approach via sparse representation, in which the image patches from different sources of each target candidate were concatenated into a one-dimensional vector that is then sparsely represented in the target template space. Liu et al. [15] designed a similarity induced by joint sparse representation to construct the likelihood function of particle filter tracker so that the visible and thermal spectrum images can be fused for object tracking. These methods are time-consuming, and ignore the local information in grayscale-thermal tracking. In this work, we follow this direction and introduce more robust sparse representation method for real-time object tracking.

3   Bayesian Filtering for Object Tracking

We first review the Bayesian filtering based tracking method. Let $\mathbf{Z}_t = [\mathbf{z}_1, \mathbf{z}_2, ..., \mathbf{z}_t]$ denote the observation set. Given $\mathbf{Z}$ and the state variable $\mathbf{x}_t$, we can compute the optimal state $\hat{\mathbf{x}}_t$ by Maximum A Posterior (MAP) estimation,

$$\hat{\mathbf{x}}_t = \arg\max_{\mathbf{x}_{t,i}} P(\mathbf{x}_{t,i}|\mathbf{Z}_t), \qquad (1)$$

where $\mathbf{x}_{t,i}$ indicates the state of the i-th sample at time t. We factorize Eq. (1) by Bayesian rules as follows,

$$P(\mathbf{x}_t|\mathbf{Z}_t) \propto P(\mathbf{z}_t|\mathbf{x}_t) \int P(\mathbf{x}_t|\mathbf{x}_{t-1})P(\mathbf{x}_{t-1}|\mathbf{Z}_{t-1})d\mathbf{x}_{t-1}, \qquad (2)$$

where $P(\mathbf{x}_t|\mathbf{x}_{t-1})$ and $P(\mathbf{z}_t|\mathbf{x}_t)$ are the motion model and the likelihood model, respectively. We use six independent affine parameters, including deformable and translated information, to represent the variation of motion, and model the dynamic process by the Gaussian distribution,

$$P(\mathbf{x}_t|\mathbf{x}_{t-1}) = N(\mathbf{x}_t; \mathbf{x}_{t-1}, \sigma), \qquad (3)$$

where σ denotes a diagonal covariance matrix whose elements are the variations of the affine parameters. Given the motion model, we can sample a candidate set.

In this way, maximizing Eq. (1) is equal to minimize the observation likelihood term $P(\mathbf{z}_t|\mathbf{x}_t)$, which will be defined by the Laplacian sparse representation to measure the confidence of each candidate in next section.

4 Observation Model

This section introduces a general multimodal sparse representation to induce the observation likelihood of each candidate.

4.1 Laplacian Sparse Representation

Motivated by recent advances of sparse coding for object tracking [16, 28, 30], this section presents a generative sparse model for object representation.

Dictionary Construction. Given the target object template of the initial frame, we construct the multimodal dictionary $\mathbf{D}$ as follows. First, the template regions of all modalities are normalized to 16×16 pixels, respectively. Second, we use overlapped sliding windows on the normalized regions to obtain N patches, and the i-th patch of m-th modality is converted to a vector $\mathbf{t}_i^m \in R^{G \times 1}$, where G is the size of the patch. We concatenate all modal vectors of i-th patch as its appearance representation: $\mathbf{t}_i = [t_i^1; t_i^w; ...; t_i^M]$, where M denotes the number of modalities, and our grayscale-thermal data is the special case with $M = 2$. Such representation seamlessly integrate the information from different modalities. Finally, we employ the k-means algorithm on N patches to obtain J cluster, and generate $\mathbf{D} \in R^{G \times J}$ by all cluster centers, which consists of the most representative patterns of the target object. The dictionary $\mathbf{D}$ is fixed during tracking process to avoid incorrectly updated due to tracking failures or occlusions.

Sparse Feature Model. For each patch vector $\mathbf{t}_i$ of the object template, we can generate a sparse coefficient vector $\mathbf{f}_i$ by the sparse representation model:

$$\min_{\mathbf{f}_i} \frac{1}{2}||\mathbf{t}_i - \mathbf{D}\mathbf{f}_i||_F^2 + \lambda||\mathbf{f}_i||_1, \ s.t. \ \mathbf{f}_i \succeq 0, \tag{4}$$

where λ is a balance parameter. Equation (4) solves l_1 minimization problems for all patch vectors independently, which is computationally inefficient especially when the number of patches is much larger than the number of dictionary elements. Meanwhile, the dependence among patches is ignored, which sometimes leads to unreasonable codes, i.e., similar patches have big difference in the responses of sparse representation. To this end, we employ reverse Laplacian sparse representation model as:

$$\min_{\mathbf{F}} \frac{1}{2}||\mathbf{D}_i - \mathbf{T}\mathbf{F}||_F^2 + \lambda||\mathbf{F}||_1 + \beta \ tr(\mathbf{F}\mathbf{L}\mathbf{F}), \ s.t. \ \mathbf{F} \succeq 0, \tag{5}$$

where β is a balance parameter, and $tr(\cdot)$ denotes the trace of a matrix. $\mathbf{L} = \mathbf{D} - \mathbf{B}$ is the Laplacian matrix, and $\mathbf{B}$ and $\mathbf{D}$ are the truncated feature dissimilarity matrix and the degree matrix. Herein, unlike only exploiting nearest neighbors of patches in [30], we make the best use of the relationship among patches to refine the sparse codes, and the feature distance between patch i and j is defined as:

$$\mathbf{B}_{ij} = \begin{cases} \exp(-\dfrac{||\mathbf{t}_i - \mathbf{t}_j||_2}{\sigma}), & ||\mathbf{t}_i - \mathbf{t}_j||_2 > 0.8d \\ 1, & else \end{cases}. \tag{6}$$

where σ is a parameter, and $0.8d$ is the truncated threshold, where d is the maximum distance between local patches. In this way, we can obtain a sparse feature representation of the object template: $\mathbf{T} = vec(\mathbf{F}) \in R^{J \times N}$, where $vec(\cdot)$ indicates the vectorize operator of a matrix.

To deal with occlusions, we modify the constructed feature representation to exclude the occluded patches, which have large reconstruction errors. More specifically, when the reconstruction error of a patch is larger than the predefined threshold ε_1, we set the corresponding sparse coefficients to be zero. In this way, the sparse feature representation encodes spatial local information and occlusion, making it more effective and robust.

In addition, we update the template every several frames to adapt the appearance variations and recover the object from occlusions. The new template $\mathbf{T}_i$ of i-th frame is updated by

$$\mathbf{T}_i = \mu \mathbf{T}_0 + (1 - \mu)\mathbf{T}_{i-1} , if \ O_i < \varepsilon_2, \tag{7}$$

where $\mathbf{T}_0$ is the sparse feature representation of the first frame, and μ is the updated rate. The occlusion measure O_i of the tracking result is computed by the mean value of occlusion indicators of its patches, and ε_2 is a predefined threshold. This update scheme takes both the first accurate template and the most recent tracking result into account to improve the robustness.

Model Optimization. For optimization to Eq. (5), we first convert it to a unconstrained problem as:

$$\min_{\mathbf{F}} \frac{1}{2}||\mathbf{D}_i - \mathbf{TF}||_F^2 + \lambda ||\mathbf{F}||_1 + \beta \ tr(\mathbf{FLF}) + \psi(\mathbf{F}), \tag{8}$$

where

$$\psi(\mathbf{F}_{ij}) = \begin{cases} 0, & if \ \mathbf{F}_{ij} \geq 0 \\ \infty, & else \end{cases}. \tag{9}$$

Equation (8) can be efficiently optimized by the Accelerated Proximal Gradient (APG) approach [17].

4.2 Candidate Likelihood

When new frame arrives, P candidates are randomly drawn by the motion model (Eq. (3)). For each candidate, we generate its sparse feature representation $\mathbf{Y}_p$ on the dictionary $\mathbf{D}$, and the process is the same as the object template. Notice

(a) Otcbvs   (b) Otcbvs1   (c) Torabi   (d) WalkingOcc   (e) FastCarNig   (f) Tunnel

Fig. 2. The first frames with ground truths of the aligned grayscale-thermal video sequences used in this paper. The red bounding box indicates the ground truth. Herein, the first three columns denote the video pairs selected in [7,19], and the remain columns are captured by our grayscale-thermal recording system (Color figure online).

that all candidates are generated independently, and thus we compute them in a parallel way to further improve the efficiency of the proposed method. We employ the intersection function [9] to compute the similarity between the candidate $\mathbf{Y}_p$ and the template $\mathbf{T}$ by

$$S_p = \sum_{i=1}^{J \times N} \min(\mathbf{Y}_p^j, \mathbf{T}^j). \tag{10}$$

We use the similarity S_p to measure the confidence of the p-th candidate, and the tracking result is achieved by maximizing Eq. (2), i.e., the candidate with the highest confidence.

5 Experiments

This section first applies the proposed tracking method over both public and collected grayscale-thermal video sequences and compares with other popular tracking methods. Then, we analyze the components of the proposed method to justify their respective contributions.

5.1 Evaluation Settings

To evaluate the performance of our tracker, we conduct experiments on three public grayscale-thermal video sequences [7,19] and three collected ones by us. These sequences cover several challenging scenarios in grayscale-thermal tracking, such as low-illumination, background clutter and thermal crossover, as shown in Fig. 2.

We fix the following parameters for our tracking method in all experiments. We set the Gaussian parameter σ of motion model in Eq. (3) is set to be 1, and the number of candidates to be 50. The row number G and column number J of the dictionary $\mathbf{D}$ (Eq. (4)) are set to be 36 and 10, respectively. The occlusion thresholds ε_1 and ε_2 (Eq. (7)) are set to be 0.04 and 0.8, respectively, and the object template is updated in every 5 frames with the update rate $\mu = 0.95$.

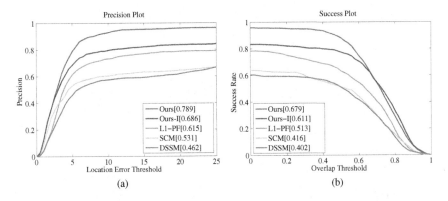

Fig. 3. The precision plots and the success plots on all video pairs are shown in (a) and (b), respectively. The precision score and the success score are presented in the legend. The compared methods include L1-PF [21], SCM [29], and DSSM [30]

Table 1. Precision Score (PS, %) and Success Score (SS, %) of the proposed tracker comparing with other grayscale and grayscale-thermal trackers. The bold fonts of results indicate the best performance.

Algorithm	Otcbvs		Otcbvs1		Torabi		WalkingOcc		FastCarNig		Tunnel		Average	
	PS	SS	PS	SS	PS	SS	PS	SS	PS	SS	PS	SS	PS	SS
Ours	**72**	**66**	88	79	62	51	57	55	**95**	66	**99**	82	**79**	**68**
Ours-I	52	47	87	**80**	73	56	**60**	**64**	54	46	98	82	69	61
SCM [29]	5	4	**100**	73	**74**	52	53	59	83	60	67	49	53	42
DSSM [30]	13	10	90	74	71	57	55	58	83	**67**	21	22	53	41
L1-PF [21]	40	33	99	78	67	**58**	15	14	73	**67**	96	78	62	51

5.2 Evaluation Metrics

We utilize two widely used metrics, precision plot and success plot, to evaluate online RGBT tracking performance. In particular, the final precision score and success score are defined as the best one of two modalities.

Precision Plot. Center Position Error (CPE) is the Euclidean distance between the center locations of the tracked object and the ground truth bounding box, and usually employed to evaluate tracking precision. However, it will be invalid when the trackers are failed [1]. To evaluate overall performance, we employ the precision score used in recent literatures [22], the percentage of frames whose output location is within the given threshold distance of ground truth. Since many targets are small, we set the threshold to be 5 pixels to obtain the representative Precision Score (PS).

Success Plot. Bounding box overlap is another effective evaluation metric. Given the output bounding box r_o and the ground truth bounding box r_g, the

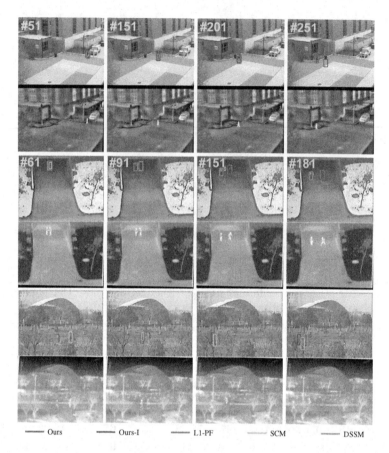

Fig. 4. Sample results of our method against other tracking methods, including L1-PF [21], SCM [29], and DSSM [30].

overlap score is defined as $S = \frac{|r_o \bigcap r_g|}{|r_o \bigcup r_g|}$, where $\bigcap$ and $\bigcup$ denote the intersection and union operators of two regions, and $|\cdot|$ indicates the number of pixels in the region. Setting a threshold t_o of overlapping area, we can calculate the success rate value, the ratio of the number of successful frames whose overlap S is larger than t_o. The success plot can be shown by the success rate at different t_o ($\in [0, 1]$). We employ the area under curve of success plots to obtain the Success Score (SS) [22].

5.3   Comparison Results

With same initialization of the targets, we compare with three sparse representation tracking algorithms, including SCM [29], DSSM [30], and L1-PF [21]. Herein, SCM and DSSM are grayscale tracking methods, and L1-PF is grayscale-thermal tracking method.

The quantitative comparison results are shown in Fig. 3 and Table 1. From Fig. 3 and Table 1, we can see that the proposed tracking algorithm substantially outperforms other grayscale and grayscale-thermal tracking algorithms. It suggests that our method can obtain robust tracking results by effectively fusing the grayscale and the thermal data in various challenging scenarios, such as illumination variation (*FastCarNig, Tunnel*) and background clutter (*Otcbvs, WalkingOcc*). In addition, our method and L1-PF achieve superior performance than SCM and DSSM, justifying the importance of complementary thermal information in visual tracking. The qualitative results is presented in Fig. 4.

5.4 Component Analysis

To justify the significance of Laplacian constraints in the proposed method, we implement one special versions for comparative analysis, i.e., *Our-I*, that set $\beta = 0$ to remove the Laplacian constraint in Eq. 5.

The evaluation results are reported in Fig. 3 and Table 1. From the evaluation results, we can conclude that the Laplacian constraints are helpful to obtain robust representation in most challenging scenarios. The worse results on *Torabi* and *WalkingOcc* of the complete version may attribute to the noise effects of low-level features.

The experiments are carried out on a desktop with an Intel i7 4.0 GHz CPU with 4 core processors and 32 GB RAM, and implemented on the platform of MATLAB without any code optimization. Our method can achieve approximately 17 frames per second.

6 Conclusion

This paper presented a generative sparse model for grayscale-thermal tracking in Bayesian filtering framework. The proposed sparse model seamlessly fused the grayscale and the thermal data for robust visual tracking, and also made the best use of local similarities to refine their codes in sparse representation for improving its robustness. The experiments demonstrated the effectiveness of the proposed method. In the future work, we will integrate the background information in our model to enhance its discriminative ability, and learn the modal reliability online to adaptively fuse different source data for more robust tracking.

Acknowledgement. This work was supported by the Development Program (863 Program) of China (No. 2014AA015104) and the Natural Science Foundation of China (No. 61472002).

References

1. Babenko, B., Yang, M.H., Belongie, S.: Robust object tracking with online multiple instance learning. IEEE Trans. Pattern Anal. Mach. Intell. **33**(7), 1619–1632 (2011)

2. Bunyak, F., Palaniappan, K., Nath, S.K., Seetharaman, G.: Geodesic active contour based fusion of visible and infrared video for persistent object tracking. In: Proceedings of IEEE Workshop on Applications of Computer Vision (2007)
3. Conaire, C.O., Connor, N.E., Cooke, E., Smeaton, A.F.: Comparison of fusion methods for thermo-visual surveillance tracking. In: Proceedings of International Conference on Information Fusion (2006)
4. Conaire, C.O., Connor, N.E., Smeaton, A.: Thermo-visual feature fusion for object tracking using multiple spatiogram trackers. Mach. Vis. Appl. **7**, 1–12 (2007)
5. Cvejic, N., Nikolov, S.G., Knowles, H.D., Loza, A., Achim, A., Bull, D.R., Canagarajah, C.N.: The effect of pixel-level fusion on object tracking in multi-sensor surveillance video. In: Proceedings of IEEE Conference on Computer Vision and Pattern Recognition (2007)
6. Danelljan, M., Khan, F.S., Felsberg, M., van de Weijer, J.: Adaptive color attributes for real-time visual tracking. In: Proceedings of IEEE Conference on Computer Vision and Pattern Recognition (2014)
7. Davis, J.W., Sharma, V.: Background-subtraction using contour-based fusion of thermal and visible imagery. Comput. Vis. Image Underst. **106**(2), 162–182 (2007)
8. Gade, R., Moeslund, T.B.: Thermal cameras and applications: a survey. Mach. Vis. Appl. **25**, 245–262 (2014)
9. Gao, S., Tsang, W.H., Chia, L.T., Zhao, P.: Local features are not lonely – laplacian sparse coding for image classification. In: Proceedings of IEEE Conference on Computer Vision and Pattern Recognition (2010)
10. Grabner, H., Leistner, C., Bischof, H.: Semi-supervised on-line boosting for robust tracking. In: Forsyth, D., Torr, P., Zisserman, A. (eds.) ECCV 2008, Part I. LNCS, vol. 5302, pp. 234–247. Springer, Heidelberg (2008)
11. Hare, S., Saffari, A., Torr, P.H.S.: Struck: structured output tracking with kernels. In: Proceedings of IEEE International Conference on Computer Vision (2011)
12. Henriques, J.F., Caseiro, R., Martins, P., Batista, J.: High-speed tracking with kernelized correlation filters. IEEE Trans. Pattern Anal. Mach. Intell. **37**(3), 583–596 (2015)
13. Leykin, A., Hammoud, R.: Pedestrian tracking by fusion of thermal-visible surveillance videos. Mach. Vis. Appl. **21**(4), 587–595 (2010)
14. Li, C., Lin, L., Zuo, W., Yan, S., Tang, J.: Sold: sub-optimal low-rank decomposition for efficient video segmentation. In: Proceedings of IEEE Conference on Computer Vision and Pattern Recognition (2015)
15. Liu, H., Sun, F.: Fusion tracking in color and infrared images using joint sparse representation. Inf. Sci. **55**(3), 590–599 (2012)
16. Mei, X., Ling, H.: Robust visual tracking using l_1 minimization. In: Proceedings of IEEE International Conference on Computer Vision (2009)
17. Parikh, N., Boyd, S.: Proximal algorithms. Found. Trends Optim. 1–96 (2013)
18. Shen, F., Shen, C., Liu, W., Shen, H.T.: Supervised discrete hashing. In: Proceedings of IEEE Conference on Computer Vision and Pattern Recognition (2015)
19. Torabi, A., Masse, G., Bilodeau, G.A.: An iterative integrated framework for thermal-visible image registration, sensor fusion, and people tracking for video surveillance applications. Comput. Vis. Image Underst. **116**(2), 210–221 (2012)
20. Walchshausal, L., Lindl, R.: Multi-sensor classification using a boosted cascade detector. In: Proceedings of IEEE Intelligent Vehicles Symposium (2007)
21. Wu, Y., Blasch, E., Chen, G., Bai, L., Ling, H.: Multiple source data fusion via sparse representation for robust visual tracking. In: Proceedings of International Conference on Information Fusion (2011)

22. Wu, Y., Lim, J., Yang, M.H.: Online object tracking: a benchmark. In: Proceedings of IEEE Conference on Computer Vision and Pattern Recognition (2013)
23. Yang, Y., Yang, Y., Huang, Z., Shen, H.T., Nie, F.: Tag localization with spatial correlations and joint group sparsity. In: Proceedings of IEEE Conference on Computer Vision and Pattern Recognition (2011)
24. Zhang, J., Ma, S., Sclaroff, S.: MEEM: robust tracking via multiple experts using entropy minimization. In: Fleet, D., Pajdla, T., Schiele, B., Tuytelaars, T. (eds.) ECCV 2014, Part VI. LNCS, vol. 8694, pp. 188–203. Springer, Heidelberg (2014)
25. Zhang, K., Zhang, L., Liu, Q., Zhang, D., Yang, M.-H.: Fast visual tracking via dense spatio-temporal context learning. In: Fleet, D., Pajdla, T., Schiele, B., Tuytelaars, T. (eds.) ECCV 2014, Part V. LNCS, vol. 8693, pp. 127–141. Springer, Heidelberg (2014)
26. Zhang, T., Ghanem, B., Liu, S., Ahuja, N.: Low-rank sparse learning for robust visual tracking. In: Fitzgibbon, A., Lazebnik, S., Perona, P., Sato, Y., Schmid, C. (eds.) ECCV 2012, Part VI. LNCS, vol. 7577, pp. 470–484. Springer, Heidelberg (2012)
27. Zhang, W., Li, C., Zheng, A., Tang, J., Luo, B.: Motion compensation based fast moving object detection in dynamic background. In: Zha, H., Chen, X., Wang, L., Miao, Q. (eds.) CCCV 2015. CCIS, vol. 547, pp. 247–256. Springer, Heidelberg (2015). doi:10.1007/978-3-662-48570-5_24
28. Zhong, W., Lu, H., Yang, M.H.: Robust object tracking via sparsity-based collaborative model. In: Proceedings of IEEE Conference on Computer Vision and Pattern Recognition (2012)
29. Zhong, W., Lu, H., Yang, M.H.: Robust object tracking via sparse collaborative appearance model. IEEE Trans. Image Process. **23**(5), 2356–2368 (2014)
30. Zhuang, B., Lu, H., Xiao, Z., Wang, D.: Visual tracking via discriminative sparse similarity map. IEEE Trans. Image Process. **23**(4), 1872–1881 (2014)

Efficient Perceptual Region Detector
Based on Object Boundary

Gang Wang[✉], Ke Gao, Yongdong Zhang, and Jintao Li

Key Lab of Intelligent Information
Processing of Chinese Academy of Sciences (CAS),
Institute of Computing Technology, CAS, Beijing 100190, China
{wanggang01,kegao,zhyd,jtli}@ict.ac.cn

Abstract. Finding nearly semantic 'visual units' which visual analysis can operate on is a long-term hard work in computer vision community. Established powerful methodologies such as SIFT, BRISK often extract numerous redundant single keypoints with little information about semantic contents. We propose a novel method called Contour-Aware Regions detector (CAR) to find representative regions in images. Inspired by the recent research conclusion of general object proposal methods that contour is important in object localization, we first alleviate the problem of super pixle overlapping multi-object regions. And then perceptual regions are generated during the merging process using the data structure similar to MSER. Extensive experiments demonstrate: (1) superpixels generated by our method significantly outperform the state-of-out methods, as measured by boundary recall and under-segmentation error. (2) our method can find less and meaningful regions in 0.125 s per image, meanwhile achieve promising repeatabiliy.

Keywords: Contour-Aware Regions (CAR) · Superpixels · MSER · Local invariant regions

1 Introduction

Representing an image as a series of features extracted from 'visual units' is a widely applied technique in computer vision applications, such as image retrieval, object recognition, object tracking and 3D scene reconstruction. The quality of 'visual units' used in these applications directly decides the upper bound of the performance, since all the following visual processes are based on it.

Ideal 'visual units' need to capture the visual-homogenous areas, have salient outer boundaries and be close to semantic objects. In this paper, we aim to find perceptual regions satisfying these conditions as much as possible. Two different types of methods, namely segmentation and local feature detector, are trying to realize this goal. In principle, segmentation is the process of partitioning the image into several regions without overlap, while all these regions collectively cover the whole image. However, the results of segmentation can not reflect the

© Springer International Publishing Switzerland 2016
Q. Tian et al. (Eds.): MMM 2016, Part II, LNCS 9517, pp. 66–78, 2016.
DOI: 10.1007/978-3-319-27674-8_7

intrinsically hierarchical structures of the image and each segment is considered
of equal importance.

The more similar work to our method is local feature detector. Over the
last decade, low-level keypoint detectors methods, as proposed by [3,4], have
tended to be the dominant strategies to find local regions. These methods are
successful to detect certain stable structures, such as corners, blobs, which are
repeatable despite the change of viewpoint, so similar patches can be matched.
However, keypoint detectors often extract a lot of redundant 'pieces' as shown
in Fig. 1(a) and (b), mainly because they focus on low-level feature. To human
visual system, an image is not just consist of a discrete collection of pixels, it is
arranged by a variety of meaningful regions. Compared to keypoints, regions have
their nature advantages: (1) They can adapt to the image structures well and
are not susceptible to the affect of background clutter. (2) Pixels surrounded by
closed contours consist of affine-invariant regions (3) They also contain shape and
scale information implicitly. One of most popular region detector is MSER [5],
which is in principle to detect gray stable regions, however such method is also
failed to find whole object region as shown in Fig. 1(c).

(a) Harris[7], 497

(b) SIFT[6], 534

(c) MSER[5], 437

(d) CAR, 1

Fig. 1. Output regions detected by Harris, SIFT, MSER, our method and their corresponding numbers. The first three methods tend to split the leopard into 'pieces',
while our method can find the whole object region.

Regions have not been as popular as features because of two inherent
difficulties. One problem is their sensitivity to segmentation errors, since the
contours are hard to detect in complex scenes. The other problem is that the
variety of region appearance in nature images is too unpredictable that it is hard

to define a common function as keypoint detectors do. This is where this paper aims to set a new milestone with CAR detector to find spatially extended and perceptually meaningful regions. The main contributions in this paper can be concluded as follows:

- This paper proposes a unified work to find perceptual-homogeneous regions in nature images. Once generated by CAR, these regions can be useful to many computer visual applications, such as image retrieval, object recognition and segmentation.
- Inspired by general object proposal methods [14], we improve SLIC [9] super-pixel method by clustering pixels considering not only in color and image plane space, but also the contour influence. The improved superpixel segmentation is fast, regular and more close to the boundaries.
- A hierarchical region tree is built with merge process using watershed algorithm [20] and an objectness function is defined to select perceptual regions.
- Extensive experiments on the Berkeley benchmark dataset [8] show that the superpixels generated by our method evidently outperform the state-of-art method, as measured by boundary recall and under-segmentation error. What's more, our method can find representative meaningful regions which exhibit promising repeatability.

The rest of this paper is organized as follows. Section 2 briefly review existing methods about superpixels generation and local detector. Section 3 describe the key stages in CAR detector. Experiments results including boundary recall, under-segmentation error and repeatability are shown in Sect. 4. At last, we conclude our paper in Sect. 5.

2   Related Work

In this section, we briefly introduce some related work and discuss the differences between these methods and ours.

2.1   Superpixel

According to the grouping strategies, these methods can be broadly divided into two categories: graph-based algorithms and Meanshift [10] like algorithms. The graph-based algorithms usually treat each pixel of the image as a normal node in a graph, and two special nodes are added as foreground and background respectively. The cost function is defined considering the similarity between each connected normal nodes and the similarity between normal nodes and special nodes. Superpixels are generated by minimizing the cost function on that graph. The well known algorithm in this category is Normalized Cuts [1]. However the complexity of these algorithms grows fast as the number of superpixels increases, because the superpixel generation scheme depends on the eigen-based solution. Due to the nature of graph-based algorithm, the methods in this category always generate irregular superpixels both in size and shape, and suffer from efficiency issue.

The Meanshift like algorithms always start from some initial seeds as cluster centers. In each iteration, each cluster center moves to areas of higher density. The iteration can be stopped when a certain number of times achieves or cluster centers converge. SLIC adopts simple k-means to group points of local area into compact and near uniform-size superpixels. However, SLIC measures the distance in color and image plane space, while neglecting the information of object boundary, so the superpixels tend to overcross the boundary like Fig. 3(c). We improve the distance measure between pixels and cluster centers by the aid of contour map. The improved methods generate superpixels with high boundary recall while still as fast as SLIC.

2.2 Local Detectors

Ever since the Harris was proposed in the late 1980s, a lot of local detectors have been proposed in literature. The well accepted best quality detector is SIFT. Searching for max response in image pyramids space makes SIFT invariant to scale change, but also makes it slow. The extending work SURF [12] speeds up SIFT using integral image. The other representative methods includes Harris, EBR [13], IBR [13]. However, regions detected by these methods are not suitable for high-level feature extraction. Perhaps the most relevant work to our method is MSER, which tries to find gray stable regions according to a global threshold δ, but our method has several important differences from MSER. Instead of using simple gray pixel as ranking unit, the processing element adopted in our method is supperpixel. What's more, in distance measure and component regions selection, our method takes full advantage of contour information, so it can find visual-homogenous regions with salient outer boundary.

2.3 General Object Proposal

In recent years, techniques for extracting general object candidate locations becomes increasingly attractive. BING [15] assumes that generic object is a well defined closed boundary. Object may be different from each other, but they share almost the same 8×8 models in gradient space. The detection can be done using linear SVM super efficiently. Edge Box [19] finds that edges provide a sparse yet informative representation of object. A box wholly contains a closed contour is more likely to contain object, however, some edges located at the center of box are not probably indicating object. According this observation, Edge Box defines a simple function that counts the number of edges that exist in the box minus those contours that overlap the boxs boundary. In this paper, we also use the research conclusions of general object proposal methods above.

3 CAR: The Method

Contours intrinsically indicate objects, the early work [16] employ contour information to do the image segmentation and recently BING [15] takes the assumption that an object should have closed outer boundary to locate objects. Inspired

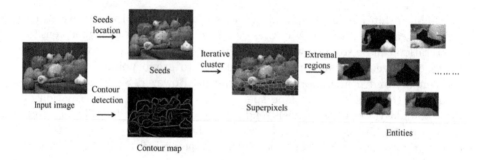

Fig. 2. Pipeline of Contour-Aware Regions detector

by these methods, this paper proposes a novel method to find meaningful regions in nature images. Figure 2 shows the pipeline of the method. Given an input image, we start to generate superpixels by cluster pixels in local area (in Sect. 3.1). A contour map is built to improve the distance measure. The contour map can be produced by as simple as Canny [21], or as complex algorithm like Structured Forests edge detector [11]. Then superpixels are merged into a hierarchical region tree using the data structure similar to MSER. The perceptual regions are selected according to objectness measure function (in Sect. 3.2).

3.1 Contour-Aware Superpixel (CAS)

Superpixel segmentation can be regarded as a cluster process. The most important matter is the distance measure between pixels and cluster center. In this part, we first describe the original SLIC, and then introduce our improved distance measure which enforces the superpixel segmentation close to boundary.

The only parameter original SLIC algorithm is the desired number of superpixel K. Given an image with N pixels, it initializes the seeds as cluster center by grid sample with equal interval $S = \sqrt{N/K}$, and further moves to lowest gradient position in 3×3 neighborhood. Next, each pixel is assigned to its nearest cluster center in local area. The distance measure D_s is explored in CIELAB color space and image plane xy space, hence each pixel p_i is presented as a 5-dimensional vector $\{l_i, a_i, b_i, x_i, y_i\}$. For a cluster c_k and a pixel p_i in $2S \times 2S$ local area centered at c_k, D_s is defined as follows:

$$
\begin{aligned}
d_{lab} &= \sqrt{(l_k - l_i)^2 + (a_k - a_i)^2 + (b_k - b_i)^2} \\
d_{xy} &= \sqrt{(x_k - x_i)^2 + (y_k - y_i)^2} \\
D_s &= d_{lab} + \frac{m}{S} d_{xy}
\end{aligned}
\tag{1}
$$

where m is assigned to 10 according to practical experiences to normalize the *lab* and *xy* distance.

However, the distance measure described above may produce superpixel overcrossing the boundary as Fig. 3(c). Neglecting boundary influence is primary

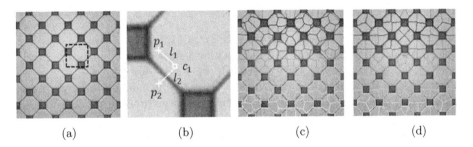

Fig. 3. (a) A grid-textured picture. (b) Patch marked by the dark box. Pixel p_1 and p_2 look very similar to center c_1, however p_2 is separated by contour in the middle, so the distance between p_2 and c_1 will be larger. (c) SLIC's result. Some superpixels cross the edges. (d) Improved result. All the segmentations adhere to contours

cause of this problem. As is shown in Fig. 3(b), though there exists a salient boundary between pixel $p2$ and cluster center c_1, the pixels p_1 and p_2 are still grouped into same cluster, since they are both similar to center c_1 as measured by Eq. (1). Thanks to recently proposed edge detector, the contour map of the image can be achieved efficiently. Let contour map $CM(p_i) \in [0\ 1]$ denote the probability of pixel p_i being on a boundary, $\phi(p_i, c_k)$ be the probability values on the path from pixel p_i to cluster center c_k in CM. We redefine distance measure D'_s as follows:

$$d_{lab|con} = \max\{d_{lab}, \omega \cdot \max\{\phi(p_i, c_k)\}\}$$
$$D'_s = d_{lab|con} + \frac{m}{S}d_{xy} \tag{2}$$

We choose $\omega = 45$ to make the contour influence comparable to color distance. With the Eq. (2), the superpixls generated by our method are more regular in shape and size, and adhere to boundary (see Fig. 3(d)) (Table 1).

3.2 Perceptual Regions Detection

Through the iterative cluster process above, the image are over-segmented into many superpixels. There still exists redundancy in these regions. In this part, we merge the superpixels into a component region tree by Union-Find operation [5]. Figure 4 shows an example of merge process. Small superpixels make up bigger component regions, and the root of component region tree is the whole image (see Fig. 4(b)). In each step, we choose two component regions which shares the smallest distance to merged. Considering the color and boundary information, the initial distance between component regions CRD is defined as distance between adjacent superpixels:

Table 1. The definitions used in Sect. 3.2

Image: I consist of a series of superpixels $SP = \{sp_1, sp_2, \cdots, sp_N\}$.
Pixel Adjacency Relation: A is a pixel level dual relation operator. Pixel $p = (x_1, y_1)$ and $q = (x_2, y_2)$ are adjacent if $|x_1 - x_2| + |y_1 - y_2| \leq 1$
Component Region: $CR = \{cr_1, cr_2, \cdots, cr_{2N-1}\}$ includes all the superpixel regions and newly generated regions during the merge process, which builds a tree structure called **Component Region Tree.**
Component Region Boundary: $\partial CR = \{q | q \notin CR \text{ and } \exists p \in CR, pAq\}$.
Component Region Boundary Recall: $BR = \sum_{x \in \partial CR} CM(x)$, it denotes how likely a component region has a close boundary.
Component Region Adjacency Relation: $\bar{A}$ is a component region level dual relation operator. Component region cr_1 and cr_2 are adjacent$(cr_1 \bar{A} cr_2)$ if $cr_1 \cap \partial CR \neq \phi$.
Component Region Shared Boundary: SB. For two adjacency component regions cr_1 and cr_2, $SB(cr_1, cr_2) = \{p \cup q | p \in cr_1, q \in cr_2 \text{ and } pAq\}$.

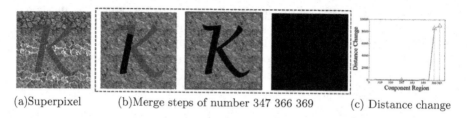

(a)Superpixel   (b)Merge steps of number 347 366 369   (c) Distance change

Fig. 4. An example of perceptual regions selection. Superpixels are merged into a component region tree and perceptual regions can be selected according to the distance change.

$$CRD(cr_i, cr_j) = \sqrt{(L(cr_i) - L(cr_j))^2 + (A(cr_i) - A(cr_j))^2 + (B(cr_i) - B(cr_j))^2}$$
$$+ T_1 \cdot SB(cr_i, cr_j) \tag{3}$$
$$L(cr_k) = \frac{\sum_{q \in cr_k} l_q}{|cr_k|}, \quad A(cr_k) = \frac{\sum_{q \in cr_k} a_q}{|cr_k|}, \quad B(cr_k) = \frac{\sum_{q \in cr_k} b_q}{|cr_k|}$$

The CRD updates during the merge process. When two component regions with different appearances are ready to merge, there will be a salient change in distances. In addition, component regions with high contour probability values on its boundary are more likely object. The distance change DC and boundary recall BR can be computed in Algorithm 1. The function **FindMin** returns the minimal value in CRD and the corresponding indexes of component regions, and **isClustered** is a boolean function denotes whether two component regions have already been merged together.

Considering the two factor, the score of component region cr_i is defined as:

$$score(cr_i) = DC(cr_i) \cdot BR(cr_i) \cdot sgn(DC(cr_i) - T_1) \cdot sgn(BR(cr_i) - T_2)$$
$$sgn(x) = \begin{cases} 1, x > 0 \\ 0, other \end{cases} \tag{4}$$

We choose $T_1 = 10, T_2 = 500, T_3 = 0.2$ in the following experiments. At last, our method outputs the component regions in descending order by non-zero $score$.

Algorithm 1. Superpixel merge process

Input: SP,CM,CRD
Output: DC,BR
Initialize: CR=SP,DC=0,BR(CR)=BR(SP)
for step $= 1$ to (N-1)
　　[minVal,m,n]=**FindMin**(CRD)
　　if isClustered(m,n)
　　　　DC(m)=minVal-DC(m),DC(n)=minVal-DC(n)
　　　　DC(N+step)=minVal
　　　　CR(N+step)=$cr_m \cup cr_n$
　　　　compute BR(N+step)
　　　　update(CRD)
　　end if
end for

4　Experiments

Our method consists of two stage: superpixel generation and perceptual region detection, so extensive experiments are performed to test the effectiveness of our method. At first, we compare CAS with four superpixel segmentation methods, namely GS [17], NC [2], TP [18], SLIC [9]. The evaluations are measured by boundary recall and under-segmentation error on the benchmark dataset BSD500 [11]. We also provide three version of CAS according to the different means of contour map generation. The fastest one is Gradient CAS (G-CAS) and the best quality is Structured Forests CAS (SF-CAS). The one between them is the Canny CAS (C-CAS). As a local feature detector, the most important matter is the repeatability under various image transformations. In Sect. 4.3, we compare our method with Harris-Affine [13], MSER [5], IBR [13], EBR [13].

　　Figure 5 provides a visual comparison of our improved method with original SLIC method, meanwhile two of perceptual regions detected by our methods are also shown. It can be seen that our method generate superpixels which are more close to the ground truth boundary, and visual-homogeneous regions with salient close boundary are well detected. We also provide comprehensive comparisons of various superpixel segmentation methods in Table 2.

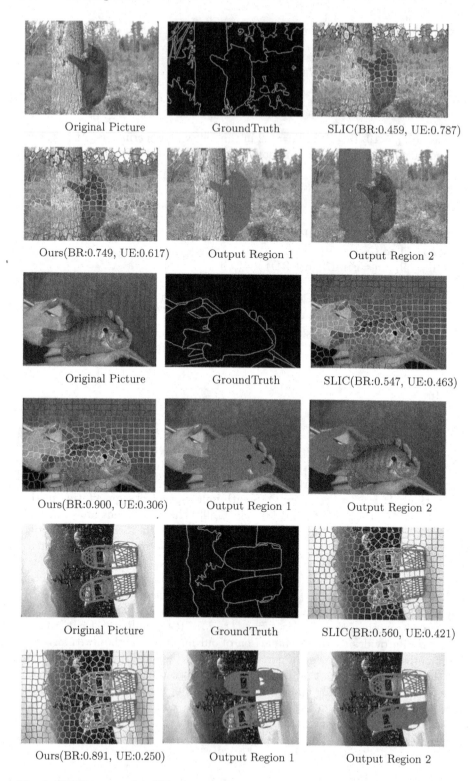

Fig. 5. SLIC's super pixle VS. Ours, and some perceptual regions generated by CAS. BR and UE are short for boundary recall and under-segment error.

4.1 Under-Segmentation Error

With respect to a known ground truth, under-segmentation error measures the error introduced by superpixel segmentation. If a certain superpixel region overlap the ground truth segmentation region more than a given threshold, the error increases according to its size. Given a known ground truth segments $g_1, g_2, \cdots, g_m$ and superpixel segments $s_1, s_2, \cdots, s_k$, the under-segmentation error UE can be computed as follows:

$$UE = \frac{1}{N}\left[\sum_{i=1}^{k} |T_i| \cdot |s_i| - N\right] \qquad (5)$$

$$T_i = \{g_j | g_j \cap s_i > B\}$$

where the operation $|\cdot|$ means the size of given set in pixels. T_i is the subset of ground truth segments with the intersection error more than a give B. In this experiment, we set $B = 0.05$ of superpixel segment s_i. Figure 6(a) give the result of under-segmentation error as a function of number of superpixel. Due to the help of contour information, our methods show obviously improved performance than original SLIC method.

4.2 Boundary Recall

A good superpixel segmentation should adhere to object boundaries. The standard boundary recall computes what the ration of ground truth boundares covered by at least one of superpixel boundaries. Figure 6(b) shows the results of boundary recall as a function of number of superpixel. Even the simple G-CAS method outperforms the SLIC with more than 40 % gain. Using the recently proposed Structure Forests edge detector, our method can reach 92.5 % boundary recall compared to SLIC's 0.68 at $k = 400$.

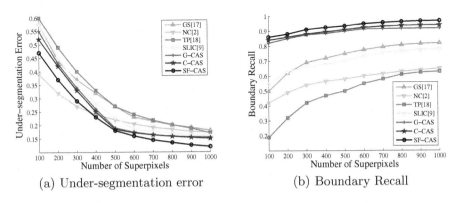

(a) Under-segmentation error   (b) Boundary Recall

Fig. 6. Under-segmentation error and boundary recall w.r.t. number of superpixels

Table 2. Performance of superpixel algorithms at $K = 400$

	GS [17]	NC [2]	TP [18]	SLIC [9]	G-CAS	C-CAS	SF-CAS
UE	0.32	0.24	0.33	0.28	0.26	0.25	**0.23**
BR	0.72	0.565	0.47	0.68	0.885	0.895	**0.925**
Speed	**0.12 s**	47 s	1.3 s	0.15 s	0.153 s	0.35 s	0.23 s

4.3 CAR Detector Repeatability

In order to test repeatability of the newly proposed method, we adopt the evaluation same to [13], Results are presented in terms of overlap area. Let R_a and R_b be two different ellipse-fitting regions being tested and H be the homography of transformation between the two regions. The overlap error ε is defined as:

$$\varepsilon = 1 - \frac{R_a \cap R_{H^T bH}}{R_a \cup R_{H^T bH}} \tag{6}$$

Region R_a and R_b are considered as corresponding if $\varepsilon < 40\,\%$.

Figure 7 gives the results of repeatability under rotation, scale change, light decreasing, blur transformations. Compared to other four methods, our methods exhibits relatively stable and high repeatability before the transformations applied are too large.

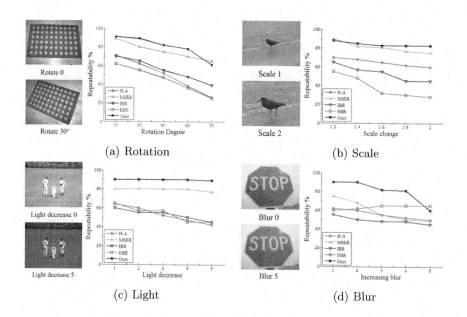

(a) Rotation   (b) Scale

(c) Light   (d) Blur

Fig. 7. Repeatability under different transforms

5   Conclusions

This paper has presented a novel efficient region detector called CAR. Inspired by the recent research result of general object proposal, we improve the SLIC superpixel method obviously. This method extracts less and meaningful regions in nature images, which can be easily employed in various computer vision applications, such as object recognition, object tracking and image retrieval.

Acknowledgments. This work was supported by National High Technology and Research Development Program of China (863 Program, 2014AA015202); the National Nature Science Foundation of China (61271428, 61273247, 61303159).

References

1. Shi, J., Malik, J.: Normalized cuts and image segmentation. TPAMI **22**(8), 888–905 (2000)
2. Mori, G.: Guiding model search using segmentation. In: ICCV, vol. 2, pp. 1417–1423. IEEE (2005)
3. Smith, S.M., Brady, J.M.: SUSANA new approach to low level image processing. IJCV **23**(1), 45–78 (1997)
4. Rosten, E., Drummond, T.W.: Machine learning for high-speed corner detection. In: Leonardis, A., Bischof, H., Pinz, A. (eds.) ECCV 2006, Part I. LNCS, vol. 3951, pp. 430–443. Springer, Heidelberg (2006)
5. Matas, J., Chum, O., Urban, M., et al.: Robust wide-baseline stereo from maximally stable extremal regions. Image Vis. Comput. **22**(10), 761–767 (2004)
6. Lowe, D.G.: Object recognition from local scale-invariant features. In: ICCV, vol.2, pp. 1150–1157 (1999)
7. Harris, C., Stephens, M.: A combined corner and edge detector. In: Alvey vision conference, vol.15, p. 50 (1988)
8. Martin, D.R., Fowlkes, C.C., Malik, J.: Learning to detect natural image boundaries using local brightness, color, and texture cues. TPAMI **26**(5), 530–549 (2004)
9. Achanta, R., Shaji, A., Smith, K., et al.: SLIC superpixels compared to state-of-the-art superpixel methods. TPAMI **34**(11), 2274–2282 (2012)
10. Comaniciu, D., Meer, P.: Mean shift: a robust approach toward feature space analysis. TPAMI **24**(5), 603–619 (2002)
11. Dollár, P., Zitnick, C.L.: Structured forests for fast edge detection. In: ICCV, pp. 1841–1848. IEEE (2013)
12. Bay, H., Ess, A., Tuytelaars, T., et al.: Speeded-up robust features (SURF). CVIU **110**(3), 346–359 (2008)
13. Mikolajczyk, K., Tuytelaars, T., Schmid, C., et al.: A comparison of affine region detectors. IJCV **65**(1–2), 43–72 (2005)
14. Zitnick, C.L., Dollár, P.: Edge boxes: locating object proposals from edges. In: Fleet, D., Pajdla, T., Schiele, B., Tuytelaars, T. (eds.) ECCV 2014, Part V. LNCS, vol. 8693, pp. 391–405. Springer, Heidelberg (2014)
15. Cheng, M.M., Zhang, Z., Lin, W.Y., et al.: BING: binarized normed gradients for objectness estimation at 300fps. In: CVPR, pp. 3286–3293 (2014)
16. Leung, T., Malik, J.: Contour continuity in region based image segmentation. In: Burkhardt, H.-J., Neumann, B. (eds.) ECCV 1998. LNCS, vol. 1406, pp. 544–559. Springer, Heidelberg (1998)

17. Felzenszwalb, P.F., Huttenlocher, D.P.: Efficient graph-based image segmentation. IJCV **59**(2), 167–181 (2004)
18. Levinshtein, A., Stere, A., Kutulakos, K.N., et al.: Turbopixels: fast superpixels using geometric flows. TPAMI **31**(12), 2290–2297 (2009)
19. Zitnick, C.L., Dollár, P.: Edge boxes: locating object proposals from edges. In: Fleet, D., Pajdla, T., Schiele, B., Tuytelaars, T. (eds.) ECCV 2014, Part V. LNCS, vol. 8693, pp. 391–405. Springer, Heidelberg (2014)
20. Vincent, L., Soille, P.: Watersheds in digital spaces: an efficient algorithm based on immersion simulations. TPAMI **13**(6), 583–598 (1991)
21. Zheng, Z., Wang, H., Teoh, E.K.: Analysis of gray level corner detection. Pattern Recogn. Lett. **20**(2), 149–162 (1999)

1D Barcode Region Detection Based on the Hough Transform and Support Vector Machine

Zhihui Wang[2], Ai Chen[1], Jianjun Li[1(✉)], Ye Yao[1],
and Zhongxuan Luo[2]

[1] School of Computer Science and Technology,
Hangzhou Dianzi University, Hangzhou 310018, China
jianjun.li@hdu.edu.cn
[2] School of Software Technology,
Dalian University of Technology, Dalian 116620, China
wangzhihui1017@gmail.com

Abstract. The barcode is widely used in logistics, identification, and other applications. Most of the research and applications now focus on how to decode the barcode. However, in a complex situation, the barcode is difficult to locate accurately. This paper presents an algorithm that can effectively detect the locations of multiple barcodes. First, image texture features are extracted by combining the local binary pattern (LBP) and gray histogram, and then a machine learning algorithm is applied to create a classifier by using the support vector machine (SVM) to extract and train the positive and negative samples. Second, the Hough transform is applied to the input image to achieve the angle invariable. Finally, our proposed method has been evaluated by the WWU Muenstar Barcode Database and the experimental results show that the proposed result has higher performance than other methods.

Keywords: Barcode · Local binary pattern · Support vector machine · Hough transform · Region detection

1 Introduction

In recent years, barcode technology has been widely used in our daily life. Indeed, barcode detection is one of today's most important technologies. By using computer image processing technology, such as image enhancement, segmentation, and recognition, we can achieve a fully automatic procedure for barcode detection. Many barcode detection algorithms have been proposed in previous research. However, these algorithms have advantages and disadvantages.

A projection method is proposed to extract the barcode region in [1]. First, the barcode image edge is extracted by the edge detector. Second, the processed image is projected in the horizontal and vertical direction. This algorithm is based on the assumption that the barcode edge is a complicating factor and is less dense than other areas of the barcode and that other areas have relatively smooth gray edge pixels. This method can achieve good performance in a simple background image. However, such

© Springer International Publishing Switzerland 2016
Q. Tian et al. (Eds.): MMM 2016, Part II, LNCS 9517, pp. 79–90, 2016.
DOI: 10.1007/978-3-319-27674-8_8

performance is difficult to achieve in a real application with a complex background. Reference [2] proposed a new algorithm. The image is first divided into 32×32 patches and then the angle of the parallel lines in each patch is determined and combined with the angle close to the block to get the bar code area. However, the algorithm cannot ensure the segmentation of various small pieces that contain only part of the barcode, which leads to angle miscalculation and incomplete extraction of the barcode from some images. Reference [3] proposed a fast barcode location method based on the discrete cosine transform (DCT) domain. First, the DCT is used to distinguish the barcode area and the other regions. Then the barcode angle and position are detected in the original image using the extracted region features. However, this algorithm is not robust because the weight matrix coefficients are not adaptive. Telkin and Coughlan [4] proposed a new algorithm, which is divided into two steps. First, the noise of the image is reduced by Gauss smoothing, the gradient amplitude of the image is calculated, and the gradient amplitude forms the edge enhanced image. The second step is to determine the barcode region by calculating the four directions of the pixel with the gradient value higher than a certain threshold. However, this algorithm is not sufficient for different kinds of barcodes that exist at the same time. Reference [5] proposed a method based on edge detection and machine learning. The Hough transform and neural network are obtained to determine the region of the barcode line. This algorithm is adequate for detection of the general image, but it has a poor effect on an image with dense text, and the result of the detection depends on the results of the Hough transform.

In this paper, we propose a method for barcode detection of camera-captured images, which includes Hough transform, local binary pattern (LBP) and color feature extraction, and support vector machine (SVM) classification. In our method, the Hough transform is used for angle correction of barcode images, and the LBP and color feature are extracted for machine learning model training and the SVM classifier. After angle correction by the Hough transform, the original image with barcode blocks is detected by the classifier. Then the blocks are merged with the extracted adjacent block region, and the isolated region is excluded. The obtained results prove that our algorithm can effectively detect the barcode region in a complex background.

The remainder of the paper is organized as follows: The optimizing algorithm is addressed in Sect. 2. The detailed simulation and results analysis are shown in Sect. 3. The conclusion is provided in the last section.

2 Proposed Method

2.1 Barcode Detection

The frame of the proposed method for detecting barcodes is shown in Fig. 1. To obtain the SVM classifier, we extract the color feature and LBP feature from image blocks and then train the features to achieve the SVM model. In a complex background, the direction of an image has an impact on detecting the barcode. First, the angle of an input image is corrected by the two-dimension Hough transform. Second, the sub-block of the corrected image is verified with the mode of the trained SVM. In the end, the classification results of each block are scanned, all isolated blocks are removed, and the neighbored area is merged to achieve the region of the barcode. The positive and negative samples (14 × 14 pixels) shown in Figs. 2 and 3 serve to extract image features by LBP and thus obtain the trained mode of SVM.

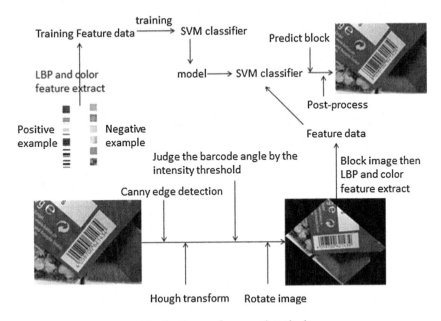

Fig. 1. Frame of proposed method

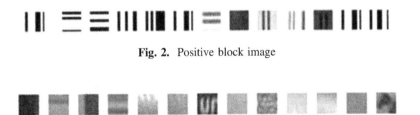

Fig. 2. Positive block image

Fig. 3. Negative block image

Local Binary Pattern Feature. The features of the positive and negative example are extracted by the LBP method. The features include a 4 × 4 × 58 dimension vector that becomes the training database. The LBP feature is defined by the local texture, which has the characteristic of low computing complexity, and is not affected by the degree of image shading. The LBP has a good effect on texture feature extraction. Indeed, the LBP has many applications in the field of machine vision. The local texture feature of a gray image is described by a joint distribution of gray value of pixels within the region, as follows:

$$T = t(g_c, g_0, \ldots, g_{p-1}) \qquad (1)$$

In Eq. 1, T represents the joint probability distribution, and g represents the gray value of the central pixel in the local area (p = 0, ..., P − 1); the gray value of the P pixel points and the image pixels are evenly distributed on the circle with the radius of the center of R. The value of P and R determines the size and range of the local field distribution, as shown in Fig. 4.

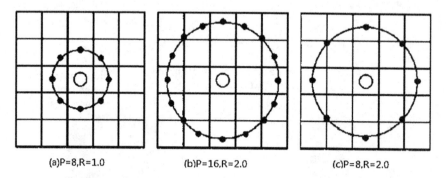

(a)P=8,R=1.0   (b)P=16,R=2.0   (c)P=8,R=2.0

Fig. 4. Circularly symmetric regions have differences (P, R)

Reference [6] introduced the derived formula:

$$\text{LBP}_{P,R} = \sum_{P=0}^{P-1} s(g_p - g_c)2^P \qquad (2)$$

Defined here:

$$S(x) = \{ \begin{smallmatrix} 1, x \geq 0 \\ 0, x < 0 \end{smallmatrix} \qquad (3)$$

By Eq. 2, the two values of the local texture model described by the LBP operator are obtained by comparing the gray values of the field pixels with the central pixel. The LBP operator is defined in the circular location and can be better defined with the invariance of the texture feature operator, in which the pixel values in the field are obtained by linear interpolation. In describing the features, the LBP has better performance in the details.

Color Feature Extraction. Since the majority of the barcode is composed of black and white lines, the color feature is extremely good for the identification of a barcode; the 256 dimension gray scale histogram method based on the statistical gray scale is used to extract the color feature of the block example. The statistics of the number of blocks in the same pixel, divided by the total number of block pixels, yields the percentage of different grays in the block, and then we obtain the 256 dimensional feature. The histogram of the horizontal coordinates is the gray t, and the vertical coordinate is the proportion of the gray scale, the gray value of the probability density of P(r), and the following relationship is established:

$$P(r) = \lim_{\Delta r \to 0} \frac{A(r + \Delta r) - A(r)}{\Delta r} \tag{4}$$

$$\int_{r\,\min}^{r\,\max} P(r)dr = 1 \tag{5}$$

In Eqs. 4 and 5, A is the area of the pixel block and r is gray value.

2.2 Support Vector Machine

The LBP feature and color feature set is trained by the SVM classifier, and then we get the model which can be used to predict the block image. The implementation process of the SVM classifier is described below.

SVM is developed by the optimal classification of linear separable conditions, and the basic idea of SVM can be illustrated by the two dimensional cases. In Fig. 5, the solid points and the hollow points are represented by the positive sample and negative sample, and then the line for classification; the other subscript number line of H represents a straight line with the nearest sample and parallel to the classification line. The distance between the two lines of H_1 and H_2 is called the gap width (margin). The significance of the optimal classification line is that the classification line can divide examples into two categories and let the margin be maximum.

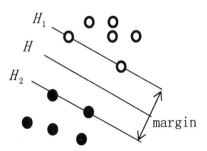

Fig. 5. Optimal classification surface

Reference [7] provides a detailed introduction. The above problems are summarized as quadratic function problems with inequality constraints. The classification function is shown in Eq. 6.

$$f(x) = \text{sgn}\{(w^*x) + b^*\} = \text{sgn}\{\sum_{i=1}^{n} a_i^* y_i(x_i x) + b^*\} \tag{6}$$

In this formula, the summations are adequate for the support vector (nonsupport vector a_i is zero). b^* is the classification threshold, which can be obtained with any support vector. The basic idea of SVM can be described as follows: First, the input space is transformed into a high dimensional space by nonlinear transformation, and then the optimal classification surface is obtained in the new space. The optimal classification surface can be used to classify the sample.

2.3   Hough Transform

Image rectifications are based on two-dimensional Hough transform. In this section, we use the Canny edge detector algorithm to get the edge map with the implementation offered by OpenCV library [8]. Then we use the Hough transform to get the angle of the barcode line. We then rotate image B to 0° or 90°, which facilitates detecting the barcode.

Angle Detection Using Hough Transform. The classical Hough transform [10] is a method for regular feature extraction; this method is used in image processing and computer vision to detect regular shapes. The Hough transform for line detection identifies a set of linear shapes L in a given image B by using a voting procedure method. The equation of a generic line $1 \in L$ in B can be defined as follows:

$$\sigma = x * \cos \theta + y * \sin \theta \tag{7}$$

where $\sigma \geq 0$, the distance of 1 from the origin, is image B and $\theta \in [0, 2\pi)$ is the angle of 1.

In this paper, we use the 2 dimensional Hough transform [5] to detect the angle of the barcode. Let the 2 dimensional Hough transform space H be the (σ, θ) plane; for any point, (x_1, y_1) can be corresponded to a sinusoid in H. If two points (x_1, y_1), (x_2, y_2) belong to the identical line l, their homologous sinusoid will be defined in a point $\sigma_1, \theta_1 \in H$. We define D_H as the accumulator matrix for the 2 dimensional Hough space $H = (\sigma, \theta)$ in image B; every element $b \in D_H$ in position (σ_b, θ_b) denotes a potential line in B having same angle θ_b and distance of axes σ_b. The goal of the angle detection phase is to change the angle of image B, and then the barcode in image B can be detected easily.

Because the density characteristics of the Hough transform of the barcode constitute a special character, we adopted the intensity characteristic to distinguish the location, which is barcode location or not. As shown in Fig. 6, the barcode line has a difference σ, so we let the element D_H be a regular grid of cells C, and every point $D_H(\sigma, \theta)$ refers to the intensity of image B. Let $C_i(x, y)$ be the value of the element of C_i in position (x, y) with $0 \leq x < vx$ and $0 \leq y < vy$ (every cell $C_i \in C$); the value is the sum of $D_H(x, y)$, as $\sum_{x=0, y=0}^{x<vx, y<vy} D_H(x, y)$. To get the angle of the barcode line in image B, we use the histogram statistics method (implemented by OpenCV) [8] to obtain the intensity Threshold T. Then we can get the map, which can describe the angle θ of the barcode line.

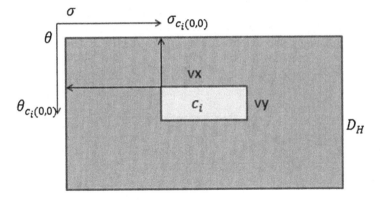

Fig. 6. Angle detection using Hough transform

$$\theta = \theta_{c_i(x,y)} \quad \square \, if \quad C_i(x,y) > T \tag{8}$$

Figure 6 denotes how the regular grid of cells C is superimposed over the Hough accumulator matrix D_H and how the elements are $C_i(x,y)$ mapped into the (σ, θ) plane.

Rotating Images. We determine the angle of the line of the barcode, then change the angle of image B to detect the barcode location. We just need to set the barcode angle, such as 0°, 90°, which is good for our detection. Our method of rotating the image is the linear spatial filtering of the OpenCV library [8]. Image B rotates to a 0° or 90° image, which will help us more easily detect the barcode. The resulting images of rotation are shown in Fig. 7.

Fig. 7. After detecting the angle of the barcode, the rotated image by the angle; the (a) row is the original image and the (b) row is the rotated image

Our method to detect the angle of the image is almost exactly the same. The result of our method is shown in Fig. 8.

Figure 8 shows that our method results in exactly the same ±5° deviation of angle as the method of [5]. Our method does not need to use the neural network, so we can cut the time for training and detecting. Thus, our method can detect the angle more quickly.

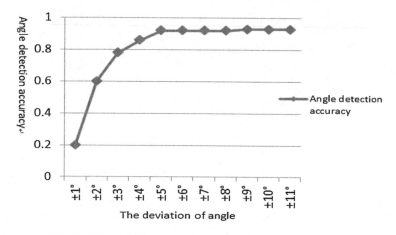

Fig. 8. The angle detection accuracy in the deviation of angle

2.4 Using the SVM Classifier to Judge Pieces of the Image

As shown in Fig. 9, after rotating image B, we divide image B into small blocks (every block is 14 × 14 pixels in this paper). Then we extract the LBP feature and color feature for each block and combine the feature data of the block. Using the SVM classifier (trained in Sect. 2.2), we judge each block of image B.

Fig. 9. Block image with a complex background

In this experiment, the size of positive and negative samples is 14 × 14 pixels. The positive samples are about 1000 blocks and the negative samples are about 2000 blocks. The positive samples contain a barcode in the test image database. The negative samples contain other elements without a barcode. Every block uses Eq. (2) to detect the LBP feature (the argument for LBP feature extraction is P = 8, R = 1.0). The color feature for every block is extracted by Eq. (4). After extracting the feature, every block has a $4 \times 4 \times 58$ dimension LBP feature vector and a 256 dimension color feature vector; we mix these features to achieve a 1184 dimension feature and enter the 1184 dimension feature in the SVM classifier; our method adopts C-SVM; the argument in SVM has the penalty factor C and RBF kernel function (coefficient k, k is the feature dimension).

In the results of the classification, '1' represents the location of the barcode; other locations are represented by '0'. The classified result is shown in Fig. 10.

$$
\begin{bmatrix}
0 & 1 & 0 & 0 & 0 & 0 & 0 & 0 \\
0 & 0 & 0 & 0 & 1 & 0 & 0 & 0 \\
0 & 0 & 0 & 0 & 0 & 0 & 0 & 0 \\
1 & 1 & 1 & 1 & 1 & 0 & 0 & 0 \\
1 & 0 & 1 & 1 & 1 & 0 & 0 & 0 \\
1 & 1 & 1 & 1 & 1 & 0 & 0 & 0 \\
1 & 1 & 1 & 1 & 1 & 0 & 1 & 0 \\
0 & 0 & 0 & 0 & 0 & 1 & 0 & 0
\end{bmatrix}
$$

Fig. 10. Classified result by SVM

2.5 Post-processing

The result distinguished by the SVM classifier is not 100 % complete; it is necessary to provide the corresponding post-processing. The post-processing steps are as follows:

1. Scan the classification results of each block in turn. Remove all isolated blocks of 1 and 0.
2. For every block, if more than half the results differ from the 4 neighborhood region, this should change the overall result.
3. Merge the unicom area and remove the small area unicom region.
4. The rectangle marked as the unicom area yields the location of the barcode.

3 Experiments and Results Analysis

3.1 Datasets

In this section, we provide the datasets used to evaluate the performance of our proposed model. In our experiments, we employ a standard 1D barcode dataset, the WWU Muenster Barcode Database [9], and the size of the image is 640 × 480 pixels. Since our method involves an SVM algorithm, we split the dataset image into training and test sets: We randomly select 50 % of the dataset's images as the training set, then artificially extract the barcode as the positive example and extract another location as the negative example; the remaining 50 % is used as the test set. The following paragraph introduces the WWU Muenster Barcode Database [9]. WWU has 1055 1D barcode images, acquired using a Nokia N95 mobile phone. All images in the database contain a non-blurred EAN or UPC-A barcode.

3.2 Result

In this section, we compare the results of our proposed method with other methods. We compare the results of before rotating the image in Fig. 11. Results for conditions to detect barcodes are shown in Fig. 12.

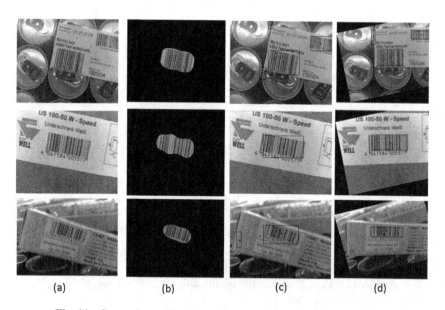

(a)   (b)   (c)   (d)

Fig. 11. Comparison of images without rotation and our rotated result

Figure 11 shows the result of detecting the barcode in the image with rotation. The (a) column is the original picture, the (b) column is ref. [5], the (c) column is our method result without rotation, and the (d) column is our result with rotation.

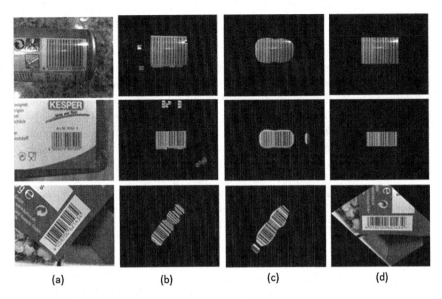

| (a) | (b) | (c) | (d) |

Fig. 12. Comparison of experimental results

Figure 12 shows the result of detecting the barcode in a complex image. The first row is the big barcode image, the second row is the small barcode image, and the third row is the rotated image. The (a) column is the original picture, the (b) column is ref. [4], the (c) column is ref. [5], and the (d) column is our method result.

As shown in Table 1, we compare the algorithm performance in the WWU Muenster Barcode Database. Our algorithm can detect the barcode in complex conditions and reduce the impact of complex backgrounds on detection results.

Table 1. Comparison of the experimental results

Dataset	Barcode detection algorithm		
	Reference [4]	Reference [5]	Our method
Muenster database [9]	90.13 %	93.23 %	95.34 %

The detection accuracy of our method is 95.34 % in the WWU Muenster Barcode Database. With a complex background, our method can detect the barcode more accurately. With the method of refs. [4, 5], the performance depends on the results of the Hough transform. In the presence of a more text intensive area, the performance of those methods will be poor. In our method, we use the LBP feature to describe the image features, which can distinguish the interference region. Our method has better robustness and can detect barcodes more accurately than the method of refs. [4, 5], which is good for barcode recognition.

4 Conclusion

In this paper, we have presented a simple method to detect the region of the barcode and proved the effectiveness of the proposed approach by using the available WWU barcode dataset. The proposed method uses LBP image feature exaction and SVM classification to locate the regions of barcodes, and the angle invariable is obtained by the Hough transform. The experimental results show that our method is precisely able to detect the regions of multiple barcodes in an image with a complex background and the angle invariable of a barcode is also verified. The accuracy of detection using our approach is higher than for approaches used in earlier research.

Acknowledgment. The work was supported by NSFC (No. 61100100), the Zhejiang Provincial Key Innovation Team on 3D Technology (No. 2011R50009).

References

1. Chen, Y., Shi, P.: Image recognition in 2D barcode. Meas. Control Technol. **25**(12), 17–19 (2006)
2. Douglas, C., Florian, H.: Locating and decoding EAN-13 barcodes from images captured by digital cameras. In: Fifth International Conference on Information, Communications and Signal Processing (ICICS 2005), Bangkok, Thailand, pp. 1556–1560 (2005)
3. Alexander, T., Douglas, C.: Locating 1-D bar codes in DCT-domain. In: IEEE International Conference on Acoustics, Speech, and Signal Processing (ICASSP 2006), Toulouse, France, pp. 14–19 (2006)
4. Tekin, E., Coughlan, J.M.: An algorithm enabling blind users to find and read barcodes. In: 2009 Workshop on Applications of Computer Vision (WACV), pp. 1–8 (2009)
5. Zamberletti, A., Gallo, I., Albertini, S., et al.: Neural 1D barcode detection using the Hough transform. IMT **10**, 157–165 (2015)
6. Ojala, T., Pietikainen, M., Harwood, D.: A comparative study of texture measures with classification based on feature distributions. Pattern Recognit. **29**, 51–59 (1996)
7. Vapanik, V.: Statistical Learning Theory. Wiley, New York (1998)
8. Bradski, G.: The OpenCV library. Dr. Dobb's J. Softw. Tools **25**(11), 120–126 (2000)
9. Wachenfeld, S., Terlunen, S., Jiang, X.: Robust 1-D barcode recognition on camera phones and mobile product information display. In: Jiang, X., Ma, M.Y., Chen, C.W. (eds.) MMP 2008. LNCS, vol. 5960, pp. 53–69. Springer, Heidelberg (2010)
10. Zamberletti, A., Gallo, I., Carullo, M., Binaghi, E.: Neural image restoration for decoding 1-D barcodes using common camera phones. In: Proceedings of International Conference on Comptuer Vision Theory and Applications (VISAPP 2010), pp. 5–11 (2010)
11. Arivazhagan, S., Ganesan, L., Kunmar, T.G.S.: Texture classification using curvelet statistical and co-occurrence features. In: Proceedings of 18th International Conference on Pattern Recognition, Hong Kong, pp. 938–941 (2006)
12. Manjunath, B.S., Chellappa, R.: Unsupervised texture segmentation using Markov random field models. IEEE Trans. Pattern Anal. Mach. Learn. **13**(5), 478–482 (1991)
13. Katona, M., Nyul, L.G.: A novel method for accurate and efficient barcode detection with morphological operations. In: 2012 Eighth International Conference on IEEE Signal Image Technology and Internet Based Systems (SITIS), pp. 307–314 (2012)

Special Session Papers

Client-Driven Strategy
of Large-Scale Scene Streaming

Laixiang Wen, Ning Xie, and Jinyuan Jia$^{(\boxtimes)}$

Tongji University, Shanghai 201804, China
{2010wenlaixiang,ningxie,jyjia}@tongji.edu.cn

Abstract. Different from the strategy of virtual scene in the stand-alone application, web version may have larger scale scene and more users online at same time. However, because of the network delay (latency) and limitation of computational power of the servers, it causes that users unable to interactively access the virtual scene fluently. Meanwhile, the large scale virtual scene cannot be successfully loaded in the entry-level client machine. In this paper, we propose a client-driven strategy of scene streaming in order to solve large-scale data transmission. The experiment demonstrates that our method enables clients to enter the scene roaming and reduce the server network load. Meanwhile, it also adapts different network architectures.

Keywords: Lightweight progressive meshes (LPM) · Web3D · Scene streaming and dynamic double layer AOI (D-DLAOI)

1 Introduction

Large-scale scene transmission is one of the key parts in web-based 3D computer graphics, where 3D scene data is downloaded from servers to clients, which charge for rendering and roaming. It is widely used in many multimedia applications, such as 3D video games and virtual reality projects. Recently, the data in video games is increasing dramatically. For example, by 2014, the most famous online virtual world game, *Second Life*, has approximately 1 million users and over 270 TB data. It is impossible to publish this game through DVD disc. Also, personal computers or smart phones cannot download entire virtual world data and run the application once.

In common, these huge amount of data is usually stored on server. Once connecting established by clients, the required parts will be sent to the clients. Thus, object selection and transmission are essential operations to guarantee users' interactive operation on the virtual world. Rest of data will be transmitted by users' requests. Many client/server (C/S) based distributed virtual environment (DVE) systems work on tasks such as large-scale scene organization, priority of data transmission, adaptation of client hardware and Internet bandwidth [1,2]. Data selection and transmission towards area of interest (AOI) [3] is not efficient, because users' trajectories of the scene roaming are rather irregular.

© Springer International Publishing Switzerland 2016
Q. Tian et al. (Eds.): MMM 2016, Part II, LNCS 9517, pp. 93–103, 2016.
DOI: 10.1007/978-3-319-27674-8_9

Server carefully processes all requests by clients so as to make load heavily. So, it is not suitable for mobile devices. P2P technique is applied to DVE in order to reduce the load in low speed of network [4,5]. It focused on discovering the optimal download source. However, scene organization graph, priority of data transmission and configuration of terminal devices have not considered yet.

In this paper, we present a 3D scene data transmission technique to stream large scale geometry scene driven by configuration of client devices in order to realize scene roaming in real time [6]. A self-adaptive method is proposed to generate hierarchy grid structure including hierarchy scene structure and its description file. The scene organization enables to speed up the procedure of sub-scene and objects selection. The description file ensures that the selection operations can be successfully completed at the client side. Meanwhile, it guarantees the extensibility of the server so as to be a pure data server without any client information maintaining. In order to further speed up object culling, we propose a *dynamic double layer AOI (D-DLAOI)* to determine the download priority of objects. D-DLAOI has double layers and four regions. For each region, it maintains a priority queue. According to density of object in D-DLAOI, the size of the inner layer is adjusted dynamically so as to control the number of request and maintain the sent request to be up-to-date. It makes sure the most needed objects will be transmitted at first (the current visible objects) and culls the most potential objects (the potential visible objects). Finally, lightweight progressive mesh (LPM) [7] is applied to achieve the stream transmission without the entire scene file under the optimal resolution.

We demonstrate our idea using various large scale scene examples, and present the comparison to evaluate the usability and efficiency of our tool.

2   Related Works

Communication Framework. Communication framework defines how computers organize together as a network. The common schemes are Client/Server (C/S), Peer-to-Peer (P2P) and Hybrid frameworks [8]. C/S is adopted by many multimedia 3D applications. For example, massively multiplayer online games (MMOGs) are built on C/S such as *ActiveWorlds*, *Second Life*, and *World of Warcraft*. This structure ensures data security and consistency. P2P media streaming has been significant progress in recent years [9]. Technically, P2P is different from C/S from the aspects of access pattern and task distribution. But, P2P data transmission is same with C/S framework, although for only small amount of data. To improve data persistence, hybrid framework is proposed to add server into P2P framework [10].

Grid Generation in 3D VE. Grid is one of the important structure for many 3D computer graphics problems such as ray tracing, collision detection and path finding. Comparing with Octree and kd-tree, 3D uniform grid is suitable for the dynamic scene. However, the computational efficiency highly depends on the grid creation strategy. Currently, the *cost evaluation* method and the *empirical model* method are two most efficient strategies. The cost evaluation method needs the classification step in advance under certain assumption [11]. The empirical model is much easier because it is based on the experience and statistics [12,13].

3D Visual Component Selection and Streaming. Visual scene streaming can be summarized into *3D geometry scene streaming (GSS)* and *image-based scene streaming (ISS)*. ISS assumes that clients do not have enough ability for 3D scene rendering. Instead, server renders and sends the rendered image frames to clients. However, when the number of users increases, processing becomes immeasurable that server has to render many different copies of contents according to many different clients' viewpoints. Applications are such as *CloudGame* [14] and *GamingAnyWhere* [15].

Alternatively, in GSS, server sends 3D geometries instead of rendered videos to clients, who render the received primitives as a visual scene. It is widely used in DVE system, where only currently visible geometries will be transmitted to clients. Wang et al. [16] applied the area of interest (AOI) to judge object visibility in local region. Constructive solid geometry (CSG) is used to construct the objects in *SecondLife*. When geometry component are contained by several visible objects, it will be sent at first as priority. By this way, the procedure of scene construction can be accelerated. Furthermore, in order to balance transmission speed and 3D model resolution, Level of detail (LOD) technique is introduced to generate 3D objects under multiple resolutions [8]. Models under low resolution will be transmitted first so as to ensure the speed of scene construction. It is simple and straightforward. However, LOD method increases Internet bandwidth consumption, because of redundancy that same object have multiple resolutions. Progressive mesh (PM) [17] is introduced to avoid this problem in P2P transmission.

The challenge is how to enhance the user experience on the visible interaction. All techniques endeavor to reduce the latency during 3D scene streaming including rapid object culling, transmission priority, resolution controlling are needed to be investigated.

3   Overview

This paper introduces a 3D scene transmission pipeline to stream large scale virtual scene which is adaptive to the client configuration. Our approach consists of three main steps:

- **Multiple-resolution 3D space adaptive grid creation.** According to the density of the objects in the scene, multi-level grids are recursively calculated in order to generate the hierarchical structure. It helps to speed up the process of object culling. Meanwhile, scene description file is also generated for client-oriented data request.
- **Visible scene determination.** D-DLAOI is proposed to determine the download priority of objects. D-DLAOI has double layers (the internal layer and the external layer). In order to avoid excessive requests by clients, the internal layer can be real-time justified according to the density of the objects.

D-DLAOI covers the user visible or potential visible region. Only objects in D-DLAOI are the candidates for transiting. According to the user's view frustum, D-DLAOI are further divided into four regions with different predefined priorities including a visible region, two potential visible regions, and a predownload region. In a particular region, the priority of object is determined by the distance to the viewpoint, and the included angle with the sight line.

– **Object resolution optimization.** After applying LPM for streaming, we need to calculate the optimal resolution for visible objects based on human visual mechanism, the configuration of the clients and network environment.

By this pipeline, server do not have to maintain any information on the client devices anymore and focuses on the operation of VE data. Hence, coupling degree is significantly reduced at server.

4   Multiple-resolution 3D Space Adaptive Grid Creation

We adaptively judge the mesh resolution from two aspects including the size of the scene and the density of objects in the scene.

In this paper, we give the method to calculate the resolution of the 2D mesh in one layer as below:

$$R_i = \lambda L_i \sqrt{\frac{N}{A}}, \quad (i \in \{X, Y\}, 0 < \lambda \leq 1), \tag{1}$$

where R_i is the resolution in the projected 2D for certain axis, i (X or Y in our system); λ is a parameter; N is the number of the objects in the given scene; A is the projected area of the bounding box of 3D grid; L_i is the length of the certain corresponding axis. In practice, bounding box of scene can be any rectangular parallelepiped but not just the cube (or regular hexahedron) with equal edge lengths. In worse case, the bad resolution will result in lower computational efficiency. In order to maintain the global optimal grid division, we need to justify the grid resolution in certain axis. In detail, once some grid resolution is less than 1, we set up it to 1, while the other is set to $\lambda\sqrt{N}$.

To further accelerate the process of object culling, we propose a hierarchical grid structure. First, we generate a one layer uniform grid for a scene who contains N objects. Then, we traverse every cell of grid to check whether it can be further divided. If so, it will be divided. This process will go on recursively until the condition is not satisfied, where the number of object in the certain cell is less than the threshold.

In practice, a large scale scene is organized by a series of sub scenes. It is rare to generate a global uniform grid. Especially in MMOGs, the grids are usually designed by scene designers. Alternatively, the algorithm processes the grid generation for each sub-scene.

Then, lightweight process is necessary to remove redundant objects in the scene. We apply LPM [7] on each non-redundant object afterwards. The hierarchical information of a scene is illustrated as Fig. 1. In this structure, there are three kinds of nodes: root node, intermediate node and leaf node.

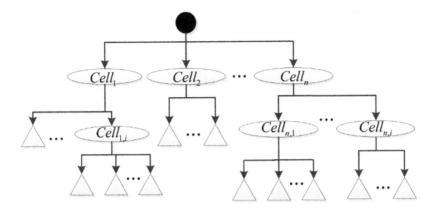

Fig. 1. Scene hierarchical structure.

- **Root node.** It is also called *scene root*. It has the highest level information including the sub-scene division, the configuration of each sub-scene (position and size), and ID of sub-scenes.
- **Intermediate node.** It is the node for a sub-scene. There are object data (position, size, pose and its ID) and the information of nested sub-scene.
- **Leaf node.** It is the object node including ID of objects and its configuration. Note that LPM can be obtained using its object ID.

Note that it has three significant characters including hierarchical structure, lightweight processing and streaming transmission. Nodes contain less amount of data, because it only has the scene descriptor file, without the object data. We later send these compact files to clients in order to assemble scene.

5   Scene Streaming Assemble Strategy

Once receiving the descriptor files for multi-levels, client enables to judge the visible scene, the priority of the object transmission, and the resolution of LOD of objects. All these are called *scene streaming assemble*. The aim is to accelerate the network transmission but not scene rendering. Therefore, AOI is applied rather than object culling, back culling, and occlusion culling.

5.1   Dynamic Double Layer AOI (D-DLAOI)

For online game, *area of interest (AOI)* is a technique to reduce communication burden even though most of the cases all-to-all communication is required within an AOI [18]. Say that AOI is the area that surrounds the player or avatar in the center. In very large scene roaming, visible objects are only who are covered by AOI in order to reduce real time computational cost of object visibility. The classic AOI is in disc shape, while the human cones of vision is sector. *Double layer AOI (DLAOI)* is then proposed to further simulate on this issue [19]. The advantage of DLAOI is to preload data in advance for rendering in order to accelerate the rendering. However, it still can not solve the issue on dense scene where contains large numbers of objects.

We propose a dynamic DLAOI (D-DLAOI) to dynamically change the size of the inner area according to the density of the objects as illustrated in Fig. 2. Similar to DLAOI, our D-DLAOI also has four partitions. As illustrated in Fig. 2(a), Q_{1a} is the *visible area*. Q_{1b} and Q_{2a} are also *potential areas*, which means that the inside objects can be observed in the next time slot. Q_{2b} is the pre-loaded area which will be loaded once no user's action exists after other three areas have been loaded. Area priority order is $Q_{1a} > Q_{1b} > Q_{2a} > Q_{2b}$. In client, once scene description file is received, the hierarchical structure will be created as illustrated in Fig. 2(b). In practice, the density of objects in the scene is not uniform. To further improve performance, we can also dynamically justify the size of Q_{1b}. In detail, we reduce the size of Q_{1b}, when $p < \varepsilon_{sh}$. Alternatively, the size of Q_{1b}, when $p > \varepsilon_{in}$, where

$$p = 1 - \frac{N_{\text{inner}}}{N_{\text{AOI}}};$$ (2)

N_{inner} is the number of the objects in the inner region of D-DLAOI; N_{AOI} is the number of the objects in AOI. ε_{sh} and ε_{in} are thresholds ($\varepsilon_{sh} = 0.01, \varepsilon_{in} = 0.99$ in our experiment).

5.2   Object Priority Determination and LOD Resolution

For designing the priority determination and LOD resolution, we follow the two common assumptions of the human vision below:

– For the viewpoint, closer to the target object;
– For the sight line, less offset from the sight line.

Therefore, it may not be helpful to improve the visual impact by loading the entire model data, whereas it will increase the amount of data transmission. By

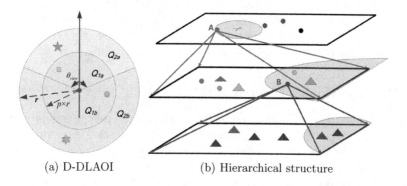

(a) D-DLAOI   (b) Hierarchical structure

Fig. 2. D-DLAOI.

doing so, users have to wait longer for scene roaming. LPM [7] is a powerful tool to distribute the optimal amount of 3D model data, p^* according to needs of LOD as below:

$$p^* = p_{\mathrm{m}}(\gamma b + \delta r + \alpha(1 - a) + \beta(1 - d)), \tag{3}$$

where p_{m} is the maximum number of grid; $b, r, a,$ and d are the normalized parameters. In detail, b means the network bandwidth; r means the rendering capabilities of the client; a is the offset angle from sight line; d is the distance from the view point to the target object. α, β, γ and δ are the coefficients ($\alpha + \beta + \gamma + \delta = 1$).

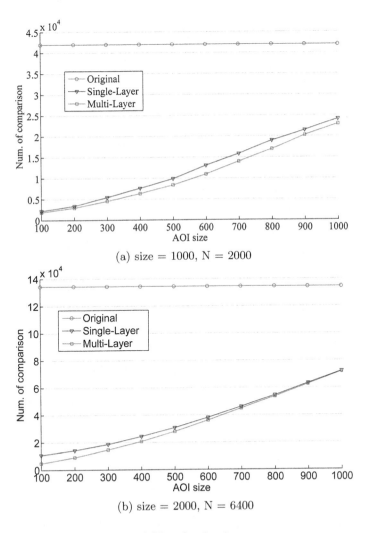

(a) size = 1000, N = 2000

(b) size = 2000, N = 6400

Fig. 3. D-DLAOI evaluation in two scenes.

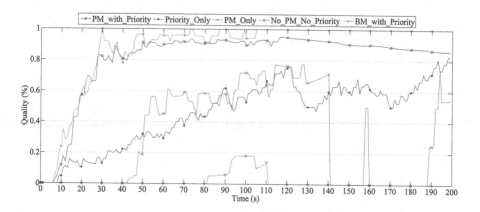

Fig. 4. The comparison on the human visual impact.

6   Experimental Results

In this section, we experimentally demonstrate the usefulness of our proposed method.

First, we evaluate our proposed D-DLAOI method through an illustrative experiment. We randomly generate two scenes in different sizes. One has 2000 objects with the size in 1000×1000. The other has 6400 objects with the size in 2000×2000. We compare three methods including the original AOI, the single layer dynamic AOI and our proposed D-DLAOI. All of them follow the same roaming trajectory. For the original AOI and the single layer dynamic AOI, the size of grid is in 100×100. For our D-DLAOI method, the shortest length of the longest axis is set to 100. In Fig. 3, we show the comparison on the number of visible object detection. Our D-DLAOI method is the best in three methods. From Fig. 3(b), with the increasing of the size of AOI, the performance of D-DLAOI is closer to the single layer method. But they are still much better than the original AOI.

Next, we evaluate quality of human visual impact. In this experiment, the network is under 10 Mbps, the size of scene is in 4000×4000 with 5000 objects, which are Stanford Bunny chosen for simple. The roaming trajectory is set as circle $((x - x_c)^2 + (y - y_c)^2 = R^2)$, where the center, (x_c, y_c) is $(2000, 2000)$; the radius, R is 1000; and the speed is $2m/s$. The measurement of the quality of human visual impact is defined as

$$\text{Quality} = \sum p_i' / \sum p_i, \tag{4}$$

where p_i is the amount of data needed for the ith object; p_i' is the received data for the ith object at this moment. The follow five methods are compared:

- **PM_with_Priority** is our proposed method in this paper;
- **Priority_Only** is to transmit the original scene data with priority;

- **PM_Only** is applied PM transmission without priority;
- **No_PM_No_Priority** is to transmit the original scene data without priority;
- **BM_with_Priority** is transmit the base mesh (BM) of the scene with priority.

The experimental results are plotted in Fig. 4, showing the human visual impact among five methods. PM_with_Priority performs the best, demonstrating that our proposed method is promising. For more realistic experiment, we also run demonstrations on the scene of forest and city as in Figs. 5 and 6.

Fig. 5. Demo of forest scene roaming.

Fig. 6. Demo of city scene roaming.

7 Conclusion and Future Work

We present a client-driven strategy of scene streaming so as to achieve real time large-scale 3D scene roaming. Our main contributions in this paper are (i) we propose a self-adaptive method to generate hierarchy grid structure including hierarchy scene structure and its description file; (ii) we propose a D-DLAOI to further speed up scene culling and object selection; (iii) we apply LPM in order to improve the human visual impact in the large scale scene roaming. The experiments demonstrated the effectiveness of our proposed approach in the realistic forest and city roaming applications.

In the future, our plan will focus on improving the performance of multiple mobile VE clients for the collaborative works. Specifically, our aim is to Accurate localization for the optimal downloading source [20] in P2P-DVE.

Acknowledgments. The research was supported jointly by National Science Foundation of China (No. 61272276), National Twelfth Five-year Plan Major Science and Technology Project (No. 2012BAC11B00-04-03), Special Research Fund of Higher College Doctorate (No. 20130072110035), Key Science and Technology Project of Jilin (No. 20140204088GX) and Tongji University Young Scholar Plan (No. 2100219052).

References

1. Hu, S.Y.: A case for 3D streaming on peer-to-peer networks. In: Proceedings of Web3D 2006, pp. 57–63 (2006)
2. Li, F.W.B., Lau, R.W.H., Kilis, D., Li, L.W.F.: Game-on-demand: an online game engine based on geometry streaming. ACM Trans. Multimedia Comput. Commun. Appl. **7**, 1–22 (2011)
3. Wang, W., Jia, J.: An incremental smlaoi algorithm for progressive downloading large scale webvr scenes. In: Proceedings of the 14th International Conference on 3D Web Technology, pp. 55–60 (2009)
4. Esch, M., Botev, J., Schloss, H., Scholtes, I.: P2P-based avatar interaction in massive multiuser virtual environments. In: International Conference on Complex, Intelligent and Software Intensive Systems, CISIS 2009, pp. 977–982 (2009)
5. Gurler, C.G., Tekalp, M.: Peer-to-peer system design for adaptive 3D video streaming. IEEE Trans. Commun. Mag. **51**, 108–114 (2013)
6. Gulliver, S.R., Ghinea, G.: Defining user perception of distributed multimedia quality. ACM Trans. Multimedia Comput. Commun. Appl. (TOMM) **2**(4), 241–257 (2006)
7. Wen, L.X., Jia, J.Y., Liang, S.: Proceedings of the 13th ACM SIGGRAPH International Conference on Virtual-Reality Continuum and its Applications in Industry, pp. 95–103 (2014)
8. Cavagna, R., Royan, J., Gioia, P., Bouville, C., Abdallah, M., Buyukkaya, E.: Signal processing: image communication. Special Issue on Advances in Three-dimensional Television and Video **24**(1–2), 115–121 (2009)
9. Hu, S.Y., Huang, T.H., Chang, S.C., Sung, W.L., Jiang, J.R., Chen, B.Y.: Flod: a framework for peer-to-peer 3d streaming. In: The 27th IEEE Conference on Computer Communications (INFOCOM) (2008)

10. Botev, J., Hohfeld, A., Schloss, H., Scholtes, I., Sturm, P., Esch, M.: The Hyper-Verse: concepts for a federated and Torrent-based '3D Web'. Int. J. Adv. Media Commun. **2**(4), 331–350 (2008)
11. Ize, T., Shirley, P., Parker, S.: Grid creation strategies for efficient ray tracing. In: IEEE Symposium on Interactive Ray Tracing, pp. 27–32 (2007)
12. Wald, I., Ize, T., Kensler, A., Knoll, A., Steven, G.: Ray tracing animated scenes using coherent grid traversal. ACM Trans. Graph. (TOG) **25**(3), 485–493 (2006)
13. Lagae, A., Dutre, P.: Compact, fast and robust grids for ray tracing. Comput. Graph. Forum **27**(4), 1235–1244 (2008)
14. Chuah, S.P., Yuen, C., Cheung, N.M.: Cloud gaming: a green solution to massive multiplayer online games. IEEE Trans. Wirel. Commun. **21**(4), 78–87 (2014)
15. Huang, C.Y., Hsu, C.H., Chang, Y.C., Chen, K.T.: GamingAnywhere: an open cloud gaming system. In: Proceedings of the 4th ACM Multimedia Systems Conference, pp. 36–47 (2013)
16. Wang, M.F., Jia, J., Zhongchu, Y.X., Zhang, C.X.: Organizing neighbors self-adaptively based on avatar interest for transmitting huge DVE scenes. In: Proceedings of the 13th ACM SIGGRAPH International Conference on Virtual-Reality Continuum and its Applications in Industry, pp. 105–111 (2014)
17. Hoppe, H.: Progressive meshes. In: Proceedings of the 23rd Annual ACM Conference on Computer Graphics and Interactive Techniques, pp. 99–108 (1996)
18. Ahmed, D.T., Shirmohammadi, S.: An auxiliary area of interest management for synchronization and load regulation in zonal P2P MMOGs. In: IEEE International Workshop on Haptic Audio visual Environments and Games, pp. 36–41 (2008)
19. Rooney, S., Bauer, D., Deydier, R.: A federated peer-to-peer network game architecture. IEEE Commun. Mag. **42**(5), 114–122 (2004)
20. Liu, E.S., Theodoropoulos, G.K.: Interest management for distributed virtual environments: a survey. ACM Comput. Surv. **46**(4), 51:1–51:42 (2014)

SELSH: A Hashing Scheme for Approximate Similarity Search with Early Stop Condition

Jie Chen, Chengkun He, Gang Hu, and Jie Shao[✉]

School of Computer Science and Engineering,
University of Electronic Science and Technology of China, Chengdu, China
{chenjie,MatthewHe,hugang}@std.uestc.edu.cn, shaojie@uestc.edu.cn

Abstract. Similarity search is a fundamental problem in various multimedia database applications. Due to the phenomenon of "curse of dimensionality", the performance of many access methods decreases significantly when the dimensionality increases. Approximate similarity search is an alternative solution, and Locality Sensitive Hashing (LSH) is the most popular method for it. Nevertheless, LSH needs to verify a large number of points to get good-enough results, which incurs plenty of I/O cost. In this paper, we propose a new scheme called SortedKey and Early stop LSH (SELSH), which extends the previous SortingKeys-LSH (SK-LSH). SELSH uses a linear order to sort all the compound hash keys. Moreover, during query processing an early stop condition and a limited page number are used to determine whether a page needs to be accessed. Our experiments demonstrate the superiority of the proposed method against two state-of-the-art methods, C2LSH and SK-LSH.

Keywords: Approximate similarity search · LSH · Sorted keys · Early stop condition

1 Introduction

With the development of Internet, the amount of multimedia data has reached an unprecedented level. There are plenty of large-scale multimedia databases nowadays. Each object stored in these databases is typically represented by a high-dimensional feature point. Similarity search is a fundamental problem in multimedia databases which aims to find the similar objects to a query object. It has many important applications, such as video retrieval, spam filter, face recognition. The ways to improve the performance of similarity search in high-dimensional space have been studied for a long time [3]. Due to the notorious "curse of dimensionality", many access methods [1,4,8] are outperformed by brute-force method of linear scan when the dimensionality is high. The challenge of finding exact nearest neighbor in high-dimensional space leads to an alternative solution: approximate similarity search, where acceptable result and high efficiency can be achieved.

© Springer International Publishing Switzerland 2016
Q. Tian et al. (Eds.): MMM 2016, Part II, LNCS 9517, pp. 104–115, 2016.
DOI: 10.1007/978-3-319-27674-8_10

Hashing method is one of the most promising and popular approximate similarity search solutions. LSH and its variants are widely adopted and well studied. Their good performance help them achieve great success in practice. LSH uses similarity-preserving hashing functions to project nearby objects in original space into the same buckets with high probability [7,9]. Generally, hashing algorithms [5,6,16] utilize m randomly-chosen hash functions to distinguish points from each other.

Several other hashing based methods have also been proposed to deal with approximate search or classification problem (such as [13,14]). Basic LSH algorithms [5,7] only use the points which share the same compound hash key with the query point by one compound hash function projection. In order to reduce the false positive, these methods create hundreds or thousands of hash tables, which cause too much space requirement. To reduce the index space occupation, Multi-Probe LSH [12] was proposed, which could find more similar objects with query via searching other near buckets. The LSB-forest [16] and C2LSH [6] have been proposed for building smaller index. However, these approaches still need a considerable space requirement.

In this paper, we propose a new scheme called SortedKey and Early stop LSH (SELSH) for approximate similarity search. Our algorithm is based on SortedKey LSH (SK-LSH) [11]. However, we add the early stop condition and a number to limit page access in SELSH. The experimental comparison results show the superiority of SELSH against the state-of-the-art hashing based approaches.

The main contributions of our work are as follows:

– We propose a new solution to answer approximate similarity search, called SELSH, which extends SK-LSH. Our method requires small I/O cost. Moreover, it returns good quality results when the parameters are appropriately set (see Sect. 4).
– We conduct experiments on three high-dimensional multimedia datasets and the results show that our method outperforms the state-of-the-art hashing methods, including SK-LSH and C2LSH, on result quality, query time and I/O cost.

The rest of this paper is organized as follows. In Sect. 2, we give the formal problem definition and the notations used in this paper. Section 3 reviews some related work. Section 4 describes our method in detail. We show the experimental results in Sect. 5. Finally, we conclude our work in Sect. 6.

2   Preliminaries

In this section, we firstly define the problem to be solved, then we introduce the notations used in this paper.

2.1   Problem Definition

The problem we deal with in this paper is that a dataset D contains n points with d dimensions. Given an approximate ratio $c > 0$, the approximate similarity search aims to find $(1 + c)$-approximate nearest neighbor which satisfies

$dist(p, q) \leq (1+c)dist(p^*, q)$. Here, p^* is the exact nearest neighbor of q, $dist(x, y)$ is distance measure between x and y.

2.2 Notations

For ease of description in following parts of this paper, we show the notations in Table 1.

Table 1. Notations

Notation	Explanation
n	The number of points in dataset
d	The dimensionality of a point
m	The dimensionality of compound hash key
L	The number of compound hash functions
k	The number of nearest neighbors
c	Approximate ratio
nq	The number of query points
Np	The maximum number of pages to load
$dist$	The Euclidean distance of two points in $\Re^d$
$hdist$	The Euclidean distance of two points in projected space
o_{min}	The point in result with minimum distance
$\mathcal{X}^2(m)$	Chisquare distribution with m freedom
$\Psi_m(p)$	Cumulative distribution function
$threshold$	A threshold of $\Psi_m(p)$

3 Related Work

To retrieve objects from multimedia databases, many indexing methods have been proposed. However, the "curse of dimensionality" causes many access methods to lose their good performance in high-dimensional space. A few methods transform the high-dimensional data to one-dimensional space [2,10]. In addition, some data approximation algorithms are proposed (e.g., VA-file [17]).

Recently, hashing based methods become popular. Among existing hashing methods, LSH is one of the most efficient approaches. The LSH is defined as follow:

Definition 1 Locality Sensitive Hashing. A hash function is called (r, cr, P_1, P_2)-sensitive if the following two conditions are satisfied at the same time for any two points p_1 and p_2.

$$\text{if} \quad p_1 \in \mathcal{B}(p_2, r), \quad \text{then} \quad P_{rH}(h(p_1) = h(p_2)) \geq P_1 \tag{1}$$

$$\text{if} \quad p_1 \notin \mathcal{B}(p_2, cr), \quad \text{then} \quad Pr_H(h(p_1) = h(p_2)) \le P_2 \tag{2}$$

$\mathcal{B}(x, y)$ means all the points within distance y to x. To make sense, $c > 1$, $P_1 \ge P_2$.

The principle of LSH is that similar objects are projected to the same hash key with high probability, and the probability of two objects with large distance have the same hash key is small.

In order to get good performance, many hash tables need to be created. As a result, the space requirement is a considerable cost. To deal with this issue, a few structures were proposed, such as LSB-forest [16], C2LSH [6], and multi-probe LSH [12]. LSB-forest represents hash keys as 1-dimensional Z-order values, and these values are indexed by B$^+$-tree. C2LSH generates many hash functions, each of which is associated with a hash table. The points frequently colliding query point are selected as candidates. Multi-probe LSH was proposed to use a few hash tables to address approximate nearest neighbor problem. It selects the points whose hash keys close to the buckets where query point's hash key stands.

4   Our Method

In this section, we describe the proposed SortedKey and Early stop LSH method in detail.

4.1   LSH Function

In order to satisfy the two conditions in Definition 1, Datar et al. [5] proposed p-stable LSH function. In Euclidean space, the 2-stable LSH function is commonly used. The function is defined as follow:

$$h(x) = \lfloor \frac{w^T \cdot x + b}{W} \rfloor \tag{3}$$

Here, w is a vector with d elements which are independently selected from the data following the Guassian distribution. b is a real number chosen uniformly from $[0, W]$, where W represents the width of LSH function.

If two points have similar hash keys, we can consider they are close to each other. However, if we just use one LSH function, it may incur many false positives in returned answers. To reduce the number of irrelevant points contained in results, m LSH functions are used to compose a compound hash function. We call the hash keys generated by compound hash function *compound hash keys*. In order to remove false positive further, usually L compound hash functions are taken into consideration.

4.2   Distance Measure, Linear Order and Early Stop Condition

Our method extends SK-LSH [11] in the sense that the distance measure and linear order used in SELSH are similar with SK-LSH.

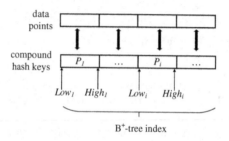

Fig. 1. Index structure

As shown in Eq. 4, the distance between two compound hash keys K_1 and K_2 consists of two parts. Assuming that each compound hash key has m dimensions. $Ln(K_1, K_2)$ denotes the number of elements which do not belong to the longest common prefix of K_1 and K_2. The difference of the first elements that differ in K_1 and K_2 is represented by $Dn(K_1, K_2)$. Usually, the parameter R is larger than $Dn(K_1, K_2)$. For example, given $K_1 = (1011021)$, $K_2 = (1010021)$, the longest common prefix of K_1 and K_2 is (101), and as a result, $Ln(K_1, K_2) = 7 - 3 = 4$, and $Dn(K_1, K_2) = |1 - 0| = 1$.

$$DH(K_1, K_2) = Ln(K_1, K_2) + Dn(K_1, K_2)/R \qquad (4)$$

Based on this distance measure, we introduce the linear order. The linear order $\leq_H$ we used is originally proposed in SK-LSH [11]. For compound hash keys K_1 whose elements are $K_{1,j}$ and K_2 whose elements are $K_{2,j}$ where $1 \leq j \leq m$, $K_1 <_H K_2$ if and only if the length of longest common prefix is smaller than m, meanwhile, the first element out of the longest common prefix of K_1 is smaller than that of K_2. When K_1 and K_2 share the same elements in corresponding positions, we use $K_1 =_H K_2$ to represent this situation. To show the linear order clearly, we use $K_1 = (1011021)$, $K_2 = (1010021)$ and $K_3 = (1011021)$, $K_4 = (1011021)$ as two examples. In the first example, the length of longest common prefix is 3 which is smaller than 7, and $0 < 1$. We can determine the relation of K_1 and K_2 is $K_2 <_H K_1$. As to the second example, K_3 and K_4 have the same elements. Thus, $K_3 =_H K_4$.

The effect of the above distance measure can be demonstrated in the experimental results of [11]. For further efficiency improvement, we introduce an early stop condition inspired by Sun et al. [15], which can be defined as $\Psi_m(\frac{c^2 \cdot hdist(K_p, K_q)}{dist(o_{min}, q)}) > threshold$. In this formula, $hdist$ is Euclidean distance of two compound hash keys. This condition provides a theoretical guarantee for approximate nearest neighbors search.

4.3 Index Structure

As described in SK-LSH [11], all the compound hash keys which are generated by one compound hash function are sorted by linear order mentioned above. Then,

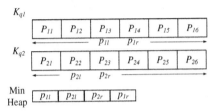

Fig. 2. Search process

we re-arrange the corresponding points according to the order of compound hash keys. In order to manage these sorted compound hash keys, we divide them into several pages. Each page contains B compound hash keys. In each page, there is a low boundary compound hash key, we represent it as *Low*, and a high boundary compound hash key, we represent it as *High*. To reduce the space requirements, we use *Low* and *High* of each page to represent this page. To organize these *Low*s and *High*s, we use a B$^+$-trees to index them. In total, we build L B$^+$-trees. Figure 1 shows an example of the index structure.

4.4   Search Process

In the search process, the page which should be accessed firstly needs to be determined carefully. To deal with this issue, we use the distance measure of compound hash key and a page [11], which is displayed in the following equation.

$$DP(q, P) = \begin{cases} DH(K_q, Low) & \text{if } K_q \text{ stands in the left page of } P \\ 0 & \text{if } K_q \text{ stands in the page } P \\ DH(K_q, High) & \text{if } K_q \text{ stands in the right page of } P \end{cases} \quad (5)$$

We retrieve L B$^+$-trees firstly. In each B$^+$-tree, we use a pointer p_r to point to the nearest page of query compound hash key, and a pointer p_l to point the left page of p_r. We call p_l left pointer which can only shift to left, and similarly, p_r right pointer whose shift direction is right. Figure 2 shows the process briefly. Algorithm 1 and Algorithm 2 demonstrate the search process. Every time, we choose the nearest page to access. In Fig. 2, Min Heap is a priority queue which keeps the page with minimum distance to query compound hash key in the first place. Before accessing the page, we should judge whether this page satisfies the early stop condition or not. If yes, we stop searching, and if not, we continue searching. The stop condition has two different cases: (1) if pointer of the page is right pointer, the low boundary of this page is taken into consideration; (2) if the pointer is left pointer, we use the high boundary. Then we delete this page from Min Heap, and insert the page which is pointed by the pointer lastly changed into Min Heap. For example, we assume that P_{13} is accessed, then we shift p_{1l} to P_{12}, and insert P_{12} into Min Heap.

Algorithm 1. Search

Input: q
Output: k nearest neighbours of q
initial $MaxHeap, MinHeap$;
for $i = 1 : L$ **do**
> compute query's hash key, $K_{q,i}$;
> find P_{il} and P_{ir};
> $MinHeap.inseart(P_{il}, P_{ir})$;

end
while $page\_count < Np$ **do**
> $P = MinHeap.extract()$;
> $verify(P, MaxHeap)$;
> $shift(P)$;
> $page\_count + +$;

end
return $MaxHeap$;

Algorithm 2. $verify(P, MaxHeap)$

Input: page P, $Maxheap$
if P's pointer is left pointer **then**
> **if** $\Psi_m(\frac{c^2 \cdot hdist(P.High, K_q)}{dist(o_{min}, q)}) > threshold$ **then**
> > verify all points in P;
> > update $MaxHeap$;
>
> **end**

end
if P's pointer is right pointer **then**
> **if** $\Psi_m(\frac{c^2 \cdot hdist(P.Low, K_q)}{dist(o_{min}, q)}) > threshold$ **then**
> > verify all points in P;
> > update $MaxHeap$;
>
> **end**

end

4.5 Complexity Analysis

We firstly analyze the space requirement of SELSH. The space cost contains two main parts. The first part comes from L B$^+$-trees, which totally occupy $O(Lnm/B)$ space. The second part comes from L copies of dataset which cost $O(Lnd)$ space. The space requirement of SELSH is $O(Ln(d + m/B))$.

Next, we discuss the I/O cost of SELSH. Retrieving a B$^+$-tree needs $O(Bd \log_B n)$ I/O cost. Loading Np needs $O(BdNp)$ I/O cost. The total I/O cost is $O(Bd(Np + L \log_B n))$. Note that this is the worse case. If the search process interrupts due to early stop condition is satisfied, the I/O cost will be smaller.

5   Experimental Results

We report our experimental results in this section.

5.1   Set up

Experimental DataSets. We conduct experiments on three real-life datasets.

- **Corel Features**[1] This dataset contains 68040 32-dimensional color histograms.
- **Audio**[2] This dataset consists of 54387 192-dimensional vectors extracted by the Marsyas library.
- **ANN_SIFT 1M**[3] There are 1 million base vectors and 10000 query vectors in this dataset. Each vector is with 128 entries.

Evaluation Metric. In our experiments, we evaluate three performance metrics.

- **ratio:** We use the ratio defined in [16] to measure the accuracy of approximate similarity search. It is defined as $\Sigma_{i=1}^{k}(dist(o_i^*)/dist(o_i))$, where o_i^* is i-th returned approximate nearest neighbor, o_i is the real i-th nearest neighbor.
- **time:** In order to make a fair comparison, we use the average query time, which is defined as $\frac{1}{nq}\Sigma_{i=1}^{nq}t_i$. Here, t_i represents the query time of the i-th query point.
- **I/O cost:** Smaller I/O cost can reduce query time and improve the performance of similarity search algorithm.

Compared Methods. We consider the state-of-the-art hashing based methods. Here, we choose C2LSH [6] and SK-LSH [11].

5.2   Selection of Appropriate Parameters

Tuning L. L is the number of compound hash functions which affects the accuracy of results. Intuitively, a larger L returns better results. To test the effect of L on *ratio* and *time*, we test L from 1 to 6. Also, we fix $Np = 10$. Figure 3(a) demonstrates the relation between L and *ratio*. As supposed, when the parameter L changes from 1 to 6, *ratio* descends. With L increasing, the decreasing speed becomes low gradually. This means that just increasing the number of L cannot make too much contributions. As L increases, longer query time is needed. The process to find out left pointers and right pointers needs to check all the index, and the heights of all the trees in our experiments are not too big. Due to these reasons given above, the curves in Fig. 3(b) do not change significantly. Considering the effects of L on *ratio* and *time*, we choose $L = 4$ in the following experiments.

---

[1] http://kdd.ics.uci.edu/databases/CorelFeatures/.
[2] http://www.cs.princeton.edu/cass/audio.tar.gz.
[3] http://corpus-texmex.irisa.fr/.

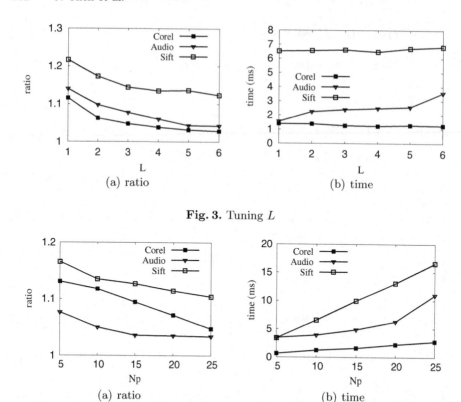

Fig. 3. Tuning L

Fig. 4. Tuning Np

Tuning Np. We use Np to represent the maximum number of the pages to be accessed. Intuitively, different Np influences the precision of our method, and the more pages we access, the answers returned will be more accurate. At the same time, a larger Np will lead to longer query time. In the experiments, we set $L = 4$, and the number of Np is from 5 to 25 at step 5. Figure 4 shows the results. As we expected, *ratio* decreases with the increase of Np in all the datasets we test. This indicates that more pages accessed lead to better accuracy of returned nearest neighbors. As to the query time, all the curves in Fig. 4(b) increase with Np nearly linearly. Obviously, Np cannot be too large. In the following experiments, Np is fixed at 20.

5.3   Comparison with SK-LSH and C2LSH

In this section, we compare the performance of our method with SK-LSH and C2LSH. As mentioned above, L is set to 4, and Np is set to 20. In order to show the performance comparison sufficiently, we vary k in the range of $\{1, 10, 20, 30, \cdots, 90, 100\}$.

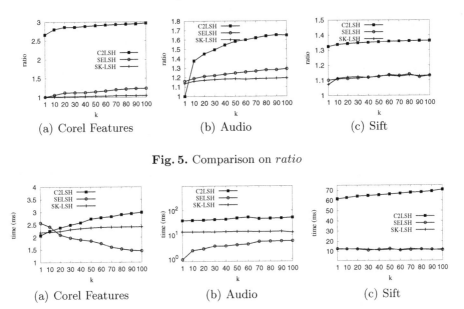

Fig. 5. Comparison on *ratio*

Fig. 6. Comparison on *time*

Firstly, we check the ratio for all compared methods on all the three datasets, which is shown in Fig. 5. SELSH and SK-LSH outperform C2LSH in each dataset. For Corel Features and Audio, SK-LSH performs slightly better than SELSH. The maximum difference between SK-LSH and SELSH is 0.1. While for Sift, SK-LSH and SELSH nearly share the same *ratio* values, which are about 1.1, and these values do not change much as k increases. With the increase of k, the *ratio* values of all the three methods ascend on Corel Features and Audio. C2LSH shows its poor performance on Audio, and *ratio* changes significantly.

Figure 6 shows the query time of three methods on different datasets. Clearly, SELSH and SK-LSH perform much better than C2LSH in almost conditions except when $k \leq 10$ for Corel Features. When dealing with Corel Features, SELSH shows a decrease trend, while SK-LSH and C2LSH have the opposite trend. This may be attributed to the early stop condition used in search process. As to Audio, SELSH and C2LSH cost longer *time* with the increase of k. SK-LSH costs nearly constant *time*. However, SELSH outperforms the other two methods. SELSH and SK-LSH cost almost the same *time* during query, and their performance is much better than C2LSH.

Considering the I/O cost, SELSH is the most efficient one among the three methods, while C2LSH is the worst. On Corel Features dataset, the I/O cost of SELSH decreases when k increases. Figure 7(a) proves the guess we make above. Although the I/O cost of SELSH shows rising trend, SELSH still outperforms SK-SLH for Audio dataset. Interestingly, on Sift dataset, SELSH and SK-LSH have the same *I/O cost*. This explains the reason why the two algorithms share

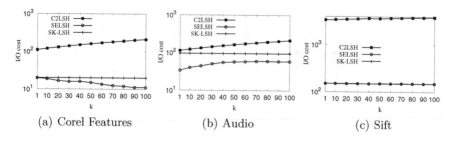

Fig. 7. Comparison on *I/O cost*

the same *time*. Due to loading a constant number of pages, SK-LSH has the same *I/O cost* in each dataset.

6   Conclusion

In this paper, we extend the hashing based method SK-LSH to propose our method SELSH. SELSH aims at addressing approximate similarity search problem based on 2-stable LSH functions. To reduce the number of random I/O, we use the distance measure and linear order of SK-LSH. In order to reduce the pages to load, we introduce the early stop condition into SELSH. Therefore, we get the answers with good-enough accuracy and smaller I/O cost. Our experiments are conducted on three datasets. The results demonstrate the superiority of SELSH against the state-of-the-art methods.

Acknowledgments. This work is partially supported by the Fundamental Research Funds for the Central Universities of China under grant No.ZYGX2014Z007.

References

1. Bentley, J.L.: Multidimensional binary search trees used for associative searching. Commun. ACM **18**(9), 509–517 (1975)
2. Berchtold, S., Böhm, C., Kriegel, H.: The pyramid-technique: towards breaking the curse of dimensionality. In: SIGMOD 1998, Proceedings ACM SIGMOD International Conference on Management of Data, June 2–4, 1998, Seattle, pp. 142–153 (1998)
3. Böhm, C., Berchtold, S., Keim, D.A.: Searching in high-dimensional spaces: index structures for improving the performance of multimedia databases. ACM Comput. Surv. **33**(3), 322–373 (2001)
4. Ciaccia, P., Patella, M., Zezula, P.: M-tree: An efficient access method for similarity search in metric spaces. In: VLDB 1997, Proceedings of 23rd International Conference on Very Large Data Bases, Athens, 25–29 August, 1997, pp. 426–435 (1997)

5. Datar, M., Immorlica, N., Indyk, P., Mirrokni, V.S.: Locality-sensitive hashing scheme based on p-stable distributions. In: Proceedings of the 20th ACM Symposium on Computational Geometry, Brooklyn, New York, 8–11 June, 2004, pp. 253–262 (2004)

6. Gan, J., Feng, J., Fang, Q., Ng, W.: Locality-sensitive hashing scheme based on dynamic collision counting. In: Proceedings of the ACM SIGMOD International Conference on Management of Data, SIGMOD 2012, Scottsdale, 20–24 May, 2012, pp. 541–552 (2012)

7. Gionis, A., Indyk, P., Motwani, R.: Similarity search in high dimensions via hashing. In: VLDB 1999, Proceedings of 25th International Conference on Very Large Data Bases, Edinburgh, 7–10 September, 1999, pp. 518–529 (1999)

8. Günther, O.: The design of the cell tree: an object-oriented index structure for geometric databases. In: Proceedings of the Fifth International Conference on Data Engineering, Los Angeles, 6–10 February, 1989, pp. 598–605 (1989)

9. Indyk, P., Motwani, R.: Approximate nearest neighbors: towards removing the curse of dimensionality. In: Proceedings of the Thirtieth Annual ACM Symposium on the Theory of Computing, Dallas, 23–26 May, 1998, pp. 604–613 (1998)

10. Jagadish, H.V., Ooi, B.C., Tan, K., Yu, C., Zhang, R.: iDistance: an adaptive b^+-tree based indexing method for nearest neighbor search. ACM Trans. Database Syst. **30**(2), 364–397 (2005)

11. Liu, Y., Cui, J., Huang, Z., Li, H., Shen, H.T.: SK-LSH: an efficient index structure for approximate nearest neighbor search. PVLDB **7**(9), 745–756 (2014)

12. Lv, Q., Josephson, W., Wang, Z., Charikar, M., Li, K.: Multi-probe LSH: efficient indexing for high-dimensional similarity search. In: Proceedings of the 33rd International Conference on Very Large Data Bases, University of Vienna, Austria, 23–27 September, 2007, pp. 950–961 (2007)

13. Shen, F., Shen, C., Shi, Q., van den Hengel, A., Tang, Z.: Inductive hashing on manifolds. In: 2013 IEEE Conference on Computer Vision and Pattern Recognition, Portland, 23–28 June, 2013, pp. 1562–1569 (2013)

14. Shen, F., Shen, C., Shi, Q., van den Hengel, A., Tang, Z., Shen, H.T.: Hashing on nonlinear manifolds. IEEE Trans. Image Process. **24**(6), 1839–1851 (2015)

15. Sun, Y., Wang, W., Qin, J., Zhang, Y., Lin, X.: SRS: solving c-approximate nearest neighbor queries in high dimensional euclidean space with a tiny index. PVLDB **8**(1), 1–12 (2014)

16. Tao, Y., Yi, K., Sheng, C., Kalnis, P.: Efficient and accurate nearest neighbor and closest pair search in high-dimensional space. ACM Trans. Database Syst., 35(3) (2010)

17. Weber, R., Böhm, K., Schek, H.: Interactive-time similarity search for large image collections using parallel va-files. In: ICDE. p. 197 (2000)

Learning Hough Transform with Latent Structures for Joint Object Detection and Pose Estimation

Hanxi Li[1,2]([✉]), Xuming He[2], Nick Barnes[2], and Mingwen Wang[1]

[1] School of Computer and Information Engineering,
Jiangxi Normal University, Nanchang, China
{mwwang,hanxi.li}@jxnu.edu.cn
[2] National ICT Australia, Canberra Research Laboratory, Canberra, Australia
{hanxi.li,xuming.he,nick.barnes}@nicta.com.au

Abstract. We present a novel max-margin Hough transform with latent structure for joint object detection and pose estimation. Our method addresses the large appearance and shape variation of objects in multiple poses by integrating three key components: First, we propose a more robust appearance model by designing a patch dictionary with complementary features; In addition, we use a group of latent components to explicitly incorporate feature selection and pooling into the Hough-based object models; Furthermore, we adopt a multiple instance learning approach to handle the lack of correspondence among training instances with noisy bounding-box labels. We design a unified objective and an efficient approximate inference that alternates the search between object location and pose space. We demonstrate the efficacy of our approach by achieving the state-of-the-art performance on two detection and two joint estimation datasets.

1 Introduction

Detecting objects and reasoning about their 3D properties from a single image is one of the key problems in visual scene understanding. The main challenge comes from the large variation in object's appearance due to different viewpoints and ambiguity in pose. Recent advances in part-based object modeling [1,2], make it possible to infer the semantic object parts and their 2D/3D configurations, from which the 3D object pose can be estimated. Nevertheless, the success of most part-based methods relies on training with a large number of labeled examples [2] and in many cases, with strong supervision (e.g., with part correspondence) [3]. It also requires designing a category-specific mixture structure to handle the multiple poses.

An alternative approach is based on feature-driven models, and in particular, the Hough transform with a patch codebook. The generalized Hough transform

Q. Tian et al. (Eds.): MMM 2016, Part II, LNCS 9517, pp. 116–129, 2016.
DOI: 10.1007/978-3-319-27674-8_11

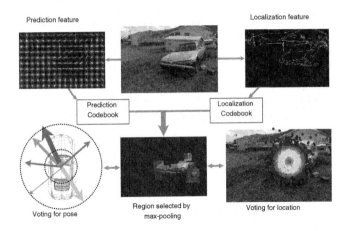

Prediction feature   Localization feature

Prediction   Localization
Codebook   Codebook

Voting for pose   Region selected by   Voting for location
                         max-pooling

Fig. 1. Illustration of the Hough transform with latent structure approach.

has been widely used for object detection [4–7] and their pose or shape estimation [8,9]. In essence, it is based on dense patch features and a nonparametric representation of feature-to-object geometric relations. The main advantage of such a representation is its simplicity, robustness towards occlusion and its intrinsic modeling capacity for objects in multiple poses [10]. Furthermore, it can be integrated with manually-designed [11] or learned patch features [12] in a flexible way.

Despite its flexibility, the traditional Hough-based approach has several limitations, which usually lead to inferior performance to more sophisticated methods such as [6]. First, it is difficult to use a single image patch feature that is stable among different instances (from multi-views) and also distinctive from background clutter. In addition, the simple additive model of patch votes ignores spatial dependency and leads poor balance in weighting sub-regions of a object. Furthermore, modeling the feature-to-object geometric relations requires well aligned training examples and thus is sensitive to the labeling of object center location, which is ambiguous and noisy for many object categories.

In this work, we propose a novel Hough-based method to address these issues in modeling objects with multiple poses. Specifically, we integrate three key components into the Hough transform framework as shown in Fig. 1. We start with a robust appearance model for the image patches, which consists of a patch vocabulary with a pair of features that complement each other. Our visual code has two parts: one is based on clustering features robustly w.r.t viewpoint change and intra-class variation, while the other is a learned discriminative feature based on a linear classifier on HOG [13]. Our second component incorporates the feature dependencies in each object instance. We explicitly add a binary latent variable for each patch feature, which indicates the relevance of corresponding feature for the target object. The latent structure allows us to reason about object support and integrates a max-pooling mechanism to balance the voting scores. Finally, we present a Multiple Instance Learning (MIL) approach to the max-margin Hough transform (M^2HT). By sampling around the original bounding-box labels, our method is capable of handling ambiguous and noisy bounding box labels. Unlike the standard MI-SVM [14], we build more informative positive bags by removing redundant instances based on spatial constraints.

We formulate a unified objective for our latent Hough transform model, which defines an additive score function on the object location and pose space. However, computing and optimizing the score function in the joint space is challenging as it requires a large number of voting elements and expensive search due to dimensionality of the object hypothesis. We propose an efficient approximate inference algorithm that decomposes the joint space into several smaller subspaces, and alternates the max-margin Hough transforms in the subspaces with shared latent variables.

We evaluate our method on two object detection benchmarks, ETHZ shape and part of VOC2007, achieving superior performance compared to the previous Hough-based approaches. In addition, we test our joint prediction method on two car detection and pose estimation datasets, and obtain the state-of-the-art results but with weaker supervision. Our main contributions are two-fold: (1) A Hough transform model with a better appearance model and latent structure for feature pooling; (2) An alternating Hough voting inference algorithm for joint object detection and pose estimation.

2   Related Work

Localizing objects and interpreting their 3D properties from images has drawn much attention recently due to progress in part-based object modeling. Most recent works are based on the deformable part model [1] and its extensions [15]. Their focus has been on detailed parsing of object parts and structure, which usually requires many training examples and/or part-level labeling of objects.

In contrast, the Generalized Hough transform [16] with a learned visual codebook is widely adopted in 3D object detection due to its simplicity and efficiency. Many early works focus on 2D object detection and localization, such as the Implicit Shape Model (ISM) [17,18]. Their Hough spaces are usually low-dimensional, limiting them to prediction of object centers or bounding boxes. Ommer et al. [19] considered joint space of location and scale, and proposed to vote with line clusters in Hough space. More recent work considers joint object detection and 3D pose estimation [20,21]. However, most of them rely on 3D object information during training stage, which is usually difficult to obtain for generic object classes.

Latent variables have been introduced into the Hough Transform for handling feature selection or objects with multiple poses. In particular, the Latent Hough Transform [6] uses a hidden variable to indicate a typical object pose and learns a pose-specific weight vector for selecting training data. However, they have to resort to a complex learning procedure to estimate a large number of parameters. Other methods exploit the back-projected support region of an object to model different poses. Razavi et al. [10] propose to use the back-projected regions as a cue for pose similarity. The voting by group method [5] introduces group assignment, correspondence and transformation, which groups patches into more distinctive parts. Nevertheless, it is challenging to optimize these variables jointly during inference.

Most Hough-based methods learn the visual codebook by clustering the image features of patches from the object regions, such as SIFT [22] or HOG [11]. Hough Forest [7] replaces the clustering with discriminative learning based on the Random Forest. The poselet [23] model exploits part annotation and 3D information to learn a spatially consistent codebook. More recently, deep network based approaches automatically learn the features for detection or pose estimation, but require a large number of labeled data [24, 25]. Unlike these methods, our codebook is learned from a small set of annotated images using a pair of complementary patch features, which integrates clustering and discriminative learning.

The voting weights of the visual code in the Hough transform can also be learned in a discriminative way. The M^2HT [4] proposes a max-margin framework for learning the code weights. Our method adapts MIL into M^2HT in a similar way to the DPM [1]. The main difference is that we make use of spatial constraints to form positive bags to increase the robustness of our algorithm.

3   Our Approach

We first briefly review Hough-based object detection and introduce our notation. The standard Hough-based detection methods assume a weakly deformable template of features, and the voting can be viewed as an implicit inference procedure in such star-shaped model.

3.1   Hough-Based Object Detection

Formally, we represent each object hypothesis from object class o as $\mathbf{y}$, which can be its location and/or pose. The object model usually consists of an codebook of image patches, denoted by $\mathbb{C} = \{C_i\}_{i=1}^K$, each of which includes an appearance feature descriptor $\mathbf{c}_i$ and a set of geometric offsets G_i of positive training patches associated with the entry. The geometric offsets G_i store the relative positions of training patches w.r.t the object hypothesis $\mathbf{y}$ in a geometric feature space. This specifies a nonparametric model of the geometric relationship between the codebook entry C_i and $\mathbf{y}$.

Given an image I, the Hough voting approaches define a scoring function $S(\mathbf{y})$ for each valid configuration $\mathbf{y}$ by summing weighted votes from local image patches. The vote of a patch is computed by considering an appearance similarity score w.r.t the codebook and a geometric consistency score w.r.t the object geometric feature $\mathbf{g_y}$. More specifically, denote each image patch $I_\mathbf{x}$ by its geometric feature $\mathbf{g_x}$ and its appearance feature $\mathbf{f_x}$,

$$S(\mathbf{y}) \propto \sum_{\mathbf{x}} \sum_{i \in N_\mathbf{x}} \omega_i p(C_i|\mathbf{f_x}) \sum_{d \in D_i} e^{\left(-\frac{\|\mathbf{g_y}-(\mathbf{g_x}+\mathbf{d})\|^2}{2\sigma_d^2}\right)}$$

$$\propto \sum_{\mathbf{x}} \sum_{i \in N_\mathbf{x}} \omega_i p(C_i|\mathbf{f_x}) K_i(\mathbf{y}, \mathbf{x}) \tag{1}$$

where $N_{\mathbf{x}}$ is the set of k-nearest neighbors in $\mathbb{C}$, ω_i is the weight for code entry C_i, $p(C_i|\mathbf{f_x})$ is the patch-to-entry matching probability, and σ_d is the standard deviation of a Gaussian kernel modeling the weak deformation of object features. We rewrite the geometric factor from C_i as $K_i(\mathbf{y}, \mathbf{x})$ for short.

For joint object detection and pose estimation, we represent the object hypothesis space with a geometrical "quadlet" $\mathbf{y} = [\mathbf{v}, s, a, \theta]$, where $\mathbf{v}$ is the 2-D location of object center, s is the scale of the object; a represents the object's aspect ratio and θ stands for the viewing-angle of the object. We also use the same space for the geometric feature $\mathbf{g_y}$.

The main weakness of Hough voting methods is the simplicity of appearance model, the k-means clustering method is purely feature driven, and the lack of feature selection and score calibration. Furthermore, it is difficult for Hough-based detectors to handle a high-dimensional hypothesis space as the required number of voting patches increases exponentially as the dimension increases.

3.2   Latent Deformable Feature Model

We propose a novel Hough-based object modeling framework to address the limitation of object representation in the existing Hough-voting methods. Our approach first introduces a new appearance model for the codebook that exploits complementary property of multiple features at different scales. In addition, to capture features more robustly, we design a group of latent variables for explicit feature selection. Therefore, we refer to our approach as the Latent deformable feature model.

Patch Model with Complementary Features. A dilemma in Hough-based modeling for multiview object detection is choosing the appearance feature used in the codebook, especially when the training dataset is small relative to the variation of object poses. A balance between selectivity/discriminativeness and stability has to be achieved. Instead of using a single type of features, we propose a hybrid appearance model that combines two complementary kinds of features: one has good recall rates (referred to as the *localization feature*), and the other has better discriminative power (referred to as the *prediction feature*).

For a patch $I_{\mathbf{x}}$, we use a localization feature $\mathbf{f_x^l}$ and prediction feature $\mathbf{f_x^p}$ pair. Each codebook $\mathbb{C}$'s entry also has a complementary feature pair $\mathbf{c}_i = (\mathbf{c}_i^l, \mathbf{w}_i^p)$. We define our appearance model in the Hough voting framework as

$$p(C_i|\mathbf{f_x}) \propto q_l(C_i|\mathbf{f_x^l})q_p(o = 1|\mathbf{f_x^p}, C_i) \tag{2}$$

where o is the indicator variable for the object class.

We first compute the localization feature codebook by the k-means clustering on the corresponding features and $q_l(C_i|f_{\mathbf{x}}^l)$ is the patch-to-entry matching probability. For each localization code $\mathbf{c}_i^l$, we further learn a discriminative code for the prediction feature. Specifically, we build our prediction code entry by learning a linear classifier from all the training patches in C_i. The linear classifier (LDA), parameterized by $\mathbf{w}_i^p$, is trained to distinguish object patches from background and to have a recall rate larger than τ_p (0.95) within C_i. Given the linear classifier, we fit a probability score by passing it through a sigmoid function: $q_p(o = 1|\mathbf{f}_\mathbf{x}^p, C_i) = 1/\left(1 + \exp(-\sigma \mathbf{w}_i^{pT} \cdot \mathbf{f}_\mathbf{x}^p)\right)$, where $\sigma, \mathbf{w}_i^p$ are learned parameters (including bias terms).

Similar to the Poselet method [23], we extend our codebook to a multiscale setting, in which $\mathbb{C}$ is learned from an image pyramid. We share the codebook from different scales but use different weights when the object score function is computed. Specifically, we extend the model in Eq. 1 as follows,

$$S(\mathbf{y}) \propto \sum_h \sum_{\mathbf{x} \in I^h} \sum_{i \in N_\mathbf{x}} \omega_i^h p(C_i|\mathbf{f}_\mathbf{x}) K_i^h(\mathbf{y}, \mathbf{x}) \tag{3}$$

where h denote different layers in the image pyramid, I^h is the image in h layer, and ω_i^h are the corresponding weights.

Latent Variables for Feature Selection. We explicitly introduce a set of latent variables for voting features, removing spurious votes belonging to background and accounting for dependency between local image patches. Specifically, a hidden variable $z_\mathbf{x}$ is associated with each image patch $I_\mathbf{x}$, indicating whether image patch $I_\mathbf{x}$ is selected for object modeling.

A. Feature Pruning with Background Modeling. For each object hypothesis, we can take the back-projection stage into consideration and introduce a constant background model ε. The Hough score can be formulated as

$$S(\mathbf{y}, \mathbf{z}) = \sum_\mathbf{x} z_\mathbf{x} \left(\sum_{i \in N_\mathbf{x}} \omega_i p(C_i|\mathbf{f}_\mathbf{x}) K_i(\mathbf{y}, \mathbf{x}) - \varepsilon \right) \tag{4}$$

where $z_\mathbf{x}$ is a binary variable that indicates whether image patch $I_\mathbf{x}$ is selected by the back-projection while the constant ε is a small threshold to remove the "trivial" voting elements in the back-projection stage. As in standard Hough-based methods, optimizing Eq. (4) can be solved with dynamic programming due to its star-shape structure. Given a hypothesis $\mathbf{y}$, the optimal $\mathbf{z}$ can be found by simply removing all the image patches $I_\mathbf{x}$ that satisfy $\sum_{i \in N_\mathbf{x}} \omega_i p(C_i|\mathbf{f}_\mathbf{x}) K_i(\mathbf{y}, \mathbf{x}) < \varepsilon$.

B. Feature Selection with Additional Max Pooling. The assumption of conditional independence between features is simplistic, and too many voting elements that are spatially close to each other usually reduce performance. We divide the object support into subregions and use max-pooling method to compute the feature votes. In particular, we define the object score as

$$S(\mathbf{y}, \mathbf{z}) = \sum_m \max_{\mathbf{x} \in \mathbf{R}_m} z_\mathbf{x} \left(\sum_{i \in N_\mathbf{x}} \omega_i p(C_i|\mathbf{f}_\mathbf{x}) K_i(\mathbf{y}, \mathbf{x}) - \varepsilon \right) \tag{5}$$

where $\mathbf{R}_m$, $m = 1, 2, \ldots, M$ are the M sub-regions of the bounding-box. For the multiscale model and patches at layer h, we evenly divide the bounding-box into M_h regions such that $M_h \propto 1/h$.

We can consider max-pooling as an additional constraint imposed on the hidden variable $\mathbf{z}$, which chooses only one feature from each subregion. With the new constraint, we can rewrite the object score (4) as follows,

$$S(\mathbf{y}, \mathbf{z}) = \sum_{\mathbf{x}} z_{\mathbf{x}} \left(\sum_{i \in N_{\mathbf{x}}} \omega_i p(C_i|\mathbf{f}_{\mathbf{x}}) K_i(\mathbf{y}, \mathbf{x}) - \varepsilon \right)$$
$$\text{s.t.} \quad \mathbf{z}^T \cdot \mathbb{1}_{\mathbf{R}_m} = 1, \quad \forall m \in \{1, 2, \ldots, M\} \tag{6}$$

where $\mathbb{1}_{\mathbf{R}_m}$ is a binary vector that indicates the m-th sub-region. In particular, $\mathbb{1}_{\mathbf{R}_m}(\mathbf{x}) = 1$ if the patch $I_{\mathbf{x}}$ is located in the sub-region $\mathbf{R}_m$ of the bounding-box.

Cascaded Voting: Pose Estimation Based on the Selected Features and Their kNNs. Given the latent variable $\mathbf{z}$, it is easy to estimate the pose (orientation) of the detected object. As the pooling process significantly reduces the number of the involved patches, we can perform the slower yet more effective (compare to visual code matching) kNN method for selected patches and then vote for the current pose according to their nearest neighbors. In specific, the score of pose $S(\hat{\mathbf{y}})$ is calculated as

$$S(\hat{\mathbf{y}}) = \sum_{\mathbf{x}} z_{\mathbf{x}} \left(\sum_{i \in k\text{NN}_{\mathbf{x}}} \omega_i \Phi(\mathbf{f}_i, \mathbf{f}_{\mathbf{x}}) K_i(\mathbf{y}, \mathbf{x}) - \varepsilon \right),$$

where $\mathbf{z}$ is estimated in Eq. 6 and $\Phi(\mathbf{f}_i, \mathbf{f}_{\mathbf{x}})$ represents a χ^2 kernel between the training sample $\mathbf{f}_i$ and test patch $\mathbf{f}_{\mathbf{x}}$. Also note that $\hat{\mathbf{y}}$ could be merely the pose angle θ, the aspect ratio a, or their combination $\hat{\mathbf{y}} = [\theta, a]$.

3.3 Multiple Instance Learning for $M^2\text{HT}$

We first consider generalizing MIL to the standard $M^2\text{HT}$ learning framework, in which no latent variables are involved. Following the notation in Eq. (1), the original $M^2\text{HT}$ method learns the codebook weight coefficient $\boldsymbol{\omega} = \{\omega_1, \cdots, \omega_K\}$ based on a max-margin criterion.

Given a training set $\{(\mathbf{f}^n, y^n)\}_{n=1}^N$, where $y^n \in \{-1, 1\}$ and $\mathbf{f}^n$ are the input features, we denote the overall response to the ith code as $A_i^n = \sum_{\mathbf{x}:i \in N_{\mathbf{x}}} p(C_i|\mathbf{f}_{\mathbf{x}}^n) K_i(\mathbf{o}_c, \mathbf{x})$, and $A^n = [A_1^n, \cdots, A_K^n]^T$. The learning objective function for object detection can be defined as follows,

$$\min_{\boldsymbol{\omega}, b} \frac{1}{2} \|\boldsymbol{\omega}\| + C \sum_{n=1}^N \max\left(0, \ 1 - y^n(A^{nT}\boldsymbol{\omega} + b)\right)$$
$$\text{s.t.} \quad \boldsymbol{\omega} \succeq 0, \tag{7}$$

We apply the max-margin multiple instance learning framework to $M^2\text{HT}$ by defining the positive and negative bags as follows. In each training image I^n, we sample a set of bounding-box hypotheses and group them into a positive bag $\mathbb{B}_n^+$ and a negative bag $\mathbb{B}_n^-$ based on the original ground-truth bounding-box labels. To do so, we first compute their PASCAL overlapping scores with the ground-truth, and assign the instances with scores larger than the threshold δ^+ to $\mathbb{B}_n^+$, and those with scores smaller than δ^- to $\mathbb{B}_n^-$, where $\delta^+ > 0.5 > \delta^-$.

Given all the positive and negative instance bags, the learning objective for the MIL M^2HT can be written as,

$$
\min_{\boldsymbol{\omega},b} \frac{1}{2}\|\boldsymbol{\omega}\| + C \sum_{n=1}^{N} \left(\max[0, 1 - \max_{j\in\mathbb{B}_n^+}(A^{jT}\boldsymbol{\omega} + b)] \right.
$$
$$
\left. + \max[0, 1 + \max_{j\in\mathbb{B}_n^-}(A^{jT}\boldsymbol{\omega} + b)] \right)
$$
$$
\text{s.t.} \quad \boldsymbol{\omega} \succeq 0, \tag{8}
$$

Note that it is straightforward to extend this objective to the latent DFMs where we replace $\max_{j\in\mathbb{B}_n^\pm}$ by $\max_{j\in\mathbb{B}_n^\pm} \max_{\mathbf{z}_j}$ in the objective.

In this work, we further exploit the spatial dependency of those instances in each bag by replacing the "max" operation by Non-Maximum-Suppression (NMS) to generate (potentially) multiple positive instances in each positive image. The non-unique outputs of NMS offer us more training instances and more appearance variation. From the perspective of learning, the multiple samples in one bag is a regularization to prevent the learner from overfitting to a single instance.

4   Experiment

We evaluate our method on four datasets, *i.e.*, the ETHZ-Shape dataset, the car dataset in the PASCAL Visual Object Classes Challenge 2007 (VOC2007) [26], the EPFL multiple-view car dataset [27], and a VOC2007 car pose estimation dataset [9]. The first two data sets are used to evaluate the object detection performance of the DFM model. For joint object detection and pose estimation, we test our cascaded voting strategy on the latter two datasets.

4.1   Experiment Settings

We use the same parameter setting in all the experiments in this paper. We choose Geometrical Blur [28] as the localization feature and HOG [13] as the prediction feature. For the parameters of GB, we exactly follow the setting employed in [28]. The HOG cell size is 8 while we extract each HOG feature from a 4×4 block of cells. We reduce the dimension of both the GB feature and HOG feature via PCA. During training, we normalize the height of the training objects to the median values of their original heights. All the codes in codebook $\mathbb{C}$ are extracted on 6-layer image pyramids generated from the training images and the scale interval is 0.8.

We perform Gaussian blurring on the Hough voting map of each visual code. The filter variance $\sigma_f = 0.02h^\star$ where $h^\star$ is the height of the object hypothesis.

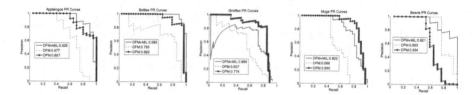

Fig. 2. The AP curves for ETHZ-Shape dataset. The curves are obtained by using the "vanilla" DFM, the DFM with MIL learner and the DPM method.

During testing, we also build an image pyramid for each test image with the same scale step. The bounding-boxes obtained in each layer are merged together and Non-Maximum-Suppression is then performed to greedily select the most promising. For parameter learning, we set the σ defined in Sect. 3.2 to 5 and the C value for multiple-instance-learning is found using cross-validation.

4.2   Object Detection

ETHZ-Shape. We first test the DFM method on the ETHZ-Shape Dataset, which is commonly used for Hough-based detectors. It contains 255 test images and involves five shape-based classes (apple logos, bottles, giraffes, mugs, and swans). For training, we follow the setting in [4], and use half the positive examples and the same number of negative examples.

To evaluate the complementary feature-pair and the multiple-instance-learning method, we illustrate the performance improvements via sequentially adding each individual component. We also report the Average Precision (AP) of detection for all the shape classes in Table 1, compared with 3 recently-proposed Hough-based detection algorithms and DPM. From the table we can see DFM achieves the 3 best performances in 5 shape categories and beats other methods in terms of average accuracy. An interesting comparison can be conducted between our method and the Group-Hough detection algorithm [5]. In their method, voting elements are first grouped together by performing alternative inference for jointly maximizing their geometrical consistency and the voting agreement. We can see that, instead of complex inference, we achieve the similar performance by using complementary feature pairs and the associated LDA-based classifiers. Another comparison is between the M^2HT and our methods. Note that after trained by the proposed MIL method, DFM outperforms the M^2HT with a large gap.

Figure 2 shows the precision-recall curves of the DPM algorithm, the "vanilla" DFM (with feature-pairs but no learning procedure) and the complete version of our method. As shown in the figure, without the proposed Multiple-Instance-Learning process, DFM already performs well but is still inferior to DPM. The MIL method increases the accuracy significantly, with the proposed MIL learning method, our DFM outperforms DPM.

Another important factor contributing to the success of DFM is the max-pooling approach. It usually captures the distinct parts of the object in detection

Table 1. Detection APs on the ETHZ-Shape dataset.

	Apple	Bottle	Giraffe	Mug	Swan	Avg.
Cluster+Verif [19]	75	65	50	75	60	65
M^2-Hough+Verif [4]	88.5	74	75	81.5	74	78.6
Group-Hough [5]	80	85	49	71	60	69
DPM [1]	89.7	89.3	**77.4**	**86.0**	60.3	81.5
DFM	67.7	73.3	55.7	58.6	85.3	75.6
DFM + MIL	**92.5**	**98.3**	68.6	82.2	**92.1**	**86.9**

while reduce the redundant evidence as well as the noise. Figure 3 demonstrates the selected regions by max-pooling. We can see that as image pyramids are employed, image patches with different scales are selected. In the same scale level, the selected patches do not cluster together and some intuitively distinct parts of the object, *e.g.*, the long neck of the giraffe, the handle of the cup and the curly neck of the swan are found by DFM.

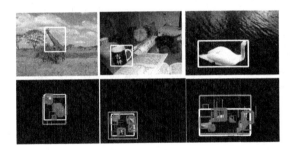

Fig. 3. Demonstration of shape detection using our DFM. The two rows shows the original test images and the corresponding detection results. In the top row, the ground-truth bounding-boxes are labeled in blue color. The detected bounding-boxes are shown in green in the bottom row. We use small blocks with different widths and colors to show the results of max-pooling inference. The wider and lighter the block is, the more important it is in determining the location of the objection (Color figure online).

To evaluate our method on larger and more challenging dataset, we perform DFM on the car dataset from PASCAL Visual Object Classes Challenge 2007. The detection APs on the VOC2007 car dataset of our method is 33.5 %, which is higher than other Hough-based method such as [6] (28 %), [29] (30 %), [7] (16.6 %), while still less than the DPM method which achieves 45 % on this test.

4.3   Car Pose Estimation

To evaluate the proposed cascaded voting strategy proposed in Sect. 3.2, DFM is used to predict bounding-boxes and viewing angles of cars on two standard viewpoint datasets, *i.e.*, the "clean" VOC2007 car dataset and the EPFL car show dataset.

We use the Leuven Car dataset as training data. There are 7 annotated viewing angles in the Leuven Car dataset, with minimum angle difference of 30°. As it only covers one side ($[0°, 90°]$, $[270°, 360°]$) of the viewing points around a car, when testing, we need to mirror the test image and predict the pose according to the score of the detected bounding-boxes. To form the training set, from each angle we randomly select 100 positive images and use the Caltech-256 dataset (remove all the car-related images) as negative samples. In other words, during training, we only have the bounding-box of the training object and rough estimation (with the resolution 30°) of the object' pose/viewing angle.

Despite the weak-supervision information in the training data, our method achieves remarkably high performance in the comparison with some state-of-the-art methods which use comprehensive annotations and sophisticated 2D to 3D transformations.

Joint Localization and Pose Estimation on EPFL Car Dataset. Another commonly used car pose dataset is the EPFL multiple-view car dataset. The EPFL car dataset was introduced in [27]. The dataset was acquired at a car show, 20 different cars were imaged every 3 to 4 degrees while the cars were rotating. There are approximately 2000 images in the dataset. Usually the first 10 cars, each corresponding to a image sequence, are used as training data. Accordingly, we use both the Leuven car dataset and the first 10 image sequences of EPFL as our training set and verify our method on the other 10 image sequences. We compare the DFM with the method proposed in [21]. Note that the comparison is fair as the authors of [21] also use extra (around 2000) car images when training their model.

Our DFM method outperforms the method proposed in [21] in terms of pose estimation and achieves a similar detection AP. In particular, our detection AP is 88.3 % (recall rate is 99.1 %), which is slightly inferior to their performance 89.54 %. However, the median error our angle prediction is 17.85°, which is around 28 % lower than their error rate 24.83°. Figure 4 illustrates the comparison between the two error histograms.

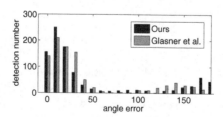

Fig. 4. The histogram of viewing angle errors for the EPFL car dataset. We compare the performance of our DFM algorithm (blue) with the 3-D voting based method (red) proposed in [21] (Color figure online).

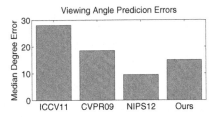

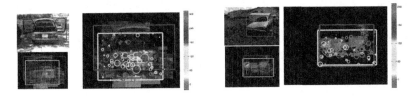

Fig. 5. The viewing angle errors on the VOC2007 car dataset. Here "ICCV2011" refers to the work [21], "CVPR09" is the work [9] while "NIPS12" stands for [15].

Fig. 6. Demonstration of car detection and pose estimation using our method. The two rows show the joint procedure for two cars, for each row, two small (from top to bottom) images are the test image with ground-truth bounding box and the detection bounding-box (green) with the max-pooling results respectively. The large image shows the pose estimations by the supporting patches. Note that the viewing angles distribute in [0, 360] and we use colors specify their values. The size of the circles in the right column indicates the importance of the supporting patches in the pose estimation (Color figure online).

Joint Detection and Pose Estimation on VOC2007 Car (subset). The Pascal data includes only coarse pose labels (frontal/rear/right/left). ArieN-achimson and Basri [9] augmented the pose labels on a subset of the 2007 test category. The subset contains 199 relatively "clean" cars and they labeled the car instances with one of 40 labels which correspond to an approximately uniform sample of viewing angles (with the step around 9°). We perform cascaded voting for both the bounding box and pose prediction. Figure 5 illustrates the median values of viewing angle error for DFM and three other methods using 3D information.

As can be seen in the Fig. 5, DFM achieves competitive results compared with those 3-D based approaches. Specifically, our method (15.0°) outperforms the approaches proposed in [9] (18.0°) and [21] (28.3°) and is comparable to the much more sophisticated method which requires 21 part annotations on training samples, [15] (9.5°). As to the detection rate, our algorithm achieves the AP 73.5 % (recall rate is 97.5 %) which is significantly higher than that reported in original work [9].

The joint-voting mechanism plays an important role in increasing the angle estimation accuracy. In Fig. 6, we can see that the constraint on the support regions makes the voting procedures in both subspaces less noisy and many patches are important for both detection and pose prediction. We show more analysis and experiment results for the joint-voting method in the supplementary part.

5   Conclusion

In this paper, we have introduced a new Hough Transform with latent structure for jointly detecting object and estimating its pose. Our method improves the traditional Hough-based object models in three aspects: First, an effective patch codebook is learned by pairing complementary features, providing a robust and discriminative appearance model; In addition, it incorporates feature pooling and selection in a principled way by modeling feature activation with latent variables; Finally, a multiple instance max margin framework is derived for learning the codebook weights in Hough Transform with noisy bounding box labels. We have demonstrated the efficacy of our approach by evaluation on four datasets, on which our method achieves state-of-the-art detection performance and superior pose estimation results.

Acknowledgement. This work is supported by National Natural Science Foundation of China (Project NO: 61503168).

References

1. Felzenszwalb, P., McAllester, D., Ramanan, D.: A discriminatively trained, multi-scale, deformable part model. In: CVPR (2008)
2. Wan, L., Eigen, D., Fergus, R.: End-to-end integration of a convolution network, deformable parts model and non-maximum suppression. In: Proceedings of the IEEE Conference on Computer Vision and Pattern Recognition, pp. 851–859 (2015)
3. Zia, M.Z., Stark, M., Schindler, K.: Explicit occlusion modeling for 3d object class representations. In: CVPR (2013)
4. Maji, S., Malik, J.: Object detection using a max-margin hough transform. In: CVPR (2009)
5. Yarlagadda, P., Monroy, A., Ommer, B.: Voting by grouping dependent parts. In: Daniilidis, K., Maragos, P., Paragios, N. (eds.) ECCV 2010, Part V. LNCS, vol. 6315, pp. 197–210. Springer, Heidelberg (2010)
6. Razavi, N., Gall, J., Kohli, P., Van Gool, L.: Latent hough transform for object detection. In: Fitzgibbon, A., Lazebnik, S., Perona, P., Sato, Y., Schmid, C. (eds.) ECCV 2012, Part III. LNCS, vol. 7574, pp. 312–325. Springer, Heidelberg (2012)
7. Gall, J., Yao, A., Razavi, N., Van Gool, L., Lempitsky, V.: Hough forests for object detection, tracking, and action recognition. TPAMI **33**(11), 2188–2202 (2011)
8. Leibe, B., Leonardis, A., Schiele, B.: Robust object detection with interleaved categorization and segmentation. IJCV **77**(1), 259–289 (2008)
9. Arie-Nachimson, M., Basri, R.: Constructing implicit 3d shape models for pose estimation. In: CVPR (2009)
10. Razavi, N., Gall, J., Van Gool, L.: Backprojection revisited: scalable multi-view object detection and similarity metrics for detections. In: Daniilidis, K., Maragos, P., Paragios, N. (eds.) ECCV 2010, Part I. LNCS, vol. 6311, pp. 620–633. Springer, Heidelberg (2010)

11. Dalal, N., Triggs, B.: Histograms of oriented gradients for human detection. In: CVPR (2005)
12. Razavian, A.S., Azizpour, H., Sullivan, J., Carlsson, S.: Cnn features off-the-shelf: an astounding baseline for recognition. In: 2014 IEEE Conference on Computer Vision and Pattern Recognition Workshops (CVPRW), pp. 512–519. IEEE (2014)
13. Dalal, N., Triggs, B.: Histograms of oriented gradients for human detection. In: CVPR, vol. 1, pp. 886–893. IEEE (2005)
14. Andrews, S., Tsochantaridis, I., Hofmann, T.: Support vector machines for multiple-instance learning. In: NIPS (2002)
15. Hejrati, M., Ramanan, D.: Analyzing 3d objects in cluttered images. In: NIPS (2012)
16. Ballard, D.: Generalizing the hough transform to detect arbitrary shapes. PR **13**(2), 111–122 (1981)
17. Leibe, B., Leonardis, A., Schiele, B.: Combined object categorization and segmentation with an implicit shape model. In: Workshop on Statistical Learning in Computer Vision, ECCV (2004)
18. Leibe, B., Leonardis, A., Schiele, B.: Robust object detection with interleaved categorization and segmentation. IJCV **77**(1–3), 259–289 (2008)
19. Ommer, B., Malik, J.: Multi-scale object detection by clustering lines. In: CVPR (2009)
20. Payet, N., Todorovic, S.: From contours to 3d object detection and pose estimation. In: ICCV (2011)
21. Glasner, D., Galun, M., Alpert, S., Basri, R., Shakhnarovich, G.: Viewpoint-aware object detection and pose estimation. In: ICCV (2011)
22. Lowe, D.: Distinctive image features from scale-invariant keypoints. IJCV **60**(2), 91–110 (2004)
23. Bourdev, L., Malik, J.: Poselets: body part detectors trained using 3d human pose annotations. In: ICCV (2009)
24. Girshick, R., Donahue, J., Darrell, T., Malik, J.: Rich feature hierarchies for accurate object detection and semantic segmentation. In: 2014 IEEE Conference on Computer Vision and Pattern Recognition (CVPR), pp. 580–587. IEEE (2014)
25. Toshev, A., Szegedy, C.: Deeppose: human pose estimation via deep neural networks. In: 2014 IEEE Conference on Computer Vision and Pattern Recognition (CVPR), pp. 1653–1660. IEEE (2014)
26. Everingham, M., Van Gool, L., Williams, C., Winn, J., Zisserman, A.: The pascal visual object classes challenge 2007 (voc 2007) results (2007) 11 (2008)
27. Ozuysal, M., Lepetit, V., Fua, P.: Pose estimation for category specific multiview object localization. In: CVPR (2009)
28. Berg, A.C., Malik, J.: Geometric blur for template matching. In: CVPR (2001)
29. Zhang, Y., Chen, T.: Implicit shape kernel for discriminative learning of the hough transform detector. In: BMVC (2010)

Consensus Guided Multiple Match Removal for Geometry Verification in Image Retrieval

Hong Wu[1]([⊠]), Xing Heng[1], and Zenglin Xu[1,2]

[1] School of Computer Science and Engineering, University of Electronic Science and Technology of China, Chengdu, China
{hwu,xheng,zlxu}@uestc.edu.cn
[2] Big Data Research Center, University of Electronic Science and Technology of China, Chengdu 611731, Sichuan, China

Abstract. State of the art image retrieval methods are mostly based on the bag-of-features (BOF) representation or its variations. Despite its success, BOF ignores the geometric relationships among local features, and thus leading to limited retrieval accuracy. This limitation has resulted in the necessity of geometric verification of the initial matches detected by visual word based matching. However, this approximate matching usually leads to many-to-many mapping, and thus introduces many false matches impairing geometric verification. To address this problem, we propose a Consensus-Guided Multiple Match Removal algorithm, which selects the confident matches relying not only on the quality of the matches, but also on their geometric consistency with others. For efficiency, the geometric consistency is verified by exploring match distribution over the Hough space instead of pairwise comparison. The matches in conflict with the confident ones are then removed based on a feature mapping constrain. Our multiple match removal method can be combined with existing geometric verification methods to improve image retrieval. Finally, we evaluate the proposed method on three benchmark datasets in comparison with previous multiple match removal method. The experimental results indicate that multiple match removal is a useful step for geometric verification, and our method can outperform the previous multiple match removal method and further improve the image retrieval performance.

Keywords: Image retrieval · Multiple match removal · Geometric verification

1 Introduction

In last decades, image retrieval has attracted increasing interests. The bag-of-features (BOF) representation was introduced in image retrieval by Sivic and Zisserman [1], and has become popular due to its effectiveness and efficiency. Among BOF approaches, local features, such as SIFT [2], are widely utilized because of their viewpoint invariant characteristic. Local features are extracted

© Springer International Publishing Switzerland 2016
Q. Tian et al. (Eds.): MMM 2016, Part II, LNCS 9517, pp. 130–139, 2016.
DOI: 10.1007/978-3-319-27674-8_12

at interest points, and quantized into visual words from a visual vocabulary, which is constructed by clustering feature descriptors. Images are then represented by the frequency histograms of visual words. At the retrieval stage, images are ranked based on the cosine similarities between the query vector and image vectors, and speed-up can be further achieved by using inverted file.

Studies have indicated that BOF-based image retrieval is intrinsically a voting approach based on approximate feature matching [3]. That is, one feature from a query and one from a database image are considered match if they are assigned the same visual word and the number of matches is used to measure the similarity between the two images. By avoiding exhaustively evaluating similarities between features, the BOF approach achieves high efficiency and can be applied to large scale image retrieval. However, a major limitation of the BOF approach is that it ignores the geometric relationships among local features and thus leads to limited retrieval accuracy. To solve this problem, geometric verification [1,3,4] was proposed to improve retrieval accuracy by checking the geometric consistency between the query image and database images.

One conventional approach to geometric verification is RANSAC [5], which is able to perform careful geometry verification by iteratively estimating parameters of a geometric model (e.g. homography, affine, or similarity) from a set of matched local features. But RANSAC is computationally slow and its performance is poor when the percent of inliers falls much below 50 %. Philbin et al. [4] used the RANSAC-based [6] method to estimate affine transformation, in which a limited degree of freedom affine transformation hypothesis is generated from each single pair of corresponding local features. Matching then becomes deterministic by enumerating all hypotheses. However, this approach is also quadratic in the number of correspondences. In this paper, we use the noun match and correspondence interchangeably.

Different from RANSAC-based methods, Lowe [2] applied a Hough scheme to find consistently transformed correspondences in the parameter space, and use them to estimate a finer affine transformation. Images are similar if such a transformation can be estimated and results in a large number of inliers. To achieve high efficient geometric verification, Weak Geometry Consistency (WGC) was proposed without explicitly estimating the affine transformation [3]. WGC assumes that the correct matched features should have similar scale and rotation transformation, and uses the peak values of scale and orientation differences histograms to measure image similarity. To improve the discrimination of BOF representation, Hamming Embedding (HE) was also proposed in [3] to refine the visual word representation by adding a binary signature.

Despite the usefulness, visual-word based feature matching can often result in a combinatorial explosion of the number of correspondences and corrupted similarity measures, when one word appears many times in an image in a small vocabulary setting, under the presence of repeating structures, or with soft assignment models as shown in Philbin et al. [7]. To alleviate the problem, Jégou et al. [8] proposed a multiple match removal strategy and two weighting functions for intra- and inter-image burstiness. Their multiple match removal method keeps the best match which has the minimum hamming distance for a query feature and remove all other matches associate with this feature. The multiple match removal method in [9] was designed to preserving as many correspondences as possible. It starts from the point that has fewest matching correspondences assigned, selects the match with the highest weight, and then discard other matches in conflict with it based on one-to-one mapping constrain. Both of the methods select confident matches only depending on their goodness. However, it was observed in [10] that correct correspondences are likely to form pairwise geometric agreements with each other while incorrect ones are unlikely to form these agreements. Therefore, not only the goodness of a match but also its pairwise geometric relationship with others should be taken into consideration when selecting a confident match. The geometric verification method proposed in [10] selects confident matches from the main strongly connected cluster found by spectral method and removes conflicts based on a given mapping constrain. But their method is time consuming due to the measuring of pairwise geometry and eigenvector computation.

To address the problems of previous algorithms, we propose a Consensus Guided Multiple Match Removal algorithm, which selects the confident matches relying not only on the goodness of the matches, but also on their geometric consistency with others. For efficiency, geometric consistency is evaluated by exploring the correspondence distribution over the Hough space instead of pairwise comparison. In more detail, it employs histograms of the Hough space to find the stronger clusters, iteratively selects correspondences from them, and removes conflicts based on the one-to-one mapping constraint. Our method can be used to filter matches before geometric verification and combined with existing geometric verification methods, such as WGC. The experimental results indicate that our method can outperform previous simple multiple match removal strategy for visual word based matching and further boost the image retrieval performance.

The rest of the paper is organized as follows: Sect. 2 describes feature matching methods based on Bag of Features and Hamming Embedding. Section 3 introduces the principle of the geometric verification based on Hough voting. Section 4 introduces our Consensus Guided Multiple Match Removal algorithm. Section 5 reports the experiment setup, the results obtained and the analysis on them. Conclusions are drawn in the last section.

2   Approximate Feature Matching

Given two sets of local features from two images, the similarity between the two images can be measure by matching the local features from the two sets. Usually, two features are considered a match if their distance is smaller than a threshold and/or it passes the distance ratio test [2], but pair-wise comparison is very time consuming. Therefore, the approximate approach is used for feature matching in image retrieval, where two features are considered a match if they are assigned the same visual word. Given feature i from query image and feature j from a database image, the matching function f can be defined as

$$f(i,j) = \begin{cases} 1 & if \; q(i) = q(j) \\ 0 & otherwise \end{cases} \qquad (1)$$

where $q(i)$ and $q(j)$ are the index of visual words assigned to feature i and j respectively. The features i and j with $f(i,j) > 0$ form a match or correspondence $c = (i,j)$. The matching score can also be weighted by the square of the inverse document frequency (idf) of the visual word, $idf^2(q(i))$, as in [3,8]. The matching scores of all correspondences are accumulated to compute the similarity measure of the two images. In [3], Humming Embedding (HE) was proposed to refine the visual word $q(i)$ with a d-dimensional binary signature $b(i) = (b_1(i), ..., b_d(i))$ that encodes the location of the feature within the Voronoi cell. The binary signatures can be used to refine visual-word-based feature matching. Then the matching function f can be defined as

$$f(i,j) = \begin{cases} 1 & if \; q(i) = q(j) \; and \; h(b(i), b(j)) \le h_t \\ 0 & otherwise \end{cases} \qquad (2)$$

where h is the Hamming distance between the two signatures $b(i)$ and $b(j)$, and h_t is a fixed Hamming threshold. This means, two features are considered a match if they are not only assigned the same word but also have a small hamming distance.

When using small codebook, under the presence of repeating structures, the "burstiness" of visual word in image will happen. And the number of matches will increase dramatically. In this case, one feature may be mapped to several features in another image, and vice versa. For example, if a visual word u appear m times in a query image and n times in a database image, there will be $m \times n$ matches based on this word. Fig. 1. illustrates this phenomenon. Vertices on the left (right) represent query (database) features, and the links between vertices

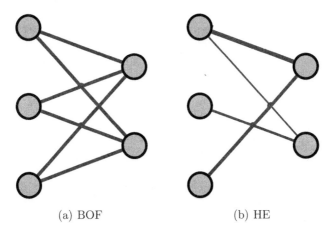

(a) BOF   (b) HE

Fig. 1. Feature mapping based on BOF or HE. Vertices on the left (right) represent query (database) features, and links represent matching between features.

represent matching between features. Then both the features and the feature mapping are represented with a bipartite graph. There are three features from a query image assigned to a visual word u and two descriptors from a database image also assigned to the same visual word. The subfigure (a) shows the feature matches generated based on visual word, and (b) shows the matches generated based on Hamming Embedding. The link width represents the strength of a match. For BOF, all the links are with same width. For HE, the matches with smaller hamming distance have larger width. In (b), there's no links between some query features and database features because of the Hamming threshold used. Obviously, many of correspondences generated by the two approaches are incorrect and will do harm to the following spatial verification.

3   Geometric Verification by Hough Voting

The affine transform is a good model for 3D projection of planar surfaces, and can be used for geometric verification. But the estimation of an affine model is costly which prevents the model from being applied to large scale image retrieval. The similarity transform is an approximation of then affine transform. Given a point i at (x_i, y_i) and a point j at (x_j, y_j), the similarity transform gives the mapping from i to j in terms of an image scaling s, an image rotation θ, and an image translation, $[t_x t_y]$:

$$\begin{bmatrix} x_j \\ y_j \end{bmatrix} = s \times \begin{bmatrix} \cos\theta & -\sin\theta \\ \sin\theta & \cos\theta \end{bmatrix} \times \begin{bmatrix} x_i \\ y_i \end{bmatrix} + \begin{bmatrix} t_x \\ t_y \end{bmatrix} \tag{3}$$

The similarity transform is more suitable for geometric verification in image retrieval, because it has only 4 parameters instead of 6 in the affine model and each of parameters can be estimated with only one correspondence by exploiting the local shape of the features. Each of the detected interest points has 4 parameters: $2D$ location $[x, y]$, scale s, and orientation θ. Given two matched features, the four parameters of their corresponding interest points can be used to approximate the four parameters of the similarity transform between the two features. More formally, the four parameters of the similarity transform can be approximated as follows. The scale s is approximated by the scale difference as

$$\hat{s} = 2^{(s_i - s_j)}, \tag{4}$$

the rotation angle θ is approximated by the orientation difference as

$$\hat{\theta} = \theta_i - \theta_j, \tag{5}$$

and the translation can be approximated as

$$\begin{bmatrix} \hat{t_x} \\ \hat{t_y} \end{bmatrix} = \begin{bmatrix} x_j \\ y_j \end{bmatrix} - \hat{s} \begin{bmatrix} \cos\hat{\theta} & -\sin\hat{\theta} \\ \sin\hat{\theta} & \cos\hat{\theta} \end{bmatrix} \begin{bmatrix} x_i \\ y_i \end{bmatrix}. \tag{6}$$

Therefore, each correspondence can vote once in the 4-dimension Hough space to predict the model location. Lowe [2] used this approach to find consistently transformed correspondences to estimate a full affine transform model. For efficiency,

WGC [3] only exploits partial geometric information, and checks for consistency of the rotation and scale hypotheses obtained from each correspondences. The peak values of scale and orientation differences histograms are considered as the number of geometrically consistent correspondences and used to measure the similarity between two images.

4   Consensus Guided Multiple Match Removal

It is presented in Sect. 2, that there will be a combinatorial explosion of the number of matches when "burstiness" of visual words occurs. In this case, many of the correspondences are outliers, or incorrect correspondences which do harm to spatial verification. A natural way to remove the redundant matches caused by visual word based matching is to first select the confident matches and then remove other matches in conflict with them based on some feature mapping constraint. In [8], for each query descriptor, the best match which has the smallest hamming distance is selected, and all the other matches associated with this query descriptors are discarded. By observing the fact that correct correspondences are likely to form pairwise geometric agreements with each other while incorrect ones are unlikely to form these agreements, Leordeanu and Hebert [10] found consistent correspondences by taking in consideration both how well the features' descriptors match and how well their pairwise geometric constraints are satisfied. They build a sparse affinity matrix of correspondences and greedily recover inliers from the main strongly connected cluster. Since this method exploits the pairwise geometric relation between correspondences, it is at least quadratic in the number of correspondences and not suitable for large scale image retrieval. To avoid pairwise comparison, our Consensus Guided Multiple Match Removal algorithm uses a histogram of the Hough space to find stronger clusters. From the stronger clusters, confident correspondences are iteratively selected and their conflict correspondences are removed.

Algorithm 1 gives a more formally description of our algorithm. The initial set of correspondences formed by Eqs. (1)(2) is input to our algorithm, and a new set of correspondences is generated after removing multiple matches from the initial set. The geometric consistence between correspondences is measured in the parameter space which is uniformly partitioned by B. If two correspondences vote in the same region or bin of the partition, we can say, they agree on their pairwise geometry. For efficiency, we use the translation (t_x, t_y) to form the parameter space in this paper. It is reasonable to think that two correspondences with similar translations are geometrically consistent under a similarity transformation. The correspondences from the denser bin have geometric agreement with more correspondences, therefore are more likely to be correct ones. First, each correspondence c is assigned to a set $H(b)$ according to the bin it belongs to $(b = \beta(c))$ in lines 2–5. Then, the largest bin is chosen in line 7, and a correspondence is chosen as the correct one from this bin in lines 9–13. If Hamming Embedding is used, a correspondence is randomly chosen from the best correspondences which has the smallest hamming distance. If BOF is used, a correspondence is randomly chosen from this bin. After the confident correspondence is chosen, all correspondences in conflict with it are erased in line 15. The confliction is defined by the mapping constrain used. In [8], the mapping constrain is that every query feature can vote only once for a database image. For a correspondence (i, j), the conflict correspondences have the form of $(i, *)$. This constrain allows two query feature to map to the same image feature. However, image matching implies that one point in one image can only have one corresponding point in another image. Therefore, we adopt the one-to-one mapping constrain in this paper. And the conflict correspondences have the form of $(i, *)$ or $(*, j)$. In implementation, we build indexes from feature Ids to correspondence Ids to make the conflicts be found more efficiently. In fact, our algorithm can accommodate different mapping constrains.

After removing multiple matches, the returned correspondences can be used for geometric verification, like fast spatial matching [4] and WGC [3].

Algorithm 1. Consensus Guided Multiple Match Removal

Require: C is a set of correspondences (c), B is a partition of the parameter space, including a set of bins (b).
Ensure: D is the set of correspondences after removing multiple match from C.
 1: **procedure** CGMMR(C, B)
 2:     **for all** $b \in B$ **do** $H(b) \leftarrow \emptyset$
 3:     **end for**
 4:     **for all** $c \in C$ **do** $H(\beta(c)) \leftarrow H(\beta(c)) \cup c$
 5:     **end for**
 6:     **while** $max(|H(b)|) > 0$ **do**
 7:         random pick b^* with $max(|H(b)|)$
 8:         **while** $|H(b^*)| > 0$ **do**
 9:             **if** HE **then**   ▷ for HE
10:                 randomly pick a $c \in BestMatch(H(b^*))$
11:             **else**   ▷ for BOF
12:                 randomly pick a $c \in H(b^*)$
13:             **end if**
14:             $D \leftarrow D \cup \{c\}$
15:             remove from all H(b) the correspondences conflict with c
16:         **end while**
17:     **end while**
18:     **return** D
19: **end procedure**

---

5   Experiments

In this section, we evaluate our Consensus Guided Multiple Matching Removal algorithm for geometry verification in image retrieval.

5.1 Datasets and Evaluation

Experiments are conducted on three publicly available datasets, commonly used in the related works, namely Oxford [4], Holiday [3], and Kentucky [11].

The Oxford building dataset contains a test set of 5K images from Flickr. There are 55 query images corresponding to 11 distinct buildings. Each query is a rectangular region delimiting the building in the image. The relevant results for a query are the other images of this building. The Holidays dataset includes 1491 images of holiday, and is divided in 500 groups, each of which represents a distinct scene or object. The first image of each group is the query, the others are relevant images for this query. The Kentucky object recognition benchmark contains 2550 different objects or scenes. Each one is represented by four images taken from four different viewpoints, resulting in totally 10200 images. Each image of the database is used as a query. The correct results are the image itself and the three other images of the same object.

The mean average precision (mAP) is used to measure the retrieval performance. For the Kentucky dataset, we also provide Kentucky Score (KS), which is the average number of relevent images in the top four results, since this performance measure is usually reported for this dataset. Feature descriptors are obtained by the Hessian-Affine detector and the SIFT descriptor, implemented in the software of Mikolajczyk [12] with the default parameters. Visual vocabulary of size 20k is learning by k-means clustering on the independent Flickr60k dataset and used for all experiments. The vocabulary and the visual features of Holiday data are downloaded from the homepage of Hervé Jégou, the author of [3].

5.2 Experimental Setup

In our experiments, Weak Geometric Constrain (WGC) [3] is used as geometric verification method for its effectiveness and efficiency. Our multiple match removal method (CGMMR) is compared to the method (MMR) used in [8] in combination with WGC, and also to WGC without multiple match removal. Our experiments are performed based on two kind of representation, Bag of Features (BOF) and Hamming Embedding (HE). In BOF, the square of the inverse document frequency (idf) of the visual word is used as the matching score. The ranking score of a database image with respect to the query image is computed as the sum of the individual matching scores divided by the L2 norm of the bag of feature vectors. MMR used in [8] is originally designed for HE. When applying it to BOF, we randomly select the confident match for each query descriptor. In HE, the signature length is set to 64 and $h_t = 24$, as done in [3,8]. The idf wight is multiplied by a Hamming based weight used in [8] as the matching score. L2 normalization is also used for computing the ranking score. Standard WGC is used in our experiments, without using orientation and scale priors. The angle and scale differences are all uniformly quantized to 16 bins. In CGMMR, the translation is used to form the parameter space, and both t_x and t_y are uniformly quantized to 16 bins.

5.3 Experimental Results

The retrieval performances of the different methods on the three dataset are listed in Table 1. The performances of our baseline are close to those reported in [3,8], and WGC and HE can improve retrieval performance as reported. From Table 1, it can be seen that removing multiple match can further improve retrieval performance for not only BOF but also HE. For BOF/WGC, MMR increases the mAP by 1 % to 2.5 %(absolute percentage points) depending on the datasets, and our CGMMR increases the mAP by 2 % to 4 % for all datasets. For HE, MMR increases the mAP by 1 % to 3 % depending on the datasets, and our CGMMR increases the mAP by 2 % to 4 %. For all cases, the methods with CGMMR outperform those with MMR.

Table 1. The mAPs of the different methods on the three datasets

Method	Kentucky		Oxford	Holidy
	KS	mAP	mAP	mAP
BOF	2.86	0.762	0.331	0.444
WGC	3.08	0.806	0.463	0.623
WGC+MMR	3.17	0.824	0.480	0.648
WGC+CGMMR	3.32	0.862	0.505	0.667
HE	3.30	0.856	0.513	0.737
HE+WGC	3.28	0.85	0.544	0.759
HE+WGC+MMR	3.34	0.862	0.579	0.771
HE+WGC+CGMMR	3.37	0.8	0.589	0.782

Table 2. The average percentages of matches removed by multiple match removal methods

Method	Kentucky	Oxford	Holidy
WGC+MMR	27.4	31.8	26.6
WGC+CGMMR	42.5	49.2	41.7
HE+WGC+MMR	9.2	11.2	7.9
HE+WGC+CGMMR	16.2	19.9	14.2

Table 2 lists the average percentages of matches removed by multiple match removal methods. It can be seen, MMR removes about 27 % to 32 % matches for BOF, about 8 % to 11 % for HE depending on the datasets. CGMMR removes 42 % to 50 % matches for BOF, about 14 % to 20 % for HE. Multiple match removal methods remove more matches for BOF than HE. This is because HE-based matching already removes some possible false matches by the hamming threshold. And CGMMR removes more matches than MMR mainly due to the one-to-one mapping constrain and also the consensus guided manner.

6   Conclusion

This paper introduces a Consensus Guided Multiple Match removal algorithm to discard possible false matches introduced by many-to-many feature mapping. Our method can be combined with existing geometric verification methods to improve image retrieval. The experimental results indicate that our method can outperform the previous multiple match removal strategy and further boost the image retrieval performance.

Acknowledgments. This work was supported by projects of NSF China (No. 61572111, 61433014, 61440036), a 973 project of China (No.2014CB340401), a 985 Project of UESTC (No.A1098531023601041) and a Basic Research Project of China Central University (No. ZYGX2014J058).

References

1. Sivic, J., Zisserman, A.: Video google: a text retrieval approach to object matching in videos. In: Ninth IEEE International Conference on computer vision, pp. 1470–1477. Nice (2003)
2. Lowe, D.: Distinctive image features from scale-invariant keypoints. Int. J. Comput. Vision **60**(2), 91–110 (2004)
3. Jegou, H., Douze, M., Schmid, C.: Hamming embedding and weak geometric consistency for large scale image search. In: Forsyth, D., Torr, P., Zisserman, A. (eds.) ECCV 2008, Part I. LNCS, vol. 5302, pp. 304–317. Springer, Heidelberg (2008)
4. Philbin, J., Chum, O., Isard, M., Sivic, J., Zisserman, A.: Object retrieval with large vocabularies and fast spatial matching. In: IEEE Conference on Computer Vision and Pattern Recognition, pp. 1–8. Minneapolis (2007)
5. Fischler, M., Bolles, R.: Random sample consensus: a paradigm for model fitting with applications to image analysis and automated cartography. Commun. ACM **24**(6), 381–395 (1981)
6. Chum, O., Matas, J., Obdrzalek, S.: Enhancing RANSAC by generalized model optimization. In: Asia Conference on Computer Vision, pp. 812–817 (2004)
7. Philbin, J., Chum, O., Sivic, J., Isard, M., Zisserman, A.: Lost in quantization: improving particular object retrieval in large scale image databases. In: IEEE Conference on Computer Vision and Pattern Recognition, pp. 1–8. Anchorage (2008)
8. Jégou, H., Douze, M., Schmid C.: On the burstiness of visual elements. In: IEEE Conference on Computer Vision and Pattern Recognition, pp. 1169–1176. Miami (2009)
9. Li, X., Larson, M., Hanjalic, A.: Pairwise geometric matching for large-scale object retrieval. In: IEEE Conference on Computer Vision and Pattern Recognition, pp. 1169–1176. Boston (2015)
10. Leordeanu, M., Hebert M.: A spectral technique for correspondence problems using pairwise constraints. In: Tenth IEEE International Conference on Computer Vision, pp. 1482–1489. Beijing (2005)
11. Nistér, D., Stewénius, H.: Scalable recognition with a vocabulary tree. In: IEEE Conference on Computer Vision and Pattern Recognition, pp. 2161–2168 (2006)
12. Mikolajczyk, K.: Binaries for affine covariant region descriptors. http://www.robots.ox.ac.uk/~vgg/research/affine/ (2007)

Locality Constrained Sparse Representation for Cat Recognition

Yu-Chen Chen[1], Shintami C. Hidayati[1,2], Wen-Huang Cheng[2], Min-Chun Hu[3], and Kai-Lung Hua[1(✉)]

[1] Department of Computer Science and Information Engineering,
National Taiwan University of Science and Technology, Taipei, Taiwan, ROC
{m10215008,d10115802,hua}@mail.ntust.edu.tw
[2] Research Center for Information Technology Innovation,
Academia Sinica, Taipei, Taiwan, ROC
whcheng@citi.sinica.edu.tw
[3] Department of Computer Science and Information Engineering,
National Cheng Kung University, Tainan, Taiwan, ROC
anita_hu@mail.ncku.edu.tw

Abstract. Cat (*Felis catus*) plays an important social role within our society and can provide considerable emotional support for their owners. Missing, swapping, theft, and false insurance claims of cat have become global problem throughout the world. Reliable cat identification is thus an essential factor in the effective management of the owned cat population. The traditional cat identification methods by permanent (e.g., tattoos, microchip, ear tips/notches, and freeze branding), semi-permanent (e.g., identification collars and ear tags), or temporary (e.g., paint/dye and radio transmitters) procedures are not robust to provide adequate level of security. Moreover, these methods might have adverse effects on the cats. Though the work on animal identification based on their phenotype appearance (face and coat patterns) has received much attention in recent years, however none of them specifically targets cat. In this paper, we therefore propose a novel biometrics method to recognize cat by exploiting their noses that are believed to be a unique identifier by cat professionals. As the pioneer of this research topic, we first collect a Cat Database that contains 700 cat nose images from 70 different cats. Based on this dataset, we design a representative dictionary with data locality constraint for cat identification. Experimental results well demonstrate the effectiveness of the proposed method compared to several state-of-the-art feature-based algorithms.

Keywords: Biometrics · Cat recognition · Sparse representation · Dictionary learning · Data locality

1 Introduction

Cats are common pets in all continents of the world, which their global population is estimated be around 600 million. According to a report of the American

© Springer International Publishing Switzerland 2016
Q. Tian et al. (Eds.): MMM 2016, Part II, LNCS 9517, pp. 140–151, 2016.
DOI: 10.1007/978-3-319-27674-8_13

Table 1. Comparison of traditional methods for cat identification [2].

| | Traditional cat identification methods | | | | | | | |
| | Permanent | | | | Semi-permanent | | Temporary | |
	Tattoo	Microchip	Ear tip/notch	Freeze brand	ID collar	Ear tag	Paint/dye	Radio transmitter
Reliability	* * *	* * * * *	* * * *	* * * * *	* *	* *	*	* *
Cost	* * *	* * * * *	* *	* * *	* *	* *	*	* * * * *
Visibility	*	-	* * *	* * * *	* * * * *	* * * * *	* * * *	-
Longevity	* * * *	* * * * *	* * * * *	* * * *	* *	* *	*	* *
Requires anesthesia	* * * *	* *	* * * * *	* * *	-	* * *	-	*
Invasiveness	* * *	* *	* * * * *	* * *	-	* * *	-	*
Risk of harm	* *	*	* * *	* *	* * * *	* * * * *	*	*
Accuracy	* * * *	* * * * *	-	* *	* * * *	* *	*	* * * *
Space-restricted	* * * * *	-	-	* * * * *	* * * *	* *	* * *	-
Uniqueness	* * * *	* * * * *	-	* *	* * * *	* * *	*	* * * * *
Instant identification	* *	-	* * * * *	* * * *	* * * *	* * * *	* * * *	*
Training required	* * * *	* * * * *	* * * * *	* * * * *	*	* * *	*	* * * * *
Database required	* * * *	* * * * *	-	* *	* *	* * *	-	* * * * *

* * * * * (Very high) * * * * (High) * * * (Medium) * * (Low) * (Very low) – (None / Not Applicable)

Pet Products Association in 2012 [1], there were about 164 million American households owned pets and almost 60 % of them owned cats. Based on this information, we notice that cat plays an important social role within our society. For any intervention that aims to manage cat population, the importance of effective identification thus should not be underestimated.

Table 1 summaries traditional methods used for cat identification. In general, permanent methods are designed to last for the lifetime of the animal, semi-permanent for months to years (but are usually lost during the cat's lifetime) and temporary methods for no longer than a few weeks. Recently, visible identification methods, such as ear notches or collars, are widely used to distinguish cats. However, wearing a mark could alter the cat's appearance, and social interaction. Furthermore, any methods for marking cats would affect their behavior due to the stress caused by the pain of tissue damage. Thus, there is clearly a need for an ideal method to identify individual cat reliably and permanently without adverse effects on the cats.

In particular, biometrics, which assign unique identity to an individual according to inherent physiological and/or behavioral characteristics, have been widely developed to overcome human identification [3–5]. Based on the success of human biometrics, we believe it's about time to develop cat biometric technology that offers reliable and effective solution for cat identification. The corresponding cat biometrics applications would contribute for not only cat identification, but also tracing, ownership assignment, and cat population management.

Recently, several research works have attempted to identify individual animal using their biometrics identifiers. Voith et al. [6] demonstrated the effectiveness of DNA analysis for identifying individual dog. However, DNA testing of an animal is costly and time-consuming. Kumar and Singh [7] addressed the problem of dog identification based on their faces. They proposed to extract a set of salient discriminatory features by holistic appearance approach (i.e., PCA, LDA, ICA, and its variants algorithm). Using animal's face as biometricsidentifiers, however, might encounter several problem due to their body dynamic, huge texture variations, and uncontrolled pose. The works of Noviyanto and Arymurthy [8], Awad et al. [9], and Tharwat et al. [10] proposed to identify cattle based on local texture features of their muzzle prints. Although these works have made great progress, local texture feature-based approaches are limited due to the lack of enough training data to capture the diverse view-point of the muzzle prints.

Cat nose print pattern is considered as a unique identifier for cat, which is similar to human fingerprints [11]. To capture cat nose print patterns, the naive approaches include using paper and ink and utilizing fingerprint scanner. However cats are not happy to be forced to put their noses on the aforementioned equipments; furthermore, cats' noses are not as flat as human fingerprints and hence even if we force the cats to act as what we ask, the resultant cat nose print pattern is of low quality. Therefore, in this work, we consider cat nose images for this pioneer cat recognition research. Besides, to the best of our knowledge, there is no public dataset for cat nose images, hence we collected a cat nose image dataset, from 70 different cats, that will be available for anyone who is interested in this related work.

Motivated by the impressive performance of image sparse representation, in this paper, we propose to employ sparse representation for describing visual information of cat nose images via the learned over-complete dictionary. Note that standard sparse representation does not preserve data locality characteristic during its encoding process, here we further consider data locality structure [13] and employ the locality regularization term for data reconstruction. With the locality regularization term, the proposed method could provide closed-form solutions for both dictionary update and sparse coding stages. The feature representations obtained based on locality-constrained sparse representation are then exploited to perform classification task via SVM framework.

The remaining of this paper is organized as follows. We present our proposed algorithm for cat recognition in Sect. 2. Experimental results are presented in Sect. 3. Finally, Sect. 4 concludes the paper and discusses some future research issues.

2   The Methodology

We now detail our proposed algorithm for recognizing cats based on the locality-constrained sparse representation of their nose images. In Sect. 2.1, we give a short overview of the proposed framework. Section 2.2 presents how we formulate the proposed feature representation. The developed supervised dictionary learning approach associated with our problem is introduced in Sect. 2.3, and the use of the extracted sparse feature representation to train a classifier for cat nose recognition is presented in Sect. 2.4.

2.1   Overview

Figure 1 shows the overview of our proposed work. Given cat nose images, we first convert them into grayscale. It has been pointed out in [18] that visual representation at a pixel level generally could not provide enough information of a useful feature, while the use of the whole image as the visual representation requires large amounts of training data to represent all possible spatial configurations. Hence, we partition each nose image into P number of patches. To obtain the representative feature of each cat nose, we exploit sparse representation of each patch. By applying sparse representation based classification approach, the choice of features becomes no longer critical because a high dimensional image vector is sparsely represented by the representative samples of its class in a low dimensional manifold [14]. Besides, in order to preserve local data structure, we enforce a data locality constraint while dictionary learning. The resultant feature of each cat nose image is the concatenation of sparse representation of its corresponding patches. Finally, a classifier is trained based on the obtained image features for our cat recognition purpose.

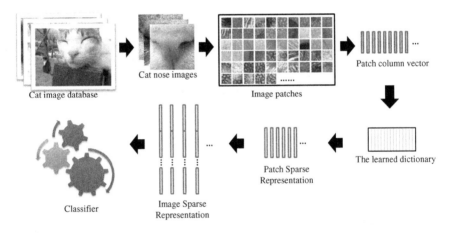

Fig. 1. The overview of our proposed cat recognition framework.

2.2   Problem Formalism for Sparse Feature Representation

Suppose there are C different cats, each cat has M nose images, and each nose image has P patches. We represent a labeled training set in our dataset as $\mathcal{D}_p = \{(\mathbf{x}_1, y_1), (\mathbf{x}_2, y_1), ..., (\mathbf{x}_N, y_C)\}$, where $\mathbf{x}_i$ is a patch of nose image with dimension d, $y_j \in \{1, 2, ..., C\}$ is the corresponding label of $\mathbf{x}_i$, and N is number of training nose image patches, that is, $N = C \times M \times P$. To obtain the sparse coefficient matrix of image patches in our training set, $\mathbf{A} = [\boldsymbol{\alpha}_1, \boldsymbol{\alpha}_2, ..., \boldsymbol{\alpha}_N] \in \mathbb{R}^{K \times N}$, we train a locality-sensitive dictionary $\mathbf{D} \in \mathbb{R}^{d \times K}$ on $\mathbf{X} = [\mathbf{x}_1, \mathbf{x}_2, ..., \mathbf{x}_N]$. Here, K is number of dictionary atoms used to encode image patches $\mathbf{X}$, and $\boldsymbol{\alpha}_i \in \mathbb{R}^{K \times 1}$

is the ith column of $\mathbf{A}$, which denotes the sparse representation of $\mathbf{x}_i$. Given $\mathbf{A}$, we then construct feature representation of each nose image by concatenating sparse representation of its patches.

2.3 Supervised Dictionary Learning Approach

In this work, motivated by [13], we enforce a locality-constraint while dictionary is learned for the cat recognition task.

More precisely, we learn a dictionary $\mathbf{D} \in \mathbb{R}^{d \times K}$ (in which $d \ll K$) by solving the following optimization problem during the training stage:

$$\min_{\mathbf{D}, \mathbf{A}} \|\mathbf{X} - \mathbf{D}\mathbf{A}\|_2^2 + \lambda \sum_{i=1}^{N} \|\boldsymbol{\ell}_i \odot \boldsymbol{\alpha}_i\|_2^2 \tag{1}$$

$$\text{s.t.}\quad \mathbf{1}^\top \boldsymbol{\alpha}_i = 1 \quad \forall i = 1, 2, ..., N,$$

where λ is the regularization parameter controlling the distances between input instances and coding results $\boldsymbol{\alpha}$, the symbol $\odot$ denotes the element-wise multiplication, and $\boldsymbol{\ell}_i \in R^{K \times 1}$ is the *locality adaptor*, in which the kth entry calculates the distance between $\mathbf{x}$ and the kth column of $\mathbf{D}$.

It is worth nothing that our dictionary learning algorithm disregards the column D to be norm-bounded by one (i.e., $\|\mathbf{d}_k\|_2 \le 1$ for $k = 1, 2, ..., K$). The two main benefits of employing this algorithm comparing to [12], which also utilized data locality into consideration but approximately solves (1) with the norm-bounded constraint on $\mathbf{D}$, are as follows. First, few constraint would yield the obtained dictionary $\mathbf{D}$ better fit the local data structure, and thus the improved representation ability can be expected. Second, the convergence rate would be faster because, based on formulation in (1), closed-form solutions for both dictionary update and sparse coding stages exist.

Sparse Coding for Updating A. In this stage, we fix $\mathbf{D}$ and minimize the objective function of (1) with respect to $\mathbf{A}$. Thus, it is equivalent to the following optimization problems

$$\min_{\boldsymbol{\alpha}_i} \|\mathbf{x}_i - \mathbf{D}\boldsymbol{\alpha}_i\|_2^2 + \lambda \|\boldsymbol{\ell}_i \odot \boldsymbol{\alpha}_i\|_2^2 \tag{2}$$

$$\text{s.t.}\quad \mathbf{1}^\top \boldsymbol{\alpha}_i = 1,$$

for i = 1, 2, ..., N. To determine the solution $\boldsymbol{\alpha}_i$ for (2), we form its Lagrange function $L(\boldsymbol{\alpha}_i, \psi)$, which is defined as

$$\|\mathbf{x}_i - \mathbf{D}\boldsymbol{\alpha}_i\|_2^2 + \lambda \|\boldsymbol{\ell}_i \odot \boldsymbol{\alpha}_i\|_2^2 + \psi(\mathbf{1}^\top \boldsymbol{\alpha}_i - 1).$$

In view of the equality $\mathbf{1}^\top \boldsymbol{\alpha}_i = 1$, the above expression can be re-formulated as

$$\boldsymbol{\alpha}_i^\top \mathbf{M}^\top \mathbf{M} \boldsymbol{\alpha}_i + \lambda \boldsymbol{\alpha}_i^\top \mathcal{L}_i \boldsymbol{\alpha}_i + \psi(\mathbf{1}^\top \boldsymbol{\alpha}_i - 1),$$

where $\mathcal{M} = \mathbf{x}_i \mathbf{1}^\top - \mathbf{D}$ and $\mathcal{L}_i = \mathrm{diag}(\ell_{i1}, \ell_{i2}, ..., \ell_{jK})^2$. Let the partial derivative of the Lagrange function with respect to $\boldsymbol{\alpha}_i$ equal to zero, i.e., $\partial L(\boldsymbol{\alpha}_i, \psi)/\partial \boldsymbol{\alpha}_i = 0$, we have

$$\boldsymbol{\rho}\boldsymbol{\alpha}_i + \psi\mathbf{1} = 0, \tag{3}$$

where $\boldsymbol{\rho} = 2(\mathcal{M}^\top\mathcal{M} + \lambda\mathcal{L}_i)$. Pre-multiplying (3) by $\mathbf{1}^\top\boldsymbol{\rho}^{-1}$ gives $\psi = -(\mathbf{1}^\top\boldsymbol{\rho}^{-1}\mathbf{1})^{-1}$. By substituting ψ into (3), we obtain the analytical closed-form solution of (2) as

$$\boldsymbol{\beta}_i = (\mathcal{M}^\top\mathcal{M} + \lambda\mathcal{L}_i)^{-1}\mathbf{1},$$
$$\boldsymbol{\alpha}_i = \boldsymbol{\beta}_i/(\mathbf{1}^\top\boldsymbol{\beta}_i).$$

Dictionary Update of D. In this stage, we fix $\mathbf{A}$ and minimize the objective function of (1) with respect to $\mathbf{D}$. Thus, it is equivalent to the following optimization problems

$$\min_{\mathbf{D}} \|\mathbf{X} - \mathbf{D}\mathbf{A}\|_2^2 + \lambda \sum_{i=1}^{N} \|\boldsymbol{\ell}_i \odot \boldsymbol{\alpha}_i\|_2^2. \tag{4}$$

Let we denote the objective function of (4) by $F(\mathbf{D})$, we have

$$F(\mathbf{D}) = \sum_{i=1}^{N} \left[\|\mathbf{x}_i - \mathbf{D}\boldsymbol{\alpha}_i\|_2^2 + \lambda \sum_{h=1}^{K} \alpha_{ih}^2 \|\mathbf{x}_i - \mathbf{d}_h\|_2^2 \right].$$

To derive the analytical solution of (4), we calculate the partial derivatives of $F(\mathbf{D})$ with respect to $\mathbf{d}_k$ for $k \in 1, 2, ..., K$ as

$$\frac{\partial F}{\partial \mathbf{d}_k} = \sum_{i=1}^{N} \left[-2\alpha_{ik}(\mathbf{x}_i - \mathbf{D}\boldsymbol{\alpha}_i) - 2\lambda\alpha_{ik}^2(\mathbf{x}_i - \mathbf{d}_k) \right]$$
$$= \sum_{i=1}^{N} \left[-2\alpha_{ik}(1 + \lambda\alpha_{ik})\mathbf{x}_i + 2\lambda\alpha_{ik}^2\mathbf{d}_k + 2\alpha_{ik} \sum_{h=1}^{K} \alpha_{ih}\mathbf{d}_h \right],$$

by setting $\partial F(\mathbf{D})/\partial \mathbf{d}_k = 0$ for $k = 1, 2, ..., K$, we have

$$\mathbf{R} \begin{bmatrix} \mathbf{d}_1^\top \\ \vdots \\ \mathbf{d}_K^\top \end{bmatrix} = \mathbf{S}$$

with

$$\mathbf{S} = \sum_{i=1}^{N} \begin{bmatrix} a_{i1}(1 + \lambda\alpha_{i1})\mathbf{x}_i^\top \\ \vdots \\ \alpha_{iK}(1 + \lambda\alpha_{iK})\mathbf{x}_i^\top \end{bmatrix}, \tag{5}$$

and

$$R \sum_{i=1}^{N} \boldsymbol{\alpha}_i\boldsymbol{\alpha}_i^\top + \lambda\mathrm{diag}(\alpha_{i1}^2, ..., \alpha_{iK}^2). \tag{6}$$

In view of (5), the closed-form solution of (4) is obtained by solving the linear system $\mathbf{R}\mathbf{D}^\top = \mathbf{S}$.

Once the iteration process converges (or a maximum number of iterations is reached), the training stage is complete and the learned dictionary will be utilized for representing the cat features.

2.4   Recognition Task

We analyze the feature representation of cat nose images obtained from previous steps and recognize their patterns using a learning prediction function f. Function f maps cat nose patterns $\mathbf{z} \in \mathbf{Z}$ to discrete cat class labels $y \in Y$. We learn f in a SVM framework, which introduces a discriminant function $\delta(\mathbf{z}, y) \in \mathbb{R}$ that measures the correctness of the association between pattern $\mathbf{z}$ and class label y, and gives a high score to those that are well-matches. More precisely, our optimal class label prediction is according to

$$f(\mathbf{z}) = \operatorname*{argmax}_{y \in Y} \delta(\mathbf{z}, y).$$

where the restriction form of the discriminant function is $\delta(\mathbf{z}, y) = \langle w, \Phi\langle \mathbf{z}, y \rangle \rangle$, and $\Phi\langle \mathbf{z}, y \rangle$ is a joint kernel map, which maps the pair $(\mathbf{z}, y)$ into a suitable feature space endowed with the dot product $\langle \cdot, \cdot \rangle$. We learn δ in a large-margin framework from a set of example pairs $\{(\mathbf{z}_1, y_1), (\mathbf{z}_2, y_2), ..., (\mathbf{z}_n, y_n)\}$ by minimizing the convex objective function

$$\min_w \frac{1}{2} \langle w, w \rangle + \gamma \sum_{l=1}^{n} \xi_l,$$

$$\text{s.t.} = \begin{cases} \forall l : \xi_l \geq 0 \\ \forall l, \forall y \neq y_l : \langle w, \sigma\Phi_l(y_l, y) \rangle \geq \Delta(y_l, y) - \xi_l \end{cases}$$

where $\sigma\Phi_l(y) = \Phi(\mathbf{z}_l, y_l) - \Phi(\mathbf{z}_l, y)$.

That optimization aims to ensure that, for each pattern $\mathbf{z}_l$, the value $\delta(\mathbf{z}_l, y_l)$ of the correct association is greater than the scores $\delta(\mathbf{z}_l, y)$ for $y \neq y_l$ of the incorrect association, by a margin which depends on a lost function Δ. We also introduce slack variables ξ_l, which measure the degree of misclassification of the data $\mathbf{z}_l$.

3   Experimental Results

3.1   Dataset and Experimental Settings

Dataset. To the best of our knowledge, there does not exist any public cat image dataset that contains cat nose images and their corresponding labels. Hence, in this work, a dataset, containing 70 different cats with 10 nose images for each cat, is collected. Note that ten nose images of each cat were taken in different angles, and the time interval between capturing each nose of a cat was

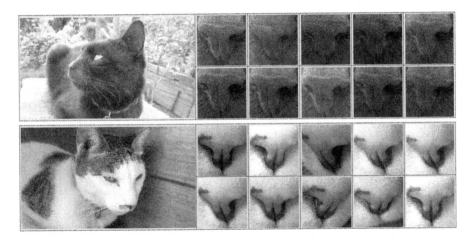

Fig. 2. An illustration of the collected cat nose database taken from two different cats. For each box, the left-most is the cat image and the rest is the corresponding nose images.

longer than five seconds. Here, we restricted the horizontal angle when taking the photo within 45° to 135° from the front face of the cat. Each nose image, normalized to the resolution of 512×512 pixels, is composed of cat nose and its nearby fur. Two set of cat nose images from two different cats are illustrated in Fig. 2. In our experiments, two nose images for each cat are randomly picked as the testing data, and the others are treated as the training data. Therefore, there are 140 (2×70) and 560 (8×70) cat nose images in testing and training set, respectively.

Experimental Settings. We partitioned each cat nose image by square non-overlapping patches at size of 4×4 pixels. In the stage of dictionary learning, we set $K = 30$ and $\lambda = 0.7$. We employed LIBSVM [19] as our library for Support Vector Machines (SVMs).

3.2   Performance Evaluation

To demonstrate its effectiveness, we compared the performance of our proposed cat recognition approach against the standard sparse representation based classification algorithm [14] and the traditional feature-based algorithms using several state-of-the-art descriptors. Seven kinds of feature descriptors were used for comparison, including color moments [15], histogram of oriented gradients (HOG) [16], Gabor wavelets [17], color moments+HOG, color moments+Gabor wavelets, and color moments+HOG+Gabor wavelets. Color moments [15] are measures that characterize mean and standard deviation of chrominance color components (Cb and Cr) in YCbCr colorspace. HOG features [16] represent local shape of an object by capturing the distribution of intensity gradients. Gabor wavelets [17] analyze the local textures. Note that "+" symbol denotes the integration of multiple features that aforementioned, for example, color moments+HOG denote the integration of color moments [15] and HOG [16]. Besides, in our experiments, 5-fold cross validation technique was employed for the performance evaluation.

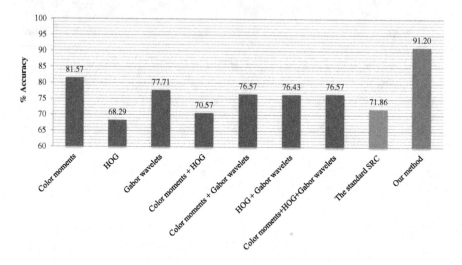

Fig. 3. Recognition performance of cat nose with different approaches.

Figure 3 reports the average recognition rate of different approaches. It can be observed that our proposed method outperformed other approaches with the average accuracy of 91.20 %. As for the standard sparse representation approach [14], which does not consider data locality, only achieved the average accuracy of 71.86 %. Hence, it's verified that the employment of the locality regularization term for data reconstruction in cat recognition task improved the performance by 19.34 %. Besides, cat recognition using color moments to characterize the content of cat nose images achieved the average accuracy of 81.57 %, which is the second best performance in this experiment. We observed that the high recognition performance of this approach was caused by the high color differences of collected cat nose images. It is noteworthy that a lot of cats share similar nose print colors, hence, when we increase the size of the collected dataset, it is foreseen that the performance of cat recognition using color features would drop dramatically. Figure 4 demonstrates one example of two set of similar color nose images for which our method works while color moments approach fails. In addition, from Fig. 3 we show that texture features employed in this evaluation did not achieve good performance compared to ours, even if they are integrated with other features, including color features.

It is obvious that the proposed method is capable of recognizing cat nose images with various angles, as illustrated in Fig. 5. This is because we focus on visual pattern information of cat nose, which contains nose print information that is unique identifier of each cat. As shown in Fig. 6, the failure cases of the proposed method are in general caused by the surrounding fur and nostril

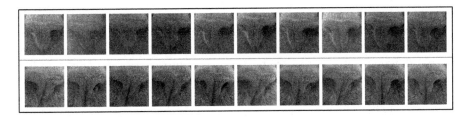

Fig. 4. Example of two set of similar color nose images for which our method works while color moments approach fails (color figure online).

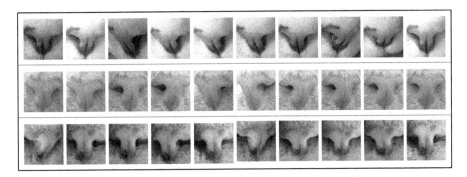

Fig. 5. Example of various angles among cat nose images.

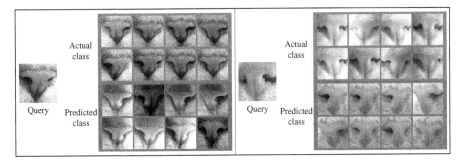

Fig. 6. Example of query images for which our method does not recognize its class correctly.

structure. Hence we expect to further improve the performance of the proposed method by extracting the cat nose images more precisely by removing the non-nose information.

4   Conclusion and Future Work

With an estimated 600 million cats living throughout the world today, reliable cat identification is a vital tool in the management of its populations. To tackle this challenge, in this paper, we presented a novel biometrics method to recognize cat by exploiting visual information of their nose. Be the frontier to study this research topic, we collected a brand new dataset that consists of 700 nose images from 70 different cats. A dictionary learning approach that employs locality regularization term, preserving local data structure, was developed to compute sparse feature of each cat. The extracted sparse features are then used to train the SVM classifier. Experiment results confirmed that our proposed method outperformed several state-of-the-art feature-based algorithms. In the future, we plan to precisely localize nose print region, which is considered as unique identifier of each cat, and extract more distinct visual information from this region.

References

1. The Humane Society of the United States (2014). Pets by the numbers. http://www.humanesociety.org/issues/pet_overpopulation/facts/pet_ownership_statistics.html
2. World Society for the Protection of Animals (2014). Identification methods for dogs and cats. http://www.icam-coalition.org/downloads/Identification%20methods%20for%20dogs%20and%20cats.pdf
3. Yang, M., Zhang, L., Yang, J., Zhang, D.: Robust sparse coding for face recognition. In: IEEE International Conference on Computer Vision and Pattern Recognition, pp. 625–632 (2011)
4. Ma, L., Wang, C., Xiao, B., Zhou, W.: Sparse representation for face recognition based on discriminative low-rank dictionary learning. In: IEEE International Conference on Computer Vision and Pattern Recognition, pp. 2586–2593 (2012)
5. Liu, F., Tang, J., Song, Y., Xiang, X., Tang, Z.: Local structure based sparse representation for face recognition with single sample per person. In: IEEE International Conference on Image Processing, pp. 713–717 (2014)
6. Voith, V.L., Ingram, E., Mitsouras, K., Irizarry, K.: Comparison of adoption agency breed identification and DNA breed identification of dogs. J. Appl. Anim. Welf. Sci. **12**, 253–262 (2009)
7. Kumar, S., Singh, S.K.: Biometric recognition for pet animal. J. Softw. Eng. Appl. **7**, 470–482 (2014)
8. Noviyanto, A., Arymurthy, A.M.: Automatic cattle identification based on muzzle photo using speed-up robust features approach. In: European Conference of Computer Science, pp. 110–114 (2012)
9. Awad, A.I., Hassanien, A.E., Zawbaa, H.M.: A cattle identification approach using live captured muzzle print images. In: Awad, A.I., Hassanien, A.E., Baba, K. (eds.) SecNet 2013. CCIS, vol. 381, pp. 143–152. Springer, Heidelberg (2013)
10. Tharwat, A., Gaber, T., Hassanien, A.E., Hassanien, H.A., Tolba, M.F.: Cattle identification using muzzle print images based on texture features approach. In: Abraham, A., Kömer, P., Snášel, V. (eds.) Proceedings of the Fourth Intern. Conf. on Innov. in Bio-Inspired Comput. and Appl. IBICA 2014. AISC, vol. 303, pp. 217–227. Springer, Heidelberg (2014)
11. Animal Bliss (2014). Cat Nose Prints as Unique as Human Finger-prints, Noseprint ID. http://www.animalbliss.com/cat-nose-prints-as-unique-as-human-fingerprints/

12. Wang, J., Yang, J., Yu, K., Lv, F., Huang, T., Gong, Y.: Locality-constrained linear coding for image classification. In: IEEE International Conference on Computer Vision and Pattern Recognition, pp. 3360–3367 (2010)

13. Wei, C.-P., Chao, Y.-W., Yeh, Y.-R., Wang, Y.-C.F.: Locality-sensitive dictionary learning for sparse representation based classification. J. Pattern Recognit. **46**, 1277–1287 (2013)

14. Wright, J., Yang, A.Y., Ganesh, A., Sastry, S.S., Yi, M.: Robust face recognition via sparse representation. IEEE Trans. Pattern Anal. Mach. Intell. **31**, 210–227 (2009)

15. Stricker, M., Orengo, M.: Similarity of color images. In: Storage Retrieval Image Video Databases, vol. 2420, pp. 381–392 (1995)

16. Dalal, N., Triggs, B.: Histograms of oriented gradients for human detection. In: IEEE International Conference on Computer Vision and Pattern Recognition, vol. 1, pp. 886–893 (2005)

17. Lee, T.S.: Image representation using 2D gabor wavelets. IEEE Trans. Pattern Anal. Mach. Intell. **18**, 959–971 (1996)

18. Singh, S., Gupta, A., Efros, A.A.: Unsupervised discovery of mid-level discriminative patches. In: European Conference Computer Vision and Pattern Recognition, pp. 73–86 (2012)

19. Chang, C.-C., Lin, C.-J.: LIBSVM: a library for support vector machines. ACM Trans. Intell. Syst. Technol. **2**, 27:1–27:27 (2011)

User Profiling by Combining Topic Modeling and Pointwise Mutual Information (TM-PMI)

Lifang Wu[1(✉)], Dan Wang[1], Cheng Guo[1],
Jianan Zhang[1], and Chang wen Chen[2]

[1] School of Electronic Information and Control Engineering,
Beijing University of Technology, Beijing 100124, China
lfwu@bjut.edu.cn, anncheng.student@sina.com,
{wangdan2013, jnzhang}@emails.bjut.edu.cn
[2] Department of Computer Science and Engineering,
State University of New York at Buffalo,
316 Davis Hall, Buffalo, NY 14260-2500, USA
chencw@buffalo.edu

Abstract. User profiling is one of the key issues in personalized recommendation systems. A content curation social network is a content-centric network; it encourages users to repin items from other users and other websites. It further permits users to arrange the pins according to their interests. It is therefore possible to estimate user interest from the pins. In this paper, we propose a user profiling approach to combining topic model and pointwise mutual information (TM-PMI). We first extract a pin's description, and then apply latent Dirichlet allocation (LDA, one of the topic modeling schemes). A three-layer hierarchical Bayesian model of user-topic-word is thus obtained. Then, a personal model is obtained by selecting a set of correlated words with constraints of word probability and PMI. The experimental results confirm the efficiency of the proposed approach.

Keywords: Topic modeling · Latent dirichlet allocation · Pointwise mutual information · User profile

1 Introduction

With the development of Web 2.0 technology, social network services (SNSs), such as blogs, Facebook, Twitter, Weibo, and Flickr are being widely used and play an important role in human lives. In 2009, Pinterest, one of the early content curation social networks (CCSNs), was launched. Pinterest is a photo sharing social networking site. With new CCSNs such as Huaban.com and Snip.it, CCSNs have attracted many social network users [1, 2].

The user-generated media on SNSs has brought about an explosion of information on SNSs. The generated information is much more than one person can deal with, leading to information overload. This has led to an increasing amount of research in the area of social recommendation. User profiling is one of the key issues in social recommendation.

Q. Tian et al. (Eds.): MMM 2016, Part II, LNCS 9517, pp. 152–161, 2016.
DOI: 10.1007/978-3-319-27674-8_14

The conventional SNSs are user-centric networks [3], which are not optimized to create comprehensive user profiles based on user-generated content. Rather, CCSNs permit users to arrange and categorize data from social media for the purpose of further consumption. For example, on Pinterest, users (called "pinners") find photos (called "pins") which are interesting in them, and manually organize these photos into boards of their own. Therefore, it is possible to extract reliable social cues based on user preference. In particular, user curation through multimedia content helps to encode multi-level content-content connections, which help to pinpoint user preference in terms of the content generated by the user. Such connections between two items include user-level connection, where the strength indicates how many users have pinned both items. For example, if two images are shared by only a few users (or bundles), the connection between them rarely suggests similar user interest. However, when shared by many users they likely indicate the same interest.

Motivated by this knowledge, we propose a scheme for modeling the user profile based on all the items repined by the user. A pin includes significant information [4], such as a description written by the user, and a link to the original item, as shown in Fig. 1. In this paper, the description information is used for user profiling. The description is text-based.

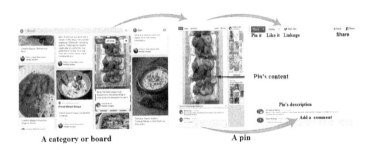

Fig. 1. A pin in Pinterest

Topic modeling [5] is an effective scheme to extract a document's hidden topics. The current popular models include Probabilistic Latent Semantic Analysis (PLSA) and Latent Dirichlet allocation (LDA). PLSA, first proposed by Hofmann [6], is a method by which a document composed of multiple topics is presented in terms of the probability in the vocabulary. Each word in the vocabulary is generated from a topic. However, PLSA is only efficient for the documents in the training set. Blei [7] extended PLSA into LDA by introducing the Dirichlet distribution. It works with both the training set and the testing set. From that point, LDA has been a popular and promising scheme of topic modeling. The improved LDA algorithms [8–10] are proposed for user profiling in SNSs.

The traditional LDA scheme combines users and terms together; users are mixtures of multiple topics. A topic has a probability distribution over a vocabulary of terms. The advantage is that it picks out coherent and semantic properties of correlated terms. However, the terms of probabilistic topics are unable to give an accurate representation of users. Accurate estimation of user preference is a challenging problem. To address this problem, we introduce pointwise mutual information (PMI).

PMI [14] is used to measure the connection between two items. If the PMI value is greater than zero, then these two items are considered related; otherwise, they are considered unrelated. In this paper, PMI is introduced to measure the correlation between a word from the topics and a specific user. The correlated words are then used to represent the user.

A user profiling scheme is proposed by combining LDA and PMI. First, LDA is applied to the descriptions of all pins from all users. A three-layer hierarchical Bayesian model of user-topic-word is then obtained. The personal model is then obtained by selecting a set of correlated words with the constraints of word probability and PMI. The results of user profiling are tested using pin recommendation results from Levenshtein distance of user profiles. The similarities between the recommended pins and the target user are then evaluated by a user study.

2 The Proposed Approach

2.1 Framework of the Proposed Approach

The framework of the proposed approach is shown in Fig. 2. First, the descriptions of all pins from all users are collected. The description is then preprocessed, including word segmentation and the building of the dictionary. Second, LDA is used to extract the probability of the words, and the word frequency vector for each user is generated. Then, the PMI between each user and a topic word is computed. Finally, user topic words are generated from LDA, using PMI and word frequency constraints. The user profile is represented as the user topic words and their probabilities. The user profiles are then tested using pin recommendation.

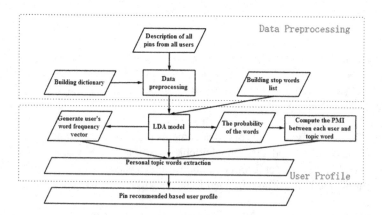

Fig. 2. Framework of the proposed approach

2.2   Data Preprocessing

In this paper, pin descriptions from Huaban.com are used, which are composed of Chinese-language documents. The first task is word segmentation. The popular Chinese word segmentation tool ICTCLAS [11] (Institute of Computing Technology, Chinese Lexical Analysis System) is employed. ICTCLAS includes word segmentation, part-of-speech tagging, and unknown word recognition; it also supports multiclass Chinese code such as GB2312, GBK, and UTF8. Following this, we set up a complementary dictionary, which includes approximately 300,000 new words. These words are frequently used in the network and are not included in the ICTCLAS dictionary. Third, generate a complementary dictionary, which includes 1433 stop words, which are filtered out.

After preprocessing, all user descriptions are represented as isolated words.

2.3   LDA from Description of User Pins

An LDA model can be represented as a three-layer hierarchical Bayesian model of user-topic-word:

- Word layer: the set of words $V = \{w_1, w_2, \ldots, w_V\}$ with frequency greater than or equal to 10.
- Topic layer: the set of topics $Z = \{z_1, z_2, \ldots, z_k, \ldots, z_K\}$, where each component is a probability distribution on the word set V, presented as $\varphi_k = \{p_{k,1}, p_{k,2}, \cdots p_{k,i}, \cdots, p_{k,V}\}$, where $p_{k,i}$ is the probability that word w_i is generated from topic z_k.
- User layer: at the word layer, a user can be represented as a word frequency vector $u = \{tf_{u,1} tf_{u,2}, \ldots, tf_{u,w_j}, \ldots, tf_{u,w_V}\}$, where, tf_{u,w_j} is the frequency of word w_j in user u; at the topic layer, a user can be represented as $\theta_u = \{p_{u,1}, p_{u,2}, \ldots, p_{u,z_k}, \cdots, p_{u,z_K}\}, p_{u,z_k}$, the probability of topic z_k for that user.

Figure 3 illustrates generation of the LDA model; we now introduce the generation process in detail:

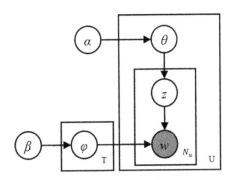

Fig. 3. Graphical model representation of the LDA model

1. For each topic $z_k, k \in \{1, 2, \ldots, K\}$:

   Draw a distribution over words $\varphi_k \sim Dir(\beta)$

2. For each user $u \in (1, 2, \ldots, U)$:
   (1) Draw a vector of topic proportions $\theta_u \sim Dir(\alpha)$
   (2) For each word in user u:
       (a) Draw a topic assignment $z_{u,n} \sim Multi(\theta_d), z_{u,n} \in \{k = 1, 2, \ldots, K\}$
       (b) Draw a word $w_{u,n} \sim Multi\left(\varphi_{z_{u,n}}\right), w_{u,n} \in \{w_1, w_2, \ldots, w_V\}$

Markov chain Monte Carlo (MCMC) is a popular probability reasoning algorithm. It is an iterative approximation algorithm to extract samples from a complex probability distribution. Gibbs sampling [12, 13] is an algorithm for obtaining a sequence of observations using MCMC. In this paper, Gibbs sampling is used to obtain the probability distribution of user-topic θ and topic-word φ.

$$P(z_{-u} = k | \{z_{-u}, w\}) \propto \exp\{\log \theta_{u,k} + \log \varphi_{k,w_u}\} \tag{1}$$

$$\theta_{u,k} = \frac{\alpha_k + n_{k|u}}{K * \alpha_k + n_{\cdot|u}} \tag{2}$$

$$\varphi_{k,w_u} = \frac{\beta + n_{w_u|k}}{\beta * V + n_{\cdot|k}} \tag{3}$$

2.4 Pointwise Mutual Information (PMI)

PMI [14] is proposed based on mutual information. It is generally used to measure the connection between two items. PMI is obtained by extracting the probability by which two events happen simultaneously. For two items x and y, their PMI could be computed as follows:

$$PMI(x,y) = \log\frac{p(x, y)}{p(x)p(y)} = \log\frac{p(x|y)}{p(x)} = \log\frac{p(y|x)}{p(y)} \tag{4}$$

From Eq. (4), we can infer the following:

- If $PMI(x, y) > 0$, then item x and item y are correlated with high probability. Larger PMI values indicate higher probability.
- If $PMI(x, y) = 0$, then x and y are isolated from each other.
- If $PMI(x, y) < 0$, then x and y are complementary to each other.

2.5   Personal Topic Words Extraction

For a specific user, personal topic words are extracted by combining the PMI and LDA models.

For a specific user u, we first order the topics by decreasing probability $\mathbf{p} = \{p_1, p_2, \ldots, p_K\}$. We select the first j topics corresponding to the first 80 % of probability [16]. Then, we sample N topic words from these topics according to the word frequency to get the word sets $w_{\mathbf{LDA}} = \{w_1, w_2, \ldots, w_N\}$. Then we compute PMI between each word and the user.

$$
\begin{aligned}
\mathrm{PMI}(\mathrm{user, word}) &= \log_2 \frac{\mathrm{p(user, word)}}{\mathrm{p(user)p(word)}} \\
&= \log_2 \frac{\mathrm{p(word|user)}}{\mathrm{p(word)}} \\
&= \log_2 \frac{\sum_{z=k}^{K} \mathrm{p(word|topic)p(topic|user)}}{\mathrm{p(word)}}
\end{aligned}
\tag{5}
$$

If PMI for a given word is greater than or equal to 0, we assume this word is correlated to the user, and the word is preserved. Otherwise, the word is discarded.

Using this method, we generate the personal topic word set $\mathbf{tw}_u$ for all users, u = 1,2,...,U. These topic words represent user preference.

2.6   Pins Recommended Based User Profile

In order to test the performance of our method, we recommend pins to a user based on the user profile model. The consistency between the recommended results and user's actual preference measures the accuracy of the user profile.

For a specific user u, we first select the related pins based on his/her topic word set $\mathbf{tw}_u$ and obtain the related pin set p_{-u}. We compute the normalized Levenshtein distance [15] between a pin that does not belong to user u. The minimum distance is used as the distance between the pin and user u. If the minimum distance is smaller than 0.8, we assume this pin is related to user u, and we can recommend this pin to user u.

3   Experiments

3.1   Dataset

The experimental dataset was crawled from a typical Chinese content curation social network, Huaban.com, which includes 34 categories (as show in Fig. 4.). In our experiments, three categories: "home", "food_drink" and "design" were randomly selected. A total of 100 users (35/30/35 for three categories, respectively) were randomly selected. From there, 633,337 pins (179,081 pins for users in "home", 225,273 pins for users in "food_drink" and 228,983 pins for users in "design") were selected.

Fig. 4. Total 34 categories in Huaban.com

3.2   Perplexity

Perplexity is a metric to measure the language generation model. It measures the prediction ability of the model with respect to a new document. The smaller the perplexity value, the higher the prediction ability. In this paper, perplexity is used to measure the performance of user profiling:

$$Perplexity(U_{test}) = exp\left\{-\frac{\sum_{u=1}^{U_t} \log p(w_u)}{\sum_{u=1}^{U_t} N_u}\right\} \tag{6}$$

$$p(w_u) = \prod_{n=1}^{N_m} \sum_{k=1}^{K} p(w_n|z_k)p(z_k|u) \tag{7}$$

Where U_{test} is the user in the testing set, U_t is the user number in the testing set, w_u is the word set of pin description from user u, $p(w_u)$ is the generated user model from pin description of user u, and N_u is the topic word number. In our experiment, 10 % of the data is used for testing. The results are shown in Fig. 5.

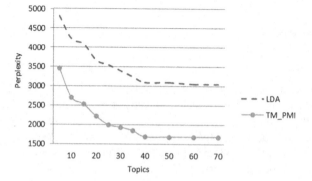

Fig. 5. Perplexity results

From Fig. 5 we can observe that the performance of user profiling is related to the topic number. Perplexity decreases as topic number increases, and Perplexity is stable when T >= 40. In the following experiment, the topic number is selected as T = 40.

3.3 User Study

This experiment is used to test the performance of algorithm-based recommendation by measuring the correlation between recommended pins and the target user.

For the 100 users in the training set, we recommended 10 pins to each user from LDA and TM-PMI, for a total of 2000 recommended pins. For each recommended pin, we selected three related pins from the corresponding user as shown in Fig. 6. The correlation between the recommended pins and the three pins from the user were tested by a user study.

The recommended pin   The related pins

Fig. 6. The recommended pin and related pins of the target user

A total of 40 subjects were selected for the user study. Each subject was randomly assigned 100 pins and required to give the correlation value (0–5) between the recommended pins and each related pin. The maximum correlation value was considered to be the correlation of the test set.

For each set, we assumed that the ground truth correlation is 5, and then based on the user study measurement, we computed the root mean squared error (RMSE) and mean absolute error (MAE) values as follows.

$$\text{MAE} = \frac{\sum_{(u,\alpha)\in T}\left|r_{u\alpha} - r'_{u\alpha}\right|}{|T|} \tag{8}$$

$$\text{RMSE} = \frac{\sqrt{\sum_{(u,\alpha)\in T}\left(r_{u\alpha} - r'_{u\alpha}\right)^2}}{|T|} \tag{9}$$

Where $r_{u\alpha}$ is the ground truth, which is generally 5; $r'_{u\alpha}$ is the rating from the user study; $|T|$ is the testing set.

3.4 Influence of the Number of Topic Words on the Result

In Sect. 2.5, some words are obtained from the former 80 % topics to represent the user. The number of the topic words is N. In order to test the influence of N on the recommendation results, the compared experiments are conducted with $N = 20, 30$, and 40 respectively. We also compare our algorithm with LDA. The experimental results are shown in Fig. 7. From Fig. 7, we can see that our algorithm is better than LDA with $N = 20, 30$, and 40. And our algorithm gets the best results with $N = 40$.

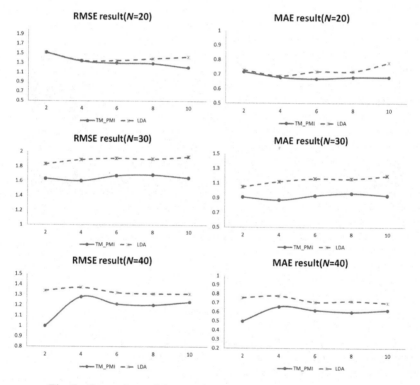

Fig. 7. Comparison of the experimental results ($N = 20, 30$ and 40)

4 Conclusion and Future Work

In this paper, we proposed a user profiling approach, combining LDA, for obtaining a three-layer hierarchical Bayesian model of user-topic-word, and pointwise mutual information to model a user profile. The model was tested using perplexity and user study of the recommended pins. The experimental results showed that the proposed algorithm performs slightly better than the traditional LDA. In this paper, only user description was used for user profiling. In future work, multimodal information will be introduced for user profiling.

References

1. Hall, C., Zarro, M.: Social curation on the website Pinterest.com. Proc. Am. Soc. Inf. Sci. Technol. **49**(1), 1–9 (2012)
2. Gilbert, E., Bakhshi, S., Chang, S., Terveen, L.: "I need to try this"? a statistical overview of pinterest. In: CHI, pp. 2427–2436. ACM (2013)
3. Geng, X., Zhang, H., Song, Z., Yang, Y., Luan, H., Chua, T.: One of a kind: user profiling by social curation. In: Proceedings of the ACM International Conference on Multimedia, Orlando, pp. 567–576. ACM (2014)
4. Bernardini, C., Silverston, T., Festor, O.: A Pin is worth a thousand words: characterization of publications in Pinterest, pp. 322–327. IEEE (2014)
5. Blei, D., Carin, L., Dunson, D.: Probabilistic topic models: a focus on graphical model design and applications to document and image analysis. IEEE Signal Process. Mag. **27**(6), 55–65 (2010)
6. Hofmann, T.: Probabilistic latent semantic analysis. In: Proceedings of the Fifteenth Conference on Uncertainty in Artificial Intelligence, Stockholm, pp. 289–296. Morgan Kaufmann Publishers Inc., (1999)
7. Blei, D.M., Ng, A.Y., Jordan, M.I.: Latent Dirichlet allocation. J. Mach. Learn. Res. **3**(4–5), 993–1022 (2003). doi:10.1162/jmlr.2003.3.4-5.993
8. Rosen-Zvi, M., Griffiths, T., Steyvers, M., Smyth, P.: The author-topic model for authors and documents. In: Proceedings of the 20th Conference on Uncertainty in Artificial Intelligence, Banff, pp. 487–494. AUAI Press (2004)
9. McCallum, A., Wang, X., Corrada-Emmanuel, A.: Topic and role discovery in social networks with experiments on enron and academic email. J. Artif. Intell. Res. **30**, 249–272 (2007)
10. Weng, J., Lim, E., Jiang, J., He, Q.: TwitterRank: finding topic-sensitive influential twitterers. In: Proceedings of the Third ACM International Conference on Web Search and Data Mining, New York, pp. 261–270. ACM (2010)
11. Zhang, H., Yu, H., Xiong, D., Liu, Q.: HHMM-based Chinese lexical analyzer ICTCLAS. In: Proceedings of the Second SIGHAN Workshop on Chinese Language Processing, Vol. 17, Sapporo, Association for Computational Linguistics, pp. 184–187 (2003)
12. Griffiths, T.L., Steyvers, M.: Finding scientific topics. Proc. Natl. Acad. Sci. USA **1011**, 5228–5235 (2004). doi:10.1073/pnas.0307752101
13. Heinrich, G.: Parameter estimation for text analysis. Technical Note Version 2.4, vsonix. Technical Report (2008)
14. https://en.wikipedia.org/wiki/Pointwise_mutual_information
15. https://en.wikipedia.org/wiki/Levenshtein_distance
16. https://en.wikipedia.org/wiki/Pareto_principle

Image Retrieval Using Color-Aware Tag on Progressive Image Search and Recommendation System

Shih-Yu Ku[1], Kai-Hsiang Chen[2]($^{(\boxtimes)}$), Jen-Wei Huang[2], and Yu Tsao[3]

[1] Department of Computer Science and Engineering,
Yuan Ze University, Taoyuan, Taiwan
aommshe@gmail.com
[2] Department of Electrical Engineering,
National Cheng Kung University, Tainan, Taiwan
kevin761015@gmail.com, jwhuang@mail.ncku.edu.tw
[3] Academia Sinica, Research Center for Information Technology Innovation,
Taipei, Taiwan
yu.tsao@citi.sinica.edu.tw

Abstract. In recent years, a large amount of different types of multimedia applications has been produced and propagated in our daily life. Accordingly, multimedia retrieval has become an important task. This paper proposes a novel Color-Aware Tag (CAT) algorithm for effective image retrieval. The proposed CAT algorithm can be divided into two phases: the offline and the online phases. In the offline phase, CAT prepares a set of color histograms for each specific tag. In the online phase, the tags along with the prepared color histograms are used to facilitate the image retrieval process. We evaluated the CAT algorithm on an image retrieval task by using a Progressive Image Search And Recommendation (PISAR) system. The Waking And Sleeping (WAS) algorithm was also incorporated to improve the performance of the CAT algorithm. Experimental results demonstrate that the CAT algorithm can effectively improve the performance of the image retrieval process.

Keywords: Image retrieval · Tag · Color · Recommendation system

1 Introduction

Due to the rapid proliferation of digital recording devices and progress in communication technology, users can easily generate and distribute photographs anywhere and anytime. Many social medium repositories have been established to provide safe storage and effective management for the shared photographs. Popular examples include Flickr, Instagram, and Facebook. In addition to the posted materials, a huge amount of metadata has been created and accumulated. These metadata generally provide important information for various user

© Springer International Publishing Switzerland 2016
Q. Tian et al. (Eds.): MMM 2016, Part II, LNCS 9517, pp. 162–173, 2016.
DOI: 10.1007/978-3-319-27674-8_15

demands, such as image classification, retrieval, and recommendation. By effectively utilizing the metadata, we can enhance the retrieval performance or satisfy particular demands.

In this study, we investigate the information of color and tags which have been used before to achieve the better capability of image retrieval. The tag of an image reflects the users' preference notations, situations, or highlighted elements, and thus, carries important information. According to Wu [1] reported that only 40 % of the tags in Flickr have high quality. A high-quality tag is a tag that is really meaningful and does not have any noise information. Noise tags and relatively low-quality tags can seriously deteriorate the retrieval accuracy. Therefore, when the information of tag has been used, a filtering process may be preferred to maintain the high quality of tags.

Furthermore, color is the most explicit element to represent information contain in image and can provide the information of emotion, location, and time. Although the color information is useful, pre-processing is required to remove unimportant parts. Because of an image can be separated into foreground part and background part. Foreground part should be the main object of an image. Therefore, we can remove the background part due to it is not important. Then, the major color of an image can be extracted to increase the accuracy of the retrieval result.

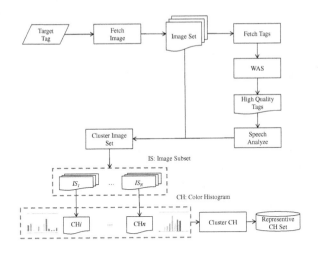

Fig. 1. Offline phase of the proposed CAT algorithm.

In this paper, we proposed a novel Color-Aware Tag, CAT, algorithm, which aims to utilize color information of an image to enhance the retrieval performance. Then, the Progressive Image Search And Recommendation system, PISAR [4], has been used to constructing the baseline image retrieval system. PISAR retrieves a set of candidate images on the basis of the user input query term, and a score has been computed for the query term and each of the retrieved

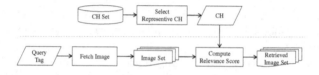

Fig. 2. Online phase of the proposed CAT algorithm.

images. Finally, the scores are ranked to determine the most suitable image. More recently, the Waking And Sleeping, WAS [1], algorithm was proposed to increase the tag quality for PISAR. WAS assigned tags to different states to filter out imperfect tags in order to improve the tag quality. In this study, the WAS algorithm is also adopted to enhance the performance of CAT.

The overall of our image retrieval system can be divided into two phases, the offline and the online phases. Figure 1 illustrates the offline phase. In this phase, a tag is first selected as the target tag from the database, and a group of images which contain the same tag have been fetched. The ensemble of the fetched images forms an image set. Next, the image set is segmented into several image subsets on the basis of the tag information. To this end, the WAS algorithm has n = been adopted to filter out low-quality tags, and only the high-quality tags are used for image clustering. Meanwhile, based on the basic of tag, we can extract the key information of each image to obtain a set of color histograms (CHs). Consequently, each tag in the database has its own set of CHs. In our preliminary experiments, we discovered that some CHs are rather similar. Therefore, we have to cluster them by calculating the similarity. The overall process is repeated to prepare a CH set for every tag in the database.

Figure 2 shows the online phase of our image retrieval system. In this phase, user can first input a query term, and an image set is fetched using the same procedure in the offline phase. Then, a set of CHs corresponding to the query term has been presented. The user selects the most relevant CH. Thereafter, the relevance score is computed on the basis of the selected CH and the CH of images from the image set. Finally, the scores have been ranked and used for locating a set of candidate images for the query term.

2   Related Works and Preliminaries

2.1   Related Works

Image retrieval is a technique that allows users to search images from a large digital images database. Many image retrieval methods have been proposed. One successful approach is to let users provide query terms to the retrieval system. According to these query terms, the system can calculate the similarity scores between images and query terms. After that, the system will return the retrieved image result which have the high similarity score to the user. In general, the retrieval methods can be roughly divided into two groups, image meta search and content-based methods.

Image meta search uses the association metadata to perform image retrieval, such as keywords, descriptions, or tags. The relationship between images and tags is often an issue to discussion. The relationship between images and tags can also define by using majority voting. If different users using the same tag to label similar images, we can intuitively consider that the visual content of image is reflect to the tag. Therefore, Li [6] proposed an algorithm which used the accumulating votes from visually similar neighbours to learn the relevance of tag. Then, retrieve the image according to the tag's relevance scores. But only use metadata to retrieve image will ignore the content information from image. Content-based image retrieval (CBIR) focuses on using image content information to save the problem.

CBIR is an image retrieval technique based on computer vision. The term "content" means the information derived from the image, such as color, texture, or shape. Kherfi [5] considered that understanding the user's needs is important in CBIR and relevance feedback (RF) is an effective tool to solve this problem. Author proposed a new RF framework that combined the advantages of a probabilistic formulation using positive example (PE) and negative example (NE). The framework records user interaction and combine PE and NE to learn the importance user assign to image. In CBIR, color is one of the most obvious features of an image. A specific color is often represented by a color space. Since the visual sense of humans resembles HSV [3], we use HSV color space to extract colors of images in this study. HSV contains three dimensions of features, hue, saturation, and value. Hue describes the basic property of a color. It divides a circle into 360° , and each degree is a pure color. With an increase of degree, the color changes from red to purple, and then back to red. Saturation represents the purity of color, and it takes a value from 0 % to 100 %. The higher of the saturation value, the purer of the color. Value characterizes the lightness of color, also in the range of 0 %–100 %. The color will be brighter with a larger value.

2.2   PISAR System and WAS Algorithm

PISAR is an image search and recommendation system. The system can allow a user to search for an image and recommend ponderable images. The PISAR system records the interaction between the system and the users to keep improving the search and recommend accuracies. For example, if the user inputs a query term and clicks a retrieval image from the system, a click event is formed associated with the query term and the retrieved image. In other words, the click event contains the information of both the tag and the image. The relationship between tags and images follows the concept of co-occurrence. On the basis of the association of tags and images, the performance of PISAR can keep improving over time.

It is discovered that the quality of tags and new items will influence the PISAR performance. Poor-quality tags may seriously degrade the system performance, and the number of new click events often less than the existing events. Therefore, although a new item is very meaningful, it may not be retrieved successfully. The WAS algorithm has been proposed to overcome this issue. WAS assigns tags into either waking or sleeping states. The tags in the walking state are representative of the image, while the tags in the sleeping state are assumed to carry little information about the image. When computing the similarity scores, the tags in the sleeping state are not included, and thus, the result can be improved.

WAS algorithm comprises two modules, importance filter and relationship retriever. Importance filter uses the time period and a click event to calculate the importance score between a tag and an image. The average score of the importance filter is computed first. If the score of a tag is higher than the average score, the tag will be assigned to the waking state. The tags that have a score lower than the average score are then assigned to the sleeping state candidate, which is the middle stage of the WAS algorithm. The tags belonging to the sleeping state candidate are considered to have the potential to provide usable information. Relationship retriever is used to find useful information in the sleeping state candidate by calculating the similarity between each tag of the sleeping state candidate and the waking state. Finally, the tags that have a high similarity are assigned to the waking state, and other tags are assigned to the sleeping state.

3   CAT Algorithm

PISAR uses the interaction between the user and the system, and the WAS algorithm analyzes the tag information. The proposed CAT algorithm incorporates the color information contented in images to enhance the retrieval results. The CAT algorithm can be divided into offline and online phases, which are discussed in detail below.

3.1   Offline Phase

In this phase, CAT finds the representative CHs of each tag in the database. To this end, a tag from the database is first selected as the target tag. Images with the same tag are then fetched to form an image set. We intend to find the representative CHs for the images within the same image set. However, from our preliminary experiments, we observe that the use of the entire image set may not help to find the exact and specific color information for the entire image set. Therefore, we propose to cluster the images into several subsets on the basis of the relations of tags. In order to find the tag relationships, we first fetch all the tags in the same image set and use them to form a tag set. Next, the WAS algorithm is used for determining the groups of tags. Figure 3 shows an example of clustered images. Assume that the user selected "sky" as the target tag. The images with the tag, "sky", are fetched to form an image set. Then, the WAS algorithm analyses the tag set and finds that "water", "clouds", and "sunset" have a high relation to "sky". The word class of the tags is used here to extract key information of image. The details will be presented later. After the clustering process, the image with the tag "water" will be assigned to an image subset called "sky, water". In this example, the original image set will be clustered into three image subsets.

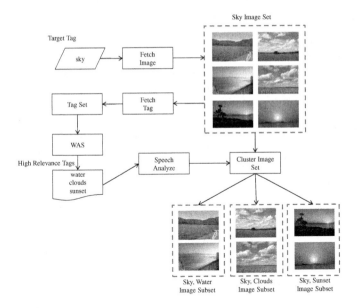

Fig. 3. Images clustering based on WAS algorithm.

3.1.1 Part-of-Speech Analysis

In this study, we use the constituent likelihood automatic word-tagging system (CLAWS) POS tagging system [2] to identify the word class of the target tag and highly related tags to extract the main target of images. There are eight word classes: noun, adjective, verb, adverb, pronoun, preposition, conjunction, and interjection. Since our method focuses on color information, we ignore word classes that do not have explicit color information, namely pronoun, preposition, conjunction, and interjection. On the basis of different types of word classes, CAT is slightly modified. In the following paragraphs, we will present the CAT algorithm for noun, adjective, verb, and adverb.

Noun. The objects in an image are nouns, for instance, nature like clouds and tree, and buildings like Taipei 101 and Tokyo Tower. We can thus consider that if a user uses a noun as the query term to search for an image, this noun is a main object in the image. Meanwhile, we also discover that some objects often occur simultaneously with the main objects. For example, assume that the main object is "sky". We can often see "clouds" in the same image, and thus, we may say that "clouds" is highly related to "sky". In our previous step, we already find these highly related tags. Other objects may be the unrelated or be the "noise" objects of the image. Therefore, we conduct background removal to remove these noise objects and extract key information.

For example, we select the "sky, clouds" image subset to explain our method. The target tag and the highly related tag of the image subset are "sky" and "clouds", respectively. We first select a correct image, which can accurately represent the target tag "sky" from the image subset. We consider "sky" and "clouds" as the main objects in the image and draw a line on them. Next, the colors on the line are fetched by the HSV color space and used for forming a color set. This color set keeps the colors of key objects, and noisy color information, which does not appear in the color set, is removed from the image. It can be observed that we only save the color of "sky" and "clouds" of the original image and remove other objects. In other words, after background removal, we can obtain key objects and thus, the key color information of the image.

Adjective and Verb. The main syntactic role of an adjective is to qualify a noun, and only a noun can follow an adjective. Given a phrase: "beautiful sunset", the color information is provided by the nouns "sunset" rather than the adjective "beautiful". Thus, a noun provides explicit color information.

In Verb, an action or an occurrence merely contains color information, such as "happen" or "be". Verb words can be divided into transitive verbs and intransitive verbs. Given a sentence expressing an action, which contains a transitive verb, such as "play basketball", we may infer the shape or color of "basketball", not "play". Because the characteristics of an adjective and a verb are similar, the process of these two word classes is the same. When the target tag is an adjective or a verb, we reserve the image subsets whose highly related tags are nouns.

Adverb. According to English grammar, four word classes can be preceded by an adverb: noun, adjective, verb, and adverb. Our method focuses on the words followed by adverbs. If an adjective is followed by an adverb, we will use the same method used in the case of an adjective as presented in Sect. 3.1.1.2 to process the adverb. For other word classes, the process is the same as that for an adjective. However, the method may fall into an infinite loop when there is an adverb followed by another adverb. To deal with such a condition, we place a constraint that the process stops when the loop repeats three times.

3.1.2   Color Histogram Estimation

With the above mentioned process, we can obtain image subsets without noise objects. This section we will extract CHs from the given image in image subset. Because the hue is a main part in the HSV color space, we use hue to prepare CHs in this study. We obtain a 360-dimensional matrix of each image. The value at a particular element in the matrix describes the total number of pixels of that degree in the image. To reduce the computational complexity, we use the color wheel to reduce the dimension. A color wheel [7] is a circular chart based on red, yellow, and blue. There are three classes of color: primary colors, secondary colors, and tertiary colors. According to these three classes of color, we can get twelve colors, including three primary colors, three secondary colors, and six tertiary colors. The twelve colors form the spectrum of colors and increase around the primary colors.

By defining the hues into twelve purest colors, we find that the interval between each color is 30°. Then, a dimensionality reduction process is performed to obtain a 12-dimensional matrix, which is used for preparing the CH of an image. In addition, we consider that each image should have different weights associated with the target tag. Therefore, we combine the Tag-Image relational score from PISAR as the weight of each image and calculate the average CH of an image set:

$$CH_i = \frac{\sum_{j=1}^{n} W_j \times I_j}{\sum_{j=1}^{n} W_j} \tag{1}$$

where CH_i denotes the i-th CH image subset, W_j represents the relational score from PISAR, I_j indicates the i-th CH of the image, and n denotes the total number of images in the image subset. Then, we can get an average CH for each image subset.

3.1.3 Color Histogram Clustering

In our preliminary experiments, we found that some CHs of the target tag are rather similar. To increase the discriminative power of different CHs, we further calculate a similarity score between each pair of CHs. We use the cosine distance:

$$Similarity(\overrightarrow{CH_p}, \overrightarrow{CH_q}) = \frac{\overrightarrow{CH_p}\overrightarrow{CH_q}}{\|CH_p\|\|CH_q\|} \tag{2}$$

$\overrightarrow{CH_p}$ and $\overrightarrow{CH_q}$ are two CHs randomly fetched from the target tag that represent colors. For example, if we calculate the similarity between the "sky, water" and the "sky, clouds" image subsets shown in Fig. 3 and note that the similarity between them is high, then we compare the relation scores of the tag "water" and "clouds" and keep the CH with the higher relation score. Next, a set of CHs is formed and used for representing the color information of the target tag.

3.2 Online Phase

As presented in the previous section, we prepared a set of CHs for each tag in the offline phase. In the online phase, when user queries a keyword to search, CH set of keyword is first retrieved. Then, one set from the entire CH set is selected by the user, and we assume that the selected CH really represents the query term. At the same time, the images that are marked with the same keyword are fetched to form an image set. This image set is the same as that prepared in the offline phase. Next, we calculate the relevance score between CH and each image in the image set.

3.2.1 Color Histogram Clustering

Figure 4 shows the process to compute the relevance scores. In the process, we first transfer the image to CH and seek the tag score (TC) from PISAR. We calculate the color similarity (CS) between the CHs of the user's choice and the target image. We also use the cosine similarity to compute the CS score as follows:

$$CS(\overrightarrow{CH_i}, \overrightarrow{CH_n^m}) = \frac{\overrightarrow{CH_i}\overrightarrow{CH_n^m}}{\|CH_i\|\|CH_n^m\|} \tag{3}$$

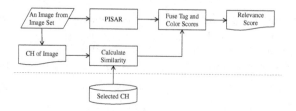

Fig. 4. Process of relevance score computation.

where $\overrightarrow{CH_i}$ denotes the i-th CH of the image, $\overrightarrow{CH_n^m}$ represents a CH based on the user's choice from the CH set, and m denotes a keyword queried by the user queries. On the basis of the TC and CS values, we define the relevance score as follows:

$$Relevance(I_i, CH_n^m) = (1 - \alpha) \times TC(I_i, T^m) + \alpha \times CS(\overrightarrow{CH_i}, \overrightarrow{CH_n^m}) \quad (4)$$

where I_i denotes an image from the image set and α represents a weight used for controlling the balance between the tag and the color scores (namely TC and CS). By ranking the relevance scores, we can retrieve images that give the highest relevance score to the user.

4   Experiment

4.1   Experimental Environment

We implemented the CAT algorithm on the PISAR system. The PISAR database contains 7,000 images and around 20,000 tags. The images were randomly downloaded from Flickr and divided into 14 categories and each category contains the same number (500) of images. To evaluate the proposed CAT algorithm, we compare the experimental results reported in [1,4]. The user first inputs a query term, and the system shows the representative CHs to the user. The user then chooses the most suitable CH corresponding to the query term. If no suitable answer is available, the user can select the "none of the above" option. With the selected CH, a set of photos are presented to be selected by the user.Our experiments included results collected from 100 users, and the average accuracies are reported in the following sections.

4.2   Optimizing the Parameters in the CAT Algorithm

To optimize the CAT algorithm, the results of 10 users were used for tuning the parameter α in Eq. (3). Since the value of α determines the weight between the tag and the color information, a larger α indicates that the color information is more emphasized for image retrieval.

   We randomly picked 10 tags as the test set to investigate the retrieval performance by using different values of α. The experimental procedure is presented

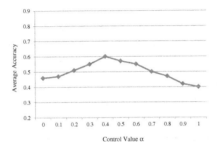

Fig. 5. Experiment of different control values of all speeches.

in Sect. 4.1, and the results are shown in Fig. 5. From the figure, we can infer that the accuracy will decrease when α is too small (smaller than 0.3) or too big (larger than 0.7), suggesting that both the tag and the color information affect the performance. From this set of results, we set α to 0.4 in the following discussion.

4.3   Example of Image Retrieval

Figure 6 show the top 5 retrieved images of different CHs. As introduced in the previous section, both the color and the tag information play important roles in image retrieval. From Fig. 6, we can see that the top 5 images between two CHs are based on color. However, we can see that the first image of the two color histograms is the same, which implies that the tag information of the first image plays an important role: although the color histogram changes, the image still ranks first. Therefore, the occurrence tells us that we have to consider both the tag and the color information at the same time.

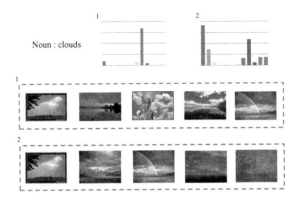

Fig. 6. Image retrieval result for the noun clouds.

4.4   Image Retrieval Result

Figure 7(a), (b), (c) and (d) illustrate the Top 5, Top 10, Top 15, and Top 20 image retrieval results of different word classes, respectively. The retrieval results obtained by the original PISAR, PISAR with WAS, and PISAR with both WAS and CAT are denoted as PISAR, WAS, and CAT, respectively. From Fig. 7(a), we conclude that the accuracy of a noun is higher than that of the other word classes. This can be attributed to the fact that our system records the user's query history and nouns are often used to search images. Meanwhile, for adjectives and verbs, CAT achieves a better performance than PISAR and WAS, confirming that color information plays an important role in these two word classes. However, the accuracy of an adverb is relatively low. A possible explanation for this could be that when dealing with an adverb, we have to search the noun preceded by an adverb. Here, the relationship between an adverb and a noun may not be very relevant and thus, decrease the retrieval accuracy. Nevertheless, CAT still outperforms PISAR and WAS for adverbs in terms of the average retrieval accuracy.

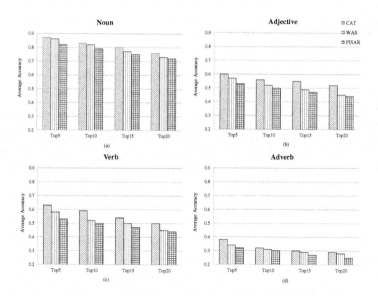

Fig. 7. Image retrieval result

5   Conclusions

In this study, we proposed a novel CAT algorithm in order to enhance the retrieval images. The CAT algorithm consists of two phases, offline and online phases. In the offline phase, the color of an image is used for calculating a set of representative CHs. The WAS algorithm is incorporated in this phase to enhance the image clustering accuracy. In the online phase, a relevance score is computed on the basis of the tag and the color information. The image that gives the highest relevance score is then retrieved and delivered to the user. We evaluated the CAT algorithm by using an image retrieval task. The original PISAR and WAS were also implemented to compare the CAT algorithm. The results show that CAT provides higher accuracy in terms of the retrieval results than both PISAR and WAS.

References

1. Du, W.-H., Rau, J.-W., Huang, J.-W., Chen, T.-S.: Improving the quality of tags using state transition on progressive image search and recommendation system. In: Proceedings of the IEEE International Conference on Systems, Man, and Cybernetics, pp. 3233–3238 (2012)
2. Garside, R., Smith, N.: A hybrid grammatical tagger: CLAWS4. In: Corpus Annotation: Linguistic Information from Computer Text Corpora, pp. 102–121 (1997)
3. Gonzalez, R.C., Woods, R.E.: Digital Image Processing. Prentice Hall Press, Upper Saddle River (2002)
4. Huang, J.-W., Tseng, C.-Y., Chen, M.-C., Chen, M.-S.: Pisar: Progressive image search and recommendation system by auto-interpretation and user behavior. In: Proceedings of the IEEE International Conference on Systems, Man, and Cybernetics, pp. 1442–1447 (2011)
5. Kherfi, M.L., Ziou, D.: Relevance feedback for CBIR: a new approach based on probabilistic feature weighting with positive and negative examples. IEEE Trans. Image Process. **15**(4), 1017–1030 (2006)
6. Li, X., Snoek, C.G., Worring, M.: Learning tag relevance by neighbor voting for social image retrieval. In: Proceedings of the 1st ACM international conference on Multimedia information retrieval, pp. 180–187 (2008)
7. Wolfrom, J.: The Magical Effects of Color. C&T Publishing, April 2010
8. Wu, L., Yang, L., Yu, N., Hua, X.-S.: Learning to tag. In: Proceedings of the 18th international conference on World wide web, pp. 361–370 (2009)

Advancing Iterative Quantization Hashing Using Isotropic Prior

Lai Li[✉], Guangcan Liu, and Qingshan Liu

Jiangsu Key Laboratory of Big Data Analysis Technology,
Nanjing University of Information Science and Technology, Nanjing 210044, China
{laili,gcliu,qsliu}@nuist.edu.cn

Abstract. It is prevalent to perform hashing on the basis of the well-known Principal Component Analysis (PCA), e.g., [1–4]. Of all those PCA-based methods, Iterative Quantization (ITQ) [1] is probably the most popular one due to its superior performance in terms of retrieval accuracy. However, the optimization problem in ITQ is severely under-deterministic, thereby the quality of the produced hash codes may be depressed. In this paper, we propose a new hashing method, termed Isotropic Iterative Quantization (IITQ), that extends the formulation of ITQ by incorporating properly the isotropic prior proposed by [3]. The optimization problem in IITQ is complicate, non-convex in nature and therefore not easy to solve. We devise a proximal method that can solve problem in a practical fashion. Extensive experiments on two benchmark datasets, CIFAR-10 [5] and 22K-LabelMe [6], show the superiorities of our IITQ over several existing methods.

Keywords: Hashing · Large-scale image retrieval · Isotropic prior · Iterative quantization

1 Introduction

With the development of the equipments for image acquisition, the volume of photos and videos we need to deal with gets bigger than ever, causing difficulties in image retrieval, image classification, and many other vision tasks related to nearest neighbor search. As a consequence, it is crucial to develop effective tools that can perform fast nearest neighbor search in large-scale databases. To this end, hashing is widely regarded as a promising approach due to its advantages of fast query speed and small storage expenses [6–12]. Various hashing methods have been proposed and investigated in the literature over the past several years, e.g., [3, 8, 13–21].

In general, hashing is a special dimension reduction task that aims to map similar data points of high dimension to adjacent binary hash codes in low dimension. As a result, it is natural to perform hashing on the basis of some

Q. Tian et al. (Eds.): MMM 2016, Part II, LNCS 9517, pp. 174–184, 2016.
DOI: 10.1007/978-3-319-27674-8_16

dimension reduction method, e.g., the well-known Principal Component Analysis (PCA) which is probably the most widely used method for dimension reduction. To perform hashing, a straightforward approach is to firstly use PCA to project high-dimensional data points to low-dimensional vectors (i.e., dimension reduction), and then convert the low-dimensional vectors to binary hash codes via quantization. However, such a naive approach cannot work well as the quantization error is often very large, usually resulting in unsatisfactory hash codes. To resolve this issue, Iterative Quantization (ITQ) [1] tries to infer an orthogonal rotation matrix that refines the projection matrix learnt by PCA such that the quantization error of mapping the data points to the binary hash codes is minimized. As has been confirmed by many reports including this paper, the performance of ITQ is superior, revealing the importance of reducing the quantization error. However, in the optimization problem of ITQ, the number of unknowns can largely exceed the number of knowns (i.e., the problem is under-deterministic), and thus ITQ may be depressed especially when code bits are relatively long. Researchers have also tried to refine ITQ by utilizing additional priors to extend the object function of ITQ. For example, Sparse Projection (SP) [4] proposes to enforce the projection matrix to have some sparsity. SP can perform better than ITQ when the code bits are extremely long, but is indeed inferior to ITQ when the code bits are of a reasonable length.

To reduce the quantization error of PCA hashing, it could be also helpful to use different projected dimensions that have an equal variance because the quantization procedure favors the case where different dimensions carry equal information. To achieve equal variance, Isotropic hashing (IsoHash) [3] proposes an objective function that minimize the diversity of the variances of different projected dimensions. Such an isotropic prior is effective, as confirmed by many studies including this work. However, since the objective function of ITQ is totally discarded, IsoHash indeed may not outperform ITQ in most cases, as shown in the experiments of this paper.

In this work, we propose to integrate ITQ and IsoHash into a unified objective function, resulting in a new hashing method termed Isotropic Iterative Quantization (IITQ). More precisely, our IITQ aims to seek an orthogonal rotation matrix to refine the initial projection matrix learnt by PCA such that both the quantization error and the diversity of variances are minimized. The optimization problem in IITQ is non-convex, complicate and cannot be solved by common optimization tools. Inspired by the success of proximal methods [22], we devise an alternating proximal algorithm that solves the optimization problem in a practical way. Extensive experiments on two benchmark datasets, CIFAR-10 [5] and 22K-LabelMe [6], show the superiorities of our IITQ over several existing methods. In summary, the contributions of this paper include:

- We propose a new hashing method called IITQ that merges the objectives of ITQ and IsoHash into a unified objective function. Experimental results show that IITQ outperforms both ITQ and IsoHash.

- It is challenging to solve the non-convex optimization problem arising from IITQ. We devise an alternating proximal algorithm that solves the problem in a practical fashion.

The rest of this paper is organized as follows. Section 2 reviews related work. Section 3 presents the details of our method and the optimization algorithm. Section 4 implements different methods to experimentally verify the effectiveness of our proposed IITQ. Section 5 concludes this paper.

2   Related Work

Traditional tree based scheme for fast nearest neighbor search, used in handling low-dimensional data, is hard to deal with the explosive growth data dimension. In sharp contrast, hashing can reasonably overcome the curse of dimensionality by mapping similar data points of high dimension to adjacent binary hash codes in low dimension, and thereby accelerating nearest neighbor search via highly efficient Hamming distances in the code space.

Existing hashing methods can be roughly divided into two categories: data-dependent [3,8,13–16] and data-independent [17–21]. Projection learning (i.e., dimension reduction) and quantization are two main steps of hashing. The computation of the projection matrix is the key to distinguish between data-dependent and data-independent. In data-independent methods, the projection matrix for dimension reduction in usually randomly generated. Such methods mainly include Locality Sensitive Hashing (LSH) [17,18], Min-wise Hashing (MinHash) [23], and their variations [19–21,24]. As the internal structure of the data is neglected, data-independent methods usually need long coding bits to obtain acceptable accuracy, thereby increasing the cost of retrieval and storage. Compared with the data-independent methods, data-dependent hashing methods learn proper projection matrix from data such that short coding bits are sufficient to produce satisfactory hash codes. As a result, this kind of methods have attracted much attention: Spectral Hashing (SH) [13] learns the hash functions based on spectral graph partitioning; Binary Reconstruction Embedding (BRE) [8] learns the hash functions through minimizing the reconstruction error between the original distances and Hamming distances; Complementary Hashing (CH) [14] employs multiple complementary hash tables to gain better results; IsoHash [3] aims to learn a projection matrix with equal variances for different dimensions; ITQ tries to seek an orthogonal rotation projection matrix to minimize quantization error; Sparse Projection (SP) [4] introduces a sparsity constrain to regularize the solution of ITQ.

3   Isotropic Iterative Quantization

In this section, we shall introduce our IITQ hashing method as well as the optimization procedure in detail.

3.1   Preliminaries

Our IITQ method is built on the basis of PCA hashing, ITQ [1] and IsoHash [3]. In this subsection, we shall brief introduce these three methods for the completeness of presentation. We shall also specify the hashing problem studied in this work for the ease of reading.

Problem Statement. Suppose we are given with a real-valued data matrix $X = [x_1, x_2, \cdots, x_n] \in \mathbb{R}^{d \times n}$, each column in which is a d-dimensional data point $x_i \in \mathbb{R}^d$. Without loss of generality, we assume that the points are zero-centered, i.e., $\sum_{i=1}^n x_i = 0$. Our goal is to learn a collection of k binary hash functions $\{h_i(\cdot)\}_{i=1}^k$ that map any data point, denoted as x, to a binary string $b \in \{-1, 1\}^k$ of length k:

$$b = [h_1(x), h_2(x), \cdots, h_k(x)] : \mathbb{R}^d \to \{-1, 1\}^k. \tag{1}$$

The similarity structure of the original data is required to be preserved. Namely, close points in the original space $\mathbb{R}^d$ should be hashed into similar binary codes in the code space $\{-1, 1\}^k$.

Note that it is NP hard to directly compute the best binary functions for a given data set. So, the same as most hashing methods, we adopt a two-stage strategy to learn $h_i(\cdot)$: a projection stage (i.e., dimension reduction) and a quantization stage. In the projection stage, k real-valued functions $\{f_i(\cdot)\}_{i=1}^k$ are learnt to map the data points in $\mathbb{R}^d$ to the vectors in $\mathbb{R}^k$. In the quantization stage, similar to most methods that use one bit to quantize each projected dimension, we adopt $h_i(x) = sgn(f_i(x)), i = 1, \cdots, k$, where $sgn(v)$ takes value 1 if $v \geq 0$ and -1 otherwise.

PCA Hashing. As we can see, hashing is indeed a special dimension reduction task, where the specificity is mainly about the binary nature of the code space. So, it would be suitable to perform hashing on the basis of some dimension reduction method, e.g., the well-known PCA which is probably the most widely used tool for dimension reduction. More precisely, one could perform PCA on X, and then use the top k eigenvectors, denoted as $\{w_i\}_{i=1}^k$, of the covariance matrix XX^T as the projection functions for dimension reduction. That is, $f_i(x) = w_i^T x, i = 1, \cdots, k$, and

$$b = [sgn(w_1^T x), sgn(w_2^T x), \cdots, sgn(w_k^T x)] \triangleq sgn(W^T x), i = 1, \cdots, n, \tag{2}$$

where $W = [w_1, w_2, \cdots, w_k] \in \mathbb{R}^{d \times k}$ is the projection matrix formed by the top k eigenvectors of XX^T.

Note that the projection matrix W learnt by PCA is also the best ortho-normal matrix for minimizing the reconstruction error of restoring the original data from its low-dimensional version, i.e., $W = \arg\min_{P \in \mathbb{R}^{d \times k}} \|X - PP^T X\|_F^2$ s.t. $P^T P = I$, where $\| \cdot \|_F$ is the Frobenius norm of a matrix and I is the identity matrix. This property is really nice. However, the above PCA hashing method cannot work well as explained in the next subsection.

Iterative Quantization. While simple, the naive PCA hashing method afore presented may not work well because the learnt projection matrix W that minimizes the reconstruction error is unnecessary minimize the quantization error, which is importance for ensuring the quality of the produced hash codes. That is to say, the quantization error could be very large and the produced hash codes may be unsatisfactory. To overcome this issue, ITQ [1] tries to refine the projection matrix W learnt by PCA by seeking an orthogonal rotation matrix that can minimize the following quantization error:

$$\min_{B,Q} \|B - Q^T V\|_F^2, \text{ s.t. } Q^T Q = \text{I}, B \in \{-1, 1\}^{k \times n}, \tag{3}$$

where $V = W^T X$ is the projected data matrix given by PCA, $B = [b_1, b_2, \cdots, b_n] \in \{-1, 1\}^{k \times n}$ is a binary code matrix whose ith column b_i is the hash code of the ith data point x_i, $Q \in \mathbb{R}^{k \times k}$ is an orthogonal matrix needs to learn, and $\|\cdot\|_F$ is the Frobenius norm of a matrix. Extensive studies have confirmed that ITQ is much better than the naive PCA hashing method [3,8,13–16]; this reveals the importance of minimizing the quantization error.

Isotropic Hashing. To improve PCA hashing, it could be also helpful to make different projected dimensions have an equal variance, because the quantization procedure favors the case where different dimensions carry equal information. With this motivation, IsoHash [3] proposes to minimize the diversity of the variances of different projected dimensions. Let the top k eigenvalues of $X X^T$ be $\{\lambda_1, \lambda_2, \cdots, \lambda_k\}$, and a be a k-dimensional vector all elements in which equal to $\sum_{i=1}^{k} \lambda_k / k$. Then IsoHash tries to learn an orthogonal rotation matrix by

$$\min_{Q \in \mathbb{R}^{k \times k}} \|diag(Q^T \Lambda Q) - diag(a)\|_F^2, \text{ s.t. } Q^T Q = \text{I}, \tag{4}$$

where $\Lambda = V V^T = W^T X X^T W$ is computed by PCA, and $diag(\cdot)$ denotes a diagonal matrix whose diagonal entries are formed from either a vector (e.g., a) or the diagonal entries of a matrix (e.g., $Q^T \Lambda Q$). It has been confirmed that IsoHash is effective, illustrating that the isotropic prior (i.e., minimizing the diversity of the variances of different projected dimensions) is useful.

3.2 Improving ITQ Using the Isotropic Prior

Observe that the problem in (3) is under-deterministic (i.e., the unknowns are more than the knowns), in this paper we propose to use the isotropic prior (4) to regularize the solution space of ITQ so as to achieve better performance. Namely, we propose a new objective function called Isotropic Iterative Quantization (IITQ) that integrates ITQ and IsoHash into a unified formulation:

$$\min_{B,Q} \frac{1}{2}\|B - Q^T V\|_F^2 + \frac{\alpha}{2}\|diag(Q^T \Lambda Q) - diag(a)\|_F^2, \tag{5}$$

$$\text{s.t. } Q^T Q = \text{I}, B \in \{-1, 1\}^{k \times n},$$

where $\alpha \geq 0$ is a parameter for balancing the weights of the two parts. This parameter is determined by cross-validation in our experiments.

Although non-convex and complicate, the optimization problem in (5) can be solved by the proximal methods [22]. Let $\beta(Q) = diag(Q^T \Lambda Q) - diag(a)$ and $g(Q) = 0.5\alpha \|\beta(Q)\|_F^2$. Then the solution (B, Q) to problem (5) can be updated via solving the following proximal problems:

$$B_{k+1} = \arg \min_{B \in \{-1,1\}^{k \times n}} \frac{1}{2} \|B - Q_k^T V\|_F^2, \tag{6}$$

$$Q_{k+1} = \arg \min_{Q^T Q = I} \frac{1}{2} \|B_{k+1} - Q^T V\|_F^2 + \langle Q - Q_k, \Delta_Q g(Q_k) \rangle + \frac{\rho_k}{2} \|Q - Q_k\|_F^2,$$

where B_k and Q_k denote the estimates at the kth iteration, $\rho_k > 0$ is a penalty parameter, and $\Delta_Q g(Q_k)$ is the gradient of the function $g(Q)$ at Q_k. By [25],

$$\Delta_Q g(Q_k) = \alpha \Lambda Q_k \beta(Q_k). \tag{7}$$

According to [22], the penalty parameter ρ_k could be set as

$$\rho_k = \kappa \alpha \|\Lambda\|^2, \tag{8}$$

where $\kappa > 1$ is a parameter ($\kappa = 10$ in this work), and $\| \cdot \|$ denote the operator norm (i.e., the largest singular value) of a matrix.

The two optimization problems in (6) both have closed-form solutions. The solution to the first problem is given by

$$B_{k+1} = sgn(Q_k^T V). \tag{9}$$

The second problem could be equivalently converted to

$$Q_{k+1} = \arg \max_{Q^T Q = I} \langle Q, V B_{k+1}^T + \rho_k Q_k - \Delta_Q g(Q_k) \rangle. \tag{10}$$

Let the singular value decomposition (SVD) of $V B_{k+1}^T + \rho_k Q_k - \Delta_Q g(Q_k)$ be $U_k \Sigma_k P_k^T$. By [26], the solution to (10) is given by

$$Q_{k+1} = U_k P_k^T. \tag{11}$$

The whole procedure for solving the problem in (5) is also summarized in Algorithm 1. According to [22], the generated sequence $\{(B_k, Q_k)\}$ is guaranteed to converge to a critical point of problem 5.

4   Experiments

We evaluate our method on two benchmark datasets, CIFAR-10 [5] and 22K-LabelMe [6], which are widely used as testbeds in large-scale image retrieval. The first dataset, CIFAR-10, contains 60K 32×32 RGB images and has been manually labeled into ten class: *airplane, automobile, bird, cat, deer, frog, horse,*

Algorithm 1. Solving the problem in (5) via alternating proximal

---

Input: Projected matrix $V \in R^{k \times n}$ computed by PCA.
Output: The rotation matrix Q.
1: Initialize $Q = \mathrm{I}$.
2: **repeat**
3:    Update the code matrix B by (9)
4:    Update the rotation matrix Q by (11)
5: **until** convergence

---

ship and truck. Each class consists of 6 K images and each image is represented by a 320-dimensional gray-scale GIST [27] feature vector. We randomly choose 59 K images as training set and the rest for testing. The second dataset we used, 22K-LabelMe, consists of 22019 images randomly sampled from the whole LabelMe database. Each image is of 32×32 and represented by a 512-dimensional GIST feature vector. The dataset is also divided into two parts: 20 K data points for training and the rest 2019 data points for testing. The definition of ground truth neighbors is a point, if it lies in the top 1 % points closed to the query in terms of the original Euclidean distance. All experiments are running on a workstation of Intel(R) Core(TM) i7-4790 CPU @3.60 GHz with 8 G memory.

In order to highlight our advantages, we compare against six state-of-the-art hashing methods, including ITQ [1], IsoHash [3], SP [4], LSH [17], SH [13], and USPLH [28]. For the ease of implementation, the retrieval results of hashing methods are determined by the Hamming ranking tactics: We calculate the Hamming distance between the query and all the data points, sort the points in an ascending order, and then consider the top K points as the retrieval results. The *precision-recall curses* and *recall curves* are the metrics used for evaluating the performance of each method.

Figures 1 and 2 show the results of comparison CIFAR-10 and 22K-LabelMe, respectively. It can be seen that our method gains better performance than the other hashing methods. In particular, our IITQ is better than both ITQ and IsoHash, revealing the superiorities of using the isotropic prior to regularize the solution of ITQ. The lack of sufficient priors accounts for the inferior performance of ITQ and IsoHash. One may have noticed that the SP method, which extends ITQ by adding a sparsity constrain, does not outperform ITQ in our setting. This is not strange, because the sparsity prior is effective only when the coding bits are very long. Since long codes always lead to low efficiency, we do not consider the cases of very long codes in our experiments.

The results of (Mean Average Precision) MAP on the two benchmark datasets are showed in Table 1. Our IITQ method achieves the best performance on both dataset. This is consistent with the results shown in Figs. 1 and 2. The corresponding training time of all methods are provided in Table 2. Due to the complicate nature of problem 5, IITQ requires more time than the others for training. Fortunately, this issue is mild because our method is indeed as fast as the others while dealing with testing samples: For a testing sample x, IITQ can

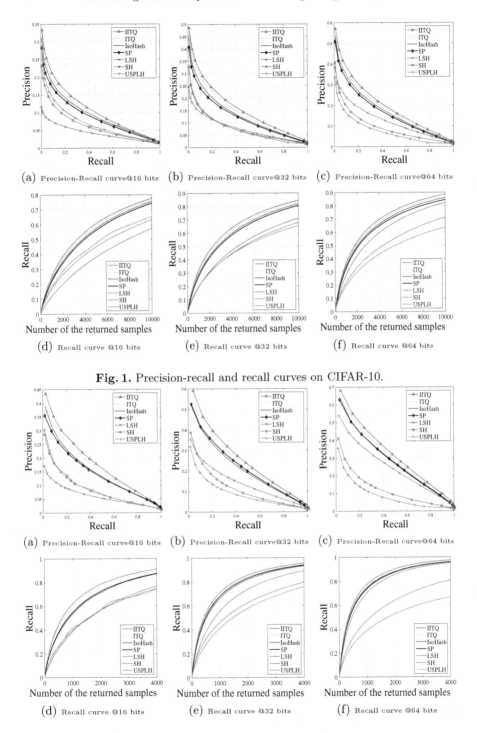

Fig. 1. Precision-recall and recall curves on CIFAR-10.

Fig. 2. Precision-recall and recall curves on 22K-LabelMe.

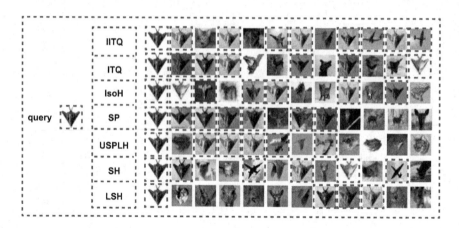

Fig. 3. A retrieval example from CIFAR-10.

Table 1. MAP on CIFAR-10 and 22 K-LabelMe.

Method	CIFAR			LabelMe		
Bits	16	32	64	16	32	64
IITQ	0.1050	0.1518	0.1909	0.1607	0.2238	0.2806
ITQ	0.0985	0.1474	0.1843	0.1490	0.2168	0.2670
IsoHash	0.0909	0.1326	0.1736	0.1322	0.1834	0.2537
SP	0.0841	0.1228	0.1608	0.1294	0.1939	0.2532
USPLH	0.0696	0.0765	0.0739	0.0808	0.0847	0.0816
SH	0.0619	0.0776	0.0932	0.0782	0.1043	0.1114
LSH	0.0470	0.0760	0.1264	0.0585	0.1218	0.1950

Table 2. Training time (in seconds) on 22K-LabelMe and CIFAR-10.

Method	LabelMe			CIFAR		
Bits	16	32	64	16	32	64
IITQ	13.26	17.96	95.13	34.71	77.87	192.34
ITQ	1.53	2.78	5.98	4.24	9.27	13.63
IsoHash	1.14	1.05	1.54	1.12	1.42	1.53
SP	8.16	9.86	15.10	15.29	18.19	24.85
USPLH	33.35	68.47	139.96	27.91	58.52	117.06
SH	0.41	0.49	0.66	0.72	0.84	1.20
LSH	0.02	0.03	0.05	0.02	0.04	0.06

calculate fast its hash code by $b = sgn(Q^T W^T x)$, where W is learnt by PCA and Q is computed by Algorithm 1.

Figure 3 shows a visual retrieval example with 32 bits. It can be seen that IITQ exhibits a better performance than the competing methods.

5   Conclusion

We studied hashing which is a key technique for addressing the challenges arising from the context of high-dimensional, massive data. In particular, we were interested in PCA-based hashing methods, e.g., ITQ and IsoHash. While effective, ITQ is not good enough due to the lack of sufficient priors while seeking the optimal rotation matrix that can minimize the quantization error. As a consequence, we proposed a new objective function, termed Isotropic Iterative Quantization (IITQ), that extends the formulation of ITQ by incorporating properly the isotropic prior proposed by IsoHash. Generally, the new objective is to minimize the quantization error as well as the diversity of the variances of different projected dimensions, leading to a complicate non-convex optimization problem. Based on proximal methods, we devised a practical algorithm to solve the optimization problem in an elegant way. Extensive experiments on two benchmark datasets, CIFAR-10 [5] and 22K-LabelMe [6], demonstrated the advantages of our IITQ over state-of-the-art methods.

Acknowledgement. This work is supported by NSFC 61502238, NSFC 61532009, BK2012045 and 15KJA520001.

References

1. Gong, Y., Lazebnik, S., Gordo, A., Perronnin, F.: Iterative quantization: a procrustean approach to learning binary codes for large-scale image retrieval. IEEE Trans. Pattern Anal. Mach. Intell. **35**(12), 2916–2929 (2013)
2. Gong, Y., Lazebnik, S.: Iterative quantization: a procrustean approach to learning binary codes. In: CVPR, pp. 817–824 (2011)
3. Kong, W., Li, W.: Isotropic hashing. In: NIPS, pp. 1655–1663 (2012)
4. Xia, Y., He, K., Kohli, P., Sun, J.: Sparse projections for high-dimensional binary codes. In: CVPR, pp. 3332–3339 (2015)
5. Krizhevsky, A., Hinton, G.: Learning multiple layers of features from tiny images (2009)
6. Torralba, A., Fergus, R., Weiss, Y.: Small codes and large image databases for recognition. In: CVPR (2008)
7. Weiss, Y., Fergus, R., Torralba, A.: Multidimensional spectral hashing. In: Fitzgibbon, A., Lazebnik, S., Perona, P., Sato, Y., Schmid, C. (eds.) ECCV 2012, Part V. LNCS, vol. 7576, pp. 340–353. Springer, Heidelberg (2012)
8. Kulis, B., Darrell, T.: Learning to hash with binary reconstructive embeddings. In: NIPS, pp. 1042–1050 (2009)
9. Kulis, B., Jain, P., Grauman, K.: Fast similarity search for learned metrics. IEEE Trans. Pattern Anal. Mach. Intell. **31**(12), 2143–2157 (2009)
10. Mu, Y., Yan, S.: Non-metric locality-sensitive hashing. In: AAAI (2010)
11. Sánchez, J., Perronnin, F.: High-dimensional signature compression for large-scale image classification. In: CVPR, pp. 1665–1672 (2011)

12. Yu, F.X., Kumar, S., Gong, Y., Chang, S.F.: Circulant binary embedding(2014)
13. Weiss, Y., Torralba, A., Fergus, R.: Spectral hashing. In: NIPS, pp. 1753–1760 (2008)
14. Xu, H., Wang, J., Li, Z., Zeng, G., Li, S., Yu, N.: Complementary hashing for approximate nearest neighbor search. In: ICCV, pp. 1631–1638 (2011)
15. Wang, J., Kumar, S., Chang, S.-F.: Semi-supervised hashing for large-scale search. IEEE Trans. Pattern Anal. Mach. Intell. 34(12), 2393–2406 (2012)
16. Mu, Y., Shen, J., Yan, S.: Weakly-supervised hashing in kernel space. In: CVPR, pp. 3344–3351 (2010)
17. Gionis, A., Indyk, P., Motwani, R.: Similarity search in high dimensions via hashing. In: VLDB, pp. 518–529 (1999)
18. Andoni, A., Indyk, P.: Near-optimal hashing algorithms for approximate nearest neighbor in high dimensions. Commun. ACM 51(1), 117–122 (2008)
19. Raginsky, M., Lazebnik, S.: Locality-sensitive binary codes from shift-invariant kernels. In: NIPS, pp. 1509–1517 (2009)
20. Datar, M., Immorlica, N., Indyk, P., Mirrokni, V.S.: Locality-sensitive hashing scheme based on p-stable distributions. In: ACM Symposium on Computational Geometry, pp. 253–262 (2004)
21. Kulis, B., Grauman, K.: Kernelized locality-sensitive hashing for scalable image search. In: ICCV, pp. 2130–2137 (2009)
22. Attouch, H., Bolte, J.: On the convergence of the proximal algorithm for nonsmooth functions involving analytic features. Math. Program. 116(1–2), 5–16 (2009)
23. Broder, A.Z., Charikar, M., Frieze, A.M., Mitzenmacher, M.: Min-wise independent permutations. J. Comput. Syst. Sci. 60, 327–336 (1998)
24. Ping Li and Arnd Christian König: Theory and applications of b-bit minwise hashing. Commun. ACM 54(8), 101–109 (2011)
25. Chu, M.T.: Constructing a hermitian matrix from its diagonal entries and eigenvalues. SIAM J. Matrix Anal. Appl 16, 207–217 (1995)
26. Gower, J.C., Dijksterhuis, G.B.: Procrustes Problems. Oxford, Oxford University Press (2004)
27. Oliva, A., Torralba, A.: Modeling the shape of the scene: a holistic representation of the spatial envelope. Int. J. Comput. Vis. 42(3), 145–175 (2001)
28. Qiao, L., Chen, S., Tan, X.: Sparsity preserving projections with applications to face recognition. Pattern Recogn. 43(1), 331–341 (2010)

An Improved RANSAC Image Stitching Algorithm Based Similarity Degree

Yule Ge[1(✉)], Chunxiao Gao[1], and GuoDong Liu[2]

[1] School of Computer Science and Technology,
Beijing Institute of Technology, Beijing, China
`skgeyule@126.com, gao_chunxiao@bit.edu.cn`
[2] North Automatic Control Technology Institute, Taiyuan, Shanxi, China
`snow_ocean@126.com`

Abstract. In terms of the deficiency in the aspects that the higher computational complexity caused by excessive iterations and the easy happened stitching dislocation caused by the difficult-to-determine parameters. In this paper, an improved RANSANC algorithm based similarity degree is proposed and is applied in image mosaic. This improved algorithm includes that sorting rough matched points by similarity degree, calculating transformation matrix, rejecting obviously wrong matched points and executing classical RANSAC algorithm. It is demonstrated by the experiments that this algorithm can effectively remove wrong matched pairs, reduce iteration times and shorten the calculation time, meanwhile ensure the accuracy of requested matrix transformation. By this method can get high quality stitching images.

Keywords: Image stitching · RANSAC · Similarity degree

1 Introduction

Sometimes the images captured by the camera will contain part of the scene people want to see. Image mosaic is a technique being used to stitch multiple images together to form a stitched image with higher resolution and large field of view. Image mosaic has been widely used in video stitching [1], biomedical image analysis [2], space exploration [3], super-resolution image reconstruction [4] and so on.

The process of image mosaic is very complicated, one important and time-consuming step is image registration which refer to calculating transformation matrix in order to make the corresponding pixels in two different images reach consensus on the coordinates. According to image similarity measurement Image registration is roughly divided into two categories [5]: kind of methods one based on gray, and others are feature-based. The first method determines the similarity between different images by means of the entire image gray values. This kind of methods has the obvious disadvantages like large amount of calculation, time consuming, and they does not make use of spatial information of the image, so that this method is now rarely used. The second kind of methods first extracted certain feature point as image registration primitive from the image. The advantage of this method is the computational

© Springer International Publishing Switzerland 2016
Q. Tian et al. (Eds.): MMM 2016, Part II, LNCS 9517, pp. 185–196, 2016.
DOI: 10.1007/978-3-319-27674-8_17

efficiency, and with affine invariant and stability, this method are more suitable for image mosaic.

There are many methods of extracting feature points, for example Harris feature point [6], Susan feature point [7], SIFT feature points [8] and Surf [9] feature points. This paper take the method of Surf. After image feature points extracted we can rely on these feature points to deduce transformation matrix. This step is usually adopt random sampling consistency (RANSAC) [10] algorithm which able to match the feature points with noise and it has good robustness. But iterative calculation is used to ensure its correctness in this algorithm so low efficiency and the parameters can only rely on empirical estimates so often leads to precision is not enough. This can be easily seen from the experimental results [11].

Considering the shortcomings of RANSAC, a large number of scholars began to try to optimize the algorithm and achieved good results. In order to reduce computation time [12] uses the relationship between the adjacent feature points to delete false matching points. However, the method removes a lot of the original matching pairs so the transformation matrix obtained by this method need to be verified. [13] assumes that all the lines between two right matching points should be roughly the same direction. So when a line intersects with many other lines which match can be considered to be false. This method is correct when there only exists translation transformation between images. But when there is rotation transformation the assumption is not set up so this method applicable scope is limited. [14] the angles between the horizontal wire and all the lines between every two matching points to be calculated. When a certain angle differ a lot with many other angles which match can be considered to be false. The essence of this method is the same as [13], but with a different method of exclusion, so the same deficiencies. [15] uses the geometric relationships between the matching points. First calculates the feature points' Delaunay triangulation on two images, then filters the points, finally selects sampling points for RANSAC by Euclidean distance. This algorithm introduced a large amount of computation lead to more time consuming.

To handle with the shortcomings of the original RANSAC and improve its performance, a more efficient and reliable improved RANSAC algorithm based similarity degree is proposed in this paper. The proposed algorithm can not only overcome the drawbacks of [13, 14] whose approaches can only be applied in the presence of a translation transformation, and compared with [15] greatly reduces the amount of calculation. Our method firstly extract SURF feature points, followed by the coarse matching using BBF, then the most important step preprocessing: purifying the coarse matching set to get a new matching set, and running the classical RANSAC on the new set for image registration, image fusion finally. The main innovation is reflected in the preprocessing unit, including sorting the coarse matching feature points, calculating preprocessing matrix for filter coarse mating set, etc. Experiments results indicates that the proposed method can effectively eliminate false matching pairs, and improves the computational efficiency, has obvious superiority.

2   The Improved RANSAC Based Similarity Degree

The innovation of this paper is that we proposed a different method to purify the coarse matching set, which process is called pretreatment. Pretreatment will delete those obvious error matching pairs so get a new set of matches with fewer error matches, and then run the classical RANSAC on the new matching set. Before we put forward details of the pretreatment method, let's first take a brief look at the image transformation matrix.

2.1   Transformation Matrix of Image Registration

To the same object shoot two images with different angles. There is always a certain part of them is overlapping. By overlapping areas we can derive the transformation relationship between two images. Through this relationship can put an image projected onto another image, thereby completing the image mosaic. Assume that a certain point $P_1(x,y)$ on an image, point $P_2(x',y')$ on another image is the corresponding points of p1 and those two points are coincide on mosaic image. We can use the homogeneous coordinate transformation matrix H with 8 parameters [16] to describe the correspondence between these two different images.

$$H = \begin{bmatrix} h_1 & h_2 & h_3 \\ h_4 & h_5 & h_6 \\ h_7 & h_8 & 1 \end{bmatrix} \tag{1}$$

Projection relationship:

$$\begin{bmatrix} x' \\ y' \\ 1 \end{bmatrix} = \begin{bmatrix} h_1 & h_2 & h_3 \\ h_4 & h_5 & h_6 \\ h_7 & h_8 & 1 \end{bmatrix} \begin{bmatrix} x \\ y \\ 1 \end{bmatrix} \tag{2}$$

can be gotten:

$$\begin{aligned} x' &= \frac{h_1 x + h_1 y + h_3}{h_7 x + h_8 y + 1} \\ y' &= \frac{h_4 x + h_5 y + h_6}{h_7 x + h_8 y + 1} \end{aligned} \tag{3}$$

From the above one pair of matching points can get two different equations. Only four pairs of correct matching points can determine all the parameters. And because there are a lot of matching pairs, so you can obtain a number of different H. So what H is the best one? You might think we can calculate final transformation matrix by using the least square method [17]. However, the question is not so simple. Because if there are many error matches in the matching set, the least square method will not run correctly. We need another way to solve this problem. RANSAC is an effective method.

2.2   Calculation Method for RANSAC Sampled

RANSAC obtains a plurality of inner point sets and transformation matrices by repeating hypothesis-verification process over and over again, and then recalculate the H in the set that contains most points. All hypothesis-verification process are performed on the coarse matching set that include a large number of points and outside point account for a large proportion. The algorithm time cost mainly lies in validating those lots of hypothesis. If we can reduce the number of hypothesis, can effectively reduce the time consumption of the algorithm. By the way, inner point refer to right matching pair and outside point refer to error matching pair in image registration.

Suppose there are M feature matching pairs in coarse matching set. Taking K to perform hypothesis-verification process to calculate a transformation matrix called a sampling. Ranging of samples can be obtained $W : [1, C_M^K]$. How many suitable sampling number W take? Because the initial matching set contain error matching pairs, if the sampling number W is too low, may each sample includes in the error pair so that the model calculated will be wrong. The optimal sampling can be achieved by take the maximum value of W. But if M is very big will result in sampling number W too many and time consuming too much, so seldom take the maximum value of W in the practical application. Through the above analysis we learn that W cannot too small nor too big. We can according to the following rules to choose W: must ensure the probability that at least there is one sampling set's all points are inside point higher than P. Practical requirements of P must be greater than 95 %. To calculate W also assumes that the ratio of the number of correct matching pairs and total number is P_1, and the ratio of the number of error matching pairs and total number is $\varepsilon = 1 - P_1$. Calculate the value of W[18]:

$$(1 - p_1^k)^W = 1 - p \tag{4}$$

This equation can be solved:

$$W = \log(1 - P) / \log(1 - (1 - \varepsilon)^K) \tag{5}$$

W will decrease with the reducing of ε.

2.3   Feature Points Matching in Coarse Matching Step

In order to find one point's matching point must seek out this point's nearest neighbor point and second nearest neighbor point. The distances between this point and those two points are nearest neighbor distance and second nearest neighbor distance. The probability that the nearest neighbor was correct match can be determined by taking the ratio of distance from the nearest neighbor to the distance of the second nearest. In order to find out this relationship Lowe [19] did a large number of experiments. As show in Fig. 1, The probability density functions for right and wrong matches are shown according to the ratio of nearest to second-nearest neighbors of each matching points. The solid line shows the PDF of this ratio for right matches, while the dotted line is for matches that were wrong. Matches that its nearest neighbor was correct match had a PDF that the ratio is focused on between 0.3 and 0.6 while its nearest neighbor was incorrect match had a PDF that the ratio was greater than 0.8.

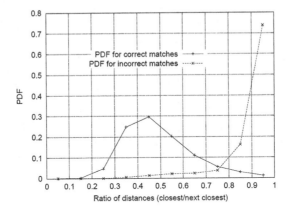

Fig. 1. The density functions for correct and incorrect matches

Table 1. The number of all matches, correct matches and their ratio

	G1	G2	G3	G4	G5	G6	G7	G8	G9	G10
M	167	366	547	458	284	416	570	335	318	470
MC	125	204	417	232	188	303	429	238	289	403
MC/M(%)	74.85	55.74	76.23	50.66	66.20	72.84	75.26	71.04	90.88	85.74
	G11	G12	G13	G14	G15	G16	G17	G18	G19	G20
M	415	470	566	1017	609	212	335	563	1169	561
MC	328	259	387	756	397	132	255	466	1006	455
MC/M(%)	79.04	55.10	68.37	74.33	65.19	62.26	76.20	82.77	86.06	81.11
	G21	G22	G23	G24	G25	G26	G27	G28	G29	G30
M	626	968	119	257	185	261	239	186	213	529
MC	561	918	70	200	148	198	183	116	164	420
MC/M(%)	89.62	94.83	58.82	77.82	80.00	75.86	76.60	62.37	78.00	79.39

3 Pretreatment

What percentage of the matching pairs is correct? Tables 1 and 2 show 30 group splicing experiments statistical results. The total number of matching pairs M, the number of the correct matching pairs MC and MC/M in coarse matching step are shown in Table 1. Analysis of experimental data: the average ratio of correct matching pairs number and the total matching pairs number is about 74.10 %. Figure 2 is a bar graph based on data drawn.

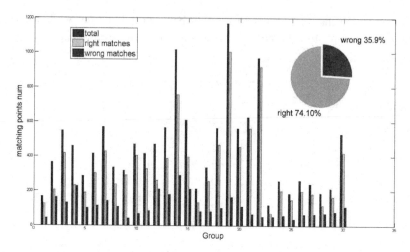

Fig. 2. Bar chart created by above data

The number of the right matching pairs which similarity degree value between 0.3 and 0.6 MCS, as well as MCS/MC, and the number of the error matching pairs which similarity degree value between 0.3 and 0.6 MES and the value of the MES/ME in coarse matching step are shown in Table 2.

Table 2. The number of correct metches which similarity value between 0.3 and 0.6 and the ratio of it and all correct matches.The number of incorrect matches which similarity value between 0.3 and 0.6 and the ratio of it and all incorrect matches.

	G1	G2	G3	G4	G5	G6	G7	G8	G9	G10
MCS	80	132	310	151	138	239	268	181	187	233
MCS/MC(%)	64	64.70	74.34	65.09	73.40	78.88	62.48	76.05	64.71	57.82
MES	11	18	12	25	9	4	15	7	4	6
MES/ME(%)	26.19	11.11	9.23	11.06	9.38	3.54	10.64	7.22	13.79	8.96
	G11	G12	G13	G14	G15	G16	G17	G18	G19	G20
MCS	221	158	236	359	234	86	152	289	411	289
MCS/MC(%)	67.38	61.01	60.98	47.49	58.94	65.15	59.61	62.02	40.85	63.52
MES	7	28	24	20	15	11	2	16	17	13
MES/ME(%)	8.05	13.27	13.41	7.66	7.08	13.75	2.5	16.49	10.43	12.26
	G21	G22	G23	G24	G25	G26	G27	G28	G29	G30
MCS	248	404	51	146	110	159	117	82	93	176
MCS/MC(%)	44.21	44.01	72.85	73	74.32	80.30	63.93	70.69	56.70	41.90
MES	6	6	6	8	7	13	11	8	8	10
MES/ME(%)	9.23	12	12.24	14.03	18.92	20.63	19.64	11.42	16.33	9.17

Figure 3 is a Line graph based on data drawn. ME represents the number of the false matches, can be determined by ME = M − MC. Analysis of experimental data: mean MCS/MC is about 63.01 %, MES/ME average is about 11.72 %. The results and the results shown in Fig. 1 roughly identical. From Tables 1, 2, Figs. 2 and 3 can draw the following conclusions:1. The number of right matches greater than the number of error matches in coarse matching phase.2. Most of the right match's similarity degree value is between 0.3 and 0.6 while a false match's similarity value is rarely in the range. From these two conclusion can be summed up most of these matches which similarity degree value between 0.3 and 0.6 are right matches. It provides a theoretical basis for this paper proposed algorithm.We can't see the correct matching points percentage of the total matching points from Lowe's experimental results.

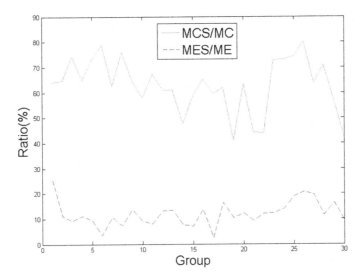

Fig. 3. Line graph created by above data

Combined with the results shown in Fig. 1 and the analysis of Calculation method for RANSAC sampled, considering using the ratio of the nearest neighbor distance and second nearest neighbor distance to improve RANSAC. We call this ratio Similarity Degree. The aim of the improvement is to get a new matching set which contains fewer outside points. In detail the process below:

The similarity degree of each matching pair can be got by using KNN [20] algorithm in rough matching step. Take the matching pairs which similarity degree value between 0.3 and 0.6 make up pretreatment set M. Assume that M contains m matching pairs. By the probability density function curves in the Fig. 1 shows the probability of similarity degree value from 0.3 to 0.6 of correct matching pairs bigger than the false matching pairs. Assuming that correct match pairs account for half of the total initial matching pairs. By probability theory knowledge can know the probability of a match is right match is higher than it is wrong match when its similarity degree is between 0.3 and 0.6. It can also to be said that the right matches are in the majority. In m matching pairs take any four don't repeat to calculate transformation matrix H_i. $P_{correct}$ is the probability of H_i is calculated by four correct matching pairs. If the number of correct matching points which similarity degree value between 0.3 and 0.6 account for large proportion of total, so $P_{correct}$ will be large. To further improve $P_{correct}$ designed following optimization strategy. First define the distance of two isomorphic matrix is calculated as:

$$dis_{A,B} = \sqrt{\sum_{\substack{i=1 \\ j=1}}^{\substack{i<=m \\ j<=n}} (A_{ij} - B_{ij})^2} \tag{6}$$

The element value in i row of the j column in the average matrix is calculated by:

$$M_{ij}^{ave} = \frac{1}{n} \sum_{k=1}^{k<=n} M_{ij}^k \tag{7}$$

Take four pairs in M to calculate H_i. Assume a total of S matrices are calculated. H_{ave} is obtained from these S matrices by using (7), then calculate the distance DIS_i between H_i and H_{ave} by using (6). Take H_i as the eliminate matrix H which minimum distance from H_{ave}.

After H is obtained then can use it to delete the error matches in initial coarse matching set. Suppose P1, P2 are a pair of matching feature points. p2' is the corresponding points of p1 converted by H. then calculate the Euclidean distance between P2 and P2'. Delete this matching feature points if the distance is greater than the threshold t'. For each pair matching points to calculate like above, then delete all pairs which distance is greater than the threshold t'. So we can get a small matching points subset than original. Specific steps are as follows:

1. Sort coarse matching set by similarity degree value in ascending order. This step can be realized in coarse matching step.
2. Taking matching pairs which similarity degree value between 0.3 and 0.6 make up set M. Then take four pairs from M to calculate H_i, and calculate the distance between H_i and H_{ave}. Take H_i as the eliminate matrix H which minimum distance from H_{ave}.
3. For each matching point calculate the Euclidean distance between its matching point and the corresponding point transformed by H. Remove matching pairs which distance are greater than t'. Getting a new matching set.
4. Running RANSAC in the new matching set.

Above is the whole process of this article pretreatment. This method eliminates the false matches so ε is small than before thus reducing the number of hypothesis and ensure the correct match remain in the new match set so also can get the highest accuracy transformation model.

4   Experimental Results and Analysis

Experiment hardware platform: Intel(R) Core(TM) i7-2600 CPU, 3.40 GHz, 8 G memory. Experiment software platform: Visual Studio 2010. Experimental images size are 816*460, shot by XiaoMi 2S. Figure 4 shows all images that will be use in which the group 1, group 2 and 22, respectively for the east gate of Beijing Institute of technology center building, Beijing institute of technology center garden and laboratories outside the window.

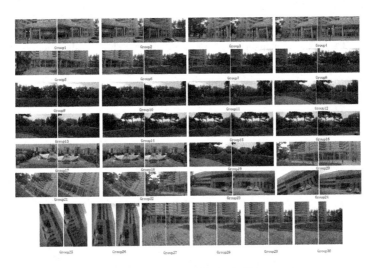

Fig. 4. Test-used images list

[15] proposed a rough matching set purification method by using Delaunay triangulation. But this method introduces a large amount of computation. Figure 6 shows the comparison of purification result between this paper and [15]. We can see that this paper's algorithm successfully purified the rough matching set, and [15] method did not remove some obvious error matching pairs.

(a) original data set   (b) new data set [14]   (c) our new data set

Fig. 5. Comparison chart of screening data set with document [4] data set under rotation transformation

When there is rotational transformation between two images [13, 14] proposed method almost removed all matches that is obviously wrong. But this paper method is still able to get the right result, as shown in Fig. 5.

Table 3 shows comparison of the number of matching pairs and running time among without using any purification method, using the method of [13, 15] and using the method proposed by this paper. As can be seen form Table 3, this paper's method improves the efficiency of RANSAC. Comparing with [13, 15] time-consuming is less. Combined with Table 3, Figs. 5, 6 show that the proposed method can effectively eliminate false matching pairs, and improves the computational efficiency, has obvious superiority.

Table 3. Comparison of proposed algorithm with other algorithm

	Number of matches before purification			Number of matches after purification			Running time/ms		
	East gate	Garden	window	East gate	Garden	window	East gate	Garden	window
RANSAC	102	275	261	–	–	–	13.76	13.06	20.58
Literature [4]	102	275	261	85	254	19	9.78	10.55	13.12
Literature [6]	102	275	261	82	232	244	706.31	14401.2	354.72
paper	102	275	261	86	260	191	7.59	10.04	9.08

(a)original data set   (b)new data set of [15]   (c)our new data set

(a)original data set   (b)new data set of [15]   (c)our new data set

Fig. 6. Comparison chart of screening data set with document [6] data set

Figures 7, 8 and 9 shows the mosaic image obtained by this paper splicing algorithm and the classical RANSAC algorithm. Those three images show stitching algorithm of this paper is feasible, it is possible to get a complete mosaic image.

(a)classical mosaic image   (b)our result

Fig. 7. Mosaic image of central teaching building east gate

(a)classical mosaic image   (b)our result

Fig. 8. Central Park mosaic image

(a) classical mosaic image   (b) our result

Fig. 9. Laboratory window mosaic image

5 Conclusions

This paper proposes an improved RANSAC algorithm based similarity degree. It first sorts matching pairs then take a part of those pairs to calculate eliminate matrix to reject error pairs. It can effectively eliminate false matching pairs, and is suitable for situations there are both translation and rotational transformation. It can improve the effectiveness of RANSAC, and the mosaic image quality is high. Experiments show that the proposed algorithm can be applied to image mosaic.

References

1. Zeng, L., Deng, D.: A selt-adaptive and real-time panoramic video mosaicing system. J. Comput. **7**(1), 218–225 (2012)
2. Li, S., Pu, F., Li, D.: Deriving emphasized course content of biomedical image processing and analysis from typical application requirements. In: International Conference on Future Biomedical Information Engineering (FBIE) (2010)
3. WeiRen, W., WangWang, L., Dong, Q.: Investigation on the development of deep space exploration. Technol. Sci. **55**(4), 1086–1091 (2012)
4. Zhu, F., Li, J., Zhu, B.: Super-resolution image reconstruction based on three-step-training neural networks. J. Syst. Eng. Electron. **21**(6), 934–940 (2010)
5. Song, Z.: Research on Image Registration Algorithm and Its Applications. Fudan University, Shanghai (2010)
6. Harris, C., Satephens, M.: A combined corner and edge detector. In: Alvey Vision Conference, Manchester, pp. 147–152 (1988)
7. Smith, S.M., Brady, J.M.: SUSAN– new approach to low level image processing. Comput. Vision **23**(1), 45–78 (1997)
8. Lowe, D.G.: Object recognition from local scale-invariant features. In: Proceedings of International Conference on Computer Vision. Corfu, Greece:[s. n.], pp. 1150–1157 (1999)
9. Bay, H., Tuytelaars, T., van Gool, L.: Speeded-up robust features (SURF). Comput. Vis. Image Underst. **3951**, 404–417 (2006)
10. Fischler, M.A., Bolles, R.C.: Random sample consensus: a paradigm for model fitting with applications to image analysis and automated cartography. ACM Graph. Image Process. (S0001-0782) **24**(6), 381–395 (1981)
11. Zhang, Y.: Research on Image and Video Mosaic Based on SURF. Xidian University, Xian (2013)
12. Zhang, Z.: An improved RANSAC algorithm for image mosaic. Measur. Control **22**(6), 1856–1858 (2014)
13. Shi, G., Xu, X., Dai, Y.: SIFT feature point matching based on improved RANSAC Algorithm. In: International Conference on Intelligent Human-Machine Systems and Cybernetics (IHMSC), pp. 474–477 (2013)
14. Yin, X.: Study on Video Mosaic for Large View. Beijing University of Chemical Technology, Beijing (2011)
15. Qianwen, F.: A RANSAC image mosaic algorithm with prerocessing. Electron. Design Eng. **21**(15), 183–187 (2013)
16. Guo, K.Y., Ye, S., Jaing, H.: An algorithm based on SURF for surveillance video mosaicing. Adv. Mater. Res. **267**, 746–751 (2011)
17. Xiangqian, G., Hongwen, K., Hongxing, C.: The least-square method in complex number domain. Prog. Nat. Sci. **16**(3), 307–312 (2006)
18. Tian, W.: Enhanced RANSAC with adaptive preverification. J. Image Graph. **14**(5), 973–977 (2000)
19. Lowe, D.G.: Distinctive image features from scale-invariant keypoints. Int. J. Comput. Vision **60**(2), 91–110 (2004)
20. Xiu-juan, L.: Research on K-nearest neighbor algorithm in classification. Sci. Technol. Inf. **31**, 81–87 (2009)

A Novel Emotional Saliency Map to Model Emotional Attention Mechanism

Xinmiao Ding[1], Lulu Huang[2], Bing Li[3(✉)], Congyan Lang[2],
Zhen Hua[1], and Yuling Wang[4]

[1] Shandong Institute of Business and Technology, Yantai, China
dingxinmiao@126.com, green8088@sina.com
[2] School of Computer and Information Technology,
Beijing Jiaotong University, Beijing 100044, China
178596166@qq.com, cylang@bjtu.edu.cn
[3] National Laboratory of Pattern Recognition,
Institute of Automation, CAS, Beijing, China
bli@nlpr.ia.ac.cn
[4] Shandong Jianzhu University, Jinan, China
anguingwyl@163.com

Abstract. Saliency map analysis provides an alternative methodology to image semantic understanding in many applications such as adaptive content delivery and region-based image retrieval. A lot of visual saliency map algorithms have been proposed during the last decades. Recent psychophysical research further reveals that visual attention can be modulated and improved by the affective significance of stimuli, called Emotional Attention, which might not only supplement but also compete with other sources of top-down control on attention. Inspired by this mechanism, we propose a novel computational emotional attention model in this paper. In particular, we present an intuitive emotional saliency map computation method by calculating Minkowski-norm of pixel's isolated saliency value and multi-scale local contrast information in color emotion space. Experimental results on diverse image datasets show that the proposed model can outperform some current state-of-the-art visual saliency map.

1 Introduction

It is well known that primate visual system employs an attention mechanism that focuses on salient parts relevant with behaviors or visual tasks. Detecting and extracting these conspicuous regions have considerable importance in many computer vision and image analysis applications, for example, automatic image cropping, adaptive image display on small devices, image/video compression, object recognition and tracking, image collection browsing, affective image classification and so on [2, 7, 14]. Inherence in this research is building attention model and developing algorithms based on the conceptual framework of saliency map [1].

An important aspect is computational visual attention model which analyzes and recognizes the distinguish objects (or regions) by using elementary properties including

Q. Tian et al. (Eds.): MMM 2016, Part II, LNCS 9517, pp. 197–206, 2016.
DOI: 10.1007/978-3-319-27674-8_18

local contrast, color difference, edge orientations and others. The research on computational visual attention model has over 10 years' history beginning with one of early papers by Itti [2]. The basic principles of most methods follow bottom-up framework involving 2 major processes: extracting low level features and computing visual saliency map. Itti et al. [2] produce a saliency map by applying the "Winner-talk-all" strategy on normalized center-surround difference of three important local features: colors, intensity and orientation. Hou et al. [3] introduce spectral residual and build up a salient map in spatial domain without requiring any prior information of the objects. Valenti et al. [4] combine color edge and curvature information to infer global information so that the salient region can be segmented from background easily. Hasel et al. [5] apply graph theory and algorithm into saliency map computation by defining Markov chain over a variety of image maps extracted from different feature vectors. Li et al. [16] construct a saliency map computational model based on tensor decomposition and reconstruction. The saliency map generator based on this model can adaptively find the most suitable features and their combinational coefficients for each pixel. Achanta et al. [6] propose a frequency-tuned approach in saliency map computation based on color and luminance features. They further point out that many existing visual saliency algorithms are essentially frequency bandpass filtering operations. Considering the structure of images, Peng et al. [15] propose a Low-rank and Structured sparse Matrix Decomposition (LSMD) model for salient object detection.

Recently, many psychological researchers indicate that emotion is of another fundamental importance in vision system and produces specific contributions in selective attention task. For instance, angry faces are more attractive than those neutral stimuli in an image [7]. Vuilleumier [8] argues that the amygdala plays a crucial role in providing both direct and indirect top-down signals on sensory pathways, which can influence the representation of emotional events. Therefore, people pay more attention to emotional stimulus so that human's attention and behaviors can be facilitated by affective influences. Anderson [9] elucidates the effect of emotional concept on the attentional selections, especially when the visual attention is missed under some specific situations. The work of Estwood [7] presents that target can be anchored from surrounding world more conveniently when it evokes some emotional values. Carretie et al. [10] further point out that any important activity related to automatic attention depends on the emotional contents.

Despite these samples evidence that there is an essential connection between emotional presentation and salient regions in an image, only few people investigate the computational model of integrating the emotion mechanism into the attention model. In this paper, we aim to rectify this imbalance by proposing, for the first time, a novel emotion attention model and corresponding emotional saliency map computation. Our solution includes several major steps. After converting image into a predefined color emotion space, we will calculate an isolated emotional saliency value for each pixel. The next stage is to get a set of emotional contrast values by applying Different-of-Gaussian (DoG) filters with different scales on an image. The final step is to compute Minkowski-norm of all of these outputs to get emotional saliency map. Experiments on two real image sets show that our proposed model outperforms some current state-of-the-art visual saliency map computations when evaluated in terms of the accuracy with which the salient regions are detected.

2 Emotional Saliency Map

2.1 Color Emotion Space

As a high-level perception, emotion is evoked by many low-level visual features. Color is inevitably among them and pixels with various colors have different emotional saliency to visual system. However, how colors affect human emotions is still a difficult issue. Fortunately, Ou et al. [11] propose a quantitative model relating the stimulus colorimetry and the emotional response to it. Specially, they present a three-dimension color emotion space, each representing color activity (CA), color weight (CW), and color heat (CH) independently. The transformation equation between color space and color emotion space is as follows:

$$
\begin{aligned}
CA &= -2.1 + 0.06\left[(L^* - 50)^2 + (a^* - 3)^2 + \left(\frac{b^* - 17}{1.4}\right)^2\right]^{1/2} \\
CW &= -1.8 + 0.04(100 - L^*) + 0.45\cos(h - 100^o) \\
CH &= -0.5 + 0.02(C^*)^{1.07}\cos(h - 50^o)
\end{aligned}
\tag{1}
$$

where (L^*, a^*, b^*) and (L^*, C^*, h) are color values in CIELAB and CIELCH color spaces respectively. Given any pixel from an image at location of X, assume that its color value is $f(X) = [R(X), G(X), B(X)]^T$ in RGB color space, we can transform it into Lab and LCH color spaces and get its color emotion values $E(X) = [CA(X), CW(X), CH(X)]^T$.

2.2 Emotional Saliency Map Computation

Now we will investigate how to define emotional saliency map in color emotion space. The proposed emotional saliency map is different from the traditional visual saliency map in two major aspects: (1) it is based on high-level color emotional features and contains more emotional semantics; (2) it takes both contrast information and isolated color emotional saliency property into account and integrate them from a general Minkowski-norm fusing framework.

Isolated Emotional Saliency. Most visual attention computational model are based on local spatial contrast without including pixels' own saliency properties [2–6]. However, some psychological researchers indicate that each pixel's color values can trigger different visual saliencies per se with no spatial features [12]. From our observation that pixels having high color activity (CA) and color heat (CH) are more emotionally salient, while color weight (CW) has less effects, we define single pixel's emotional saliency as:

$$S_0(X) = \sqrt{[CA(X) - minCA]^2 + [CH(X) - minCH]^2} \qquad (2)$$

where $minCA$ and $minCH$ are the minimal color activity value and minimal color heat value in the image respectively. The isolated emotional saliency has at least two obvious advantages: (1) the pixels in a unified color region can have similar saliency values. (2) It can provide saliency cues when no obvious contrast information exists in an image.

Contrast-Based Emotional Saliency. Another important factor of emotional saliency analysis is local contrast information which has been widely used in existing visual saliency computation schemes [2, 5, 6]. The key issue for computing local contrast is how to define the 'local range' in the Center-surround difference operation. Here, we define it in color emotion space through local Gaussian smooth operations.

$$EC(X) = \|E(X) - E^\sigma(X)\|$$
$$= \sqrt{[CA(X) - CA^\sigma(X)]^2 + [CW(X) - CW^\sigma(X)]^2 + [CH(X) - CH^\sigma(X)]^2}$$
$$(3)$$

where $E^\sigma(X) = E(X) \otimes G^\sigma$, is the convolution of the image with a Gaussian filter G^σ, with standard deviation of σ. In fact, $E^\sigma(X)$ is the local average emotion, while $EC(X)$ is the local emotional difference. We can control the local range by adjusting the value of σ. The larger σ means the larger local range. In particular, $E^\sigma(X)$ is actually the mean emotion of the whole image when we set $\sigma = \infty$.

Furthermore, to avoid noise and texture patterns affecting the final saliency map, a Gaussian smooth operation is also necessary for original image. Consequently, we propose a multi-scale local emotional contrast computation algorithm in color emotion space using Difference of Gaussian Filter (DoG). The equation of DoG is defined as:

$$DoG(\sigma_0, \sigma_1) = G^{\sigma_0} - G^{\sigma_1}$$
$$= \frac{1}{\sqrt{2\pi}} \left[\frac{1}{\sigma_0} \exp(\frac{x^2 + y^2}{2\sigma_0^2}) - \frac{1}{\sigma_1} \exp(\frac{x^2 + y^2}{2\sigma_1^2}) \right] \qquad (4)$$

where σ_0 and σ_1 are two distinct standard deviations of Gaussian Functions. Then the local emotional contrast can be gained by a convolution operation as:

$$\|E(X) \otimes DoG(\sigma_0, \sigma_1)\| = \|E^{\sigma_0}(X) - E^{\sigma_1}(X)\| \qquad (5)$$

Now we can get multi-scale local emotional contrast matrix with different values of σ_0 and σ_1 from Eq. (5). However, considering that a good saliency map not only can be free from noisy effects, but also have a high-resolution and clear boundaries of salient regions [6], σ_0 should not be too large, so we set the value of σ_0 untouched ($\sigma_0 = 3$ in this paper) but apply different values on σ_1, then a set of local contrast based emotional saliency values S_k can be computed out as

$$S_k(X) = \left\| E(X) \otimes DoG(\sigma_0, \sigma_1^k) \right\| = \left\| E^{\sigma_0}(X) - E^{\sigma_1^k}(X) \right\|,$$
$$where \quad \sigma_1^k = 2^{k-1}, k = 1...N \tag{6}$$

Emotional Saliency Map. Given pixel's isolated emotional saliency value (S_0) and contrast-based emotional saliency values $(S_k, k = 1...N)$, we integrate them into the final saliency map output, $EM^{p,N}(X)$, using a uniformed Minkowski-norm as

$$EM^{p,N}(X) = \left(\frac{\sum_{k=0}^{N} [S_k(X)]^p}{N+1} \right)^{1/p} \tag{7}$$

where $EM^{p,N}(X)$ is the final emotional saliency value of pixel X by combining $N+1$ different scales emotional values with parameter p. The value of p decides different fusing strategies:

- $p = 1$, **Simple Average.** It is essentially to get the mean output of the $N+1$ different emotional saliency values $S_k(X)$.
- $1 < p < \infty$, **Weighted Average.** Larger p means that the large saliency value among $S_k(X)$ will contribute more in this combination.
- $p = \infty$, **Max Saliency Value.** The maximal value among all $N+1$ different emotional saliency values will be set as the saliency map value of current pixel.

3 Experiments

3.1 Data Set and Error Measure

To evaluate the performance of proposed emotional saliency map, we conducted two separate experiments on two different image sets. The first one includes 500 horror images collected from Internet by ourselves. Since a horror image generally has a salient region with strong emotional stimuli, it is well suitable for emotional saliency map evaluations. The second one is Microsoft Visual Salient image set (MS) [5] which contains 5000 images with annotated visual salient regions.

Since the emotional salient regions in horror image set have not been given out, we required 7 students in our Lab to draw a bounding box around the most emotional salient region (according to their understanding of emotional saliency) in each image.

Their manually annotations are used to create a ground-truth saliency map $A = \{a(X)|a(X) \in [0,1]\}$ as

$$a(X) = \frac{1}{M}\sum_{m=1}^{M} w^m(X), \qquad (8)$$

where M is the number of labeling users and $w^m(X) \in \{0,1\}$ is a binary label given by m^{th} user to indicate whether or not the pixel X belongs to salient region.

Given $a(X)$ and the obtained emotional saliency map $EM(X)$ of an image, the Precision (*Pre*), Recall (*Rec*), and F_α measure are used to evaluate the performance of saliency map. Same as some previous work [13], α is set to be 0.5.

$$Pre = \frac{\sum_X a(X)EM(X)}{\sum_X EM(X)}, Rec = \frac{\sum_X a(X)EM(X)}{\sum_X a(X)}$$

$$F_\alpha = \frac{(1+\alpha) \times Pre \times Rec}{(\alpha \times Pre + Rec)} \qquad (9)$$

3.2 Experiments on Horror Image Set

Parameters Selection. The first experiment is to select optimal parameters N and p for proposed emotional saliency map calculation $EM^{p,N}(X)$. In finding best parameter combination, we set $N = 1, 2, \ldots, 8$ and $p = 1, 3, 5, \ldots, 19$, totally 80 parameter combinations. $F_{0.5}$ is taken to evaluate the performance of each parameter combination on the horror image set. The change of $F_{0.5}$ with N and p is shown in Fig. 1.

Figure 1 tells us that the performances with larger N are much better than these with lower N. This phenomenon indicates that combination with more different scale

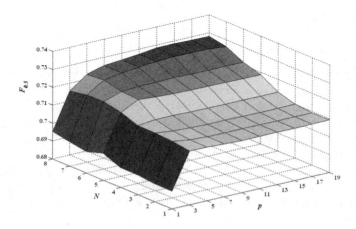

Fig. 1. $F_{0.5}$ changes with N and p on horror image set.

saliency maps can achieve much better result. However, considering both computational cost and performance, we select the parameter combination $N = 8$ and $p = 15$ (i.e. $EM^{15,8}$ in Eq. (7)), which gets the best result in Fig. 1 as the parameter setting in the following experiments. We also consider other two typical parameter combinations, $EM^{1,8}$ and $EM^{\infty,8}$ in the following experiments.

Comparison with Single Scale Emotional Saliency Map. In this subsection, we compare the combinational emotional saliency maps $EM^{15,8}$, $EM^{1,8}$ and $EM^{\infty,8}$ with the single scale saliency map $(S_0, S_2, \ldots, S_8)$ on the horror image set.

The comparison results in Fig. 2 show that the combinational emotional saliency maps have better performances than single scale saliency maps. In addition, we can find that S_0 also gives out better results than some contrast based saliency map on this image set, which indicates that isolated emotional saliency indeed plays important role in emotional attention mechanism.

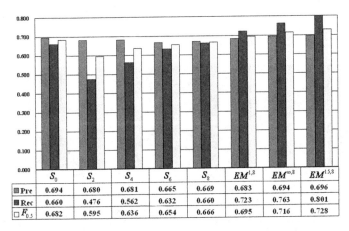

	S_0	S_2	S_4	S_6	S_8	$EM^{1,8}$	$EM^{\infty,8}$	$EM^{15,8}$
▤Pre	0.694	0.680	0.681	0.665	0.669	0.683	0.694	0.696
▮Rec	0.660	0.476	0.562	0.632	0.660	0.723	0.763	0.801
☐$F_{0.5}$	0.682	0.595	0.636	0.654	0.666	0.695	0.716	0.728

Fig. 2. Comparison with single scale emotional saliency maps

Comparison with Visual Saliency Map. We also compare our emotional saliency computation method with other existing visual saliency algorithms, Itti's method (ITTI) [2], Hou's method (HOU) [3], Graph-based visual saliency algorithm (GBVS) [5], and Frequency-tuned Salient Region Detection algorithm (FS) [6]. The precision (Pre), recall (Rec) and $F_{0.5}$ values for each method are calculated and shown in Fig. 3.

The outputs of our method $EM^{15,8}$ are $Pre = 0.696$, $Rec = 0.801$ and $F_{0.5} = 0.728$ respectively, showing that it outperforms all other visual saliency computation methods on this set. This is due to the fact that visual saliency computations are only based on local feature contrast without considering any single pixel's salient property and some emotional factors. Some saliency maps generated by different algorithms are also given as Fig. 4.

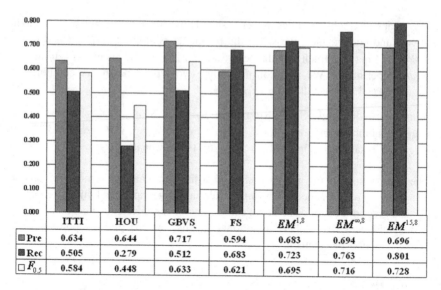

	ITTI	HOU	GBVS	FS	$EM^{1,8}$	$EM^{\infty,8}$	$EM^{15,8}$
▨ Pre	0.634	0.644	0.717	0.594	0.683	0.694	0.696
▨ Rec	0.505	0.279	0.512	0.683	0.723	0.763	0.801
▢ $F_{0.5}$	0.584	0.448	0.633	0.621	0.695	0.716	0.728

Fig. 3. Comparison with existing visual saliency algorithm on horror image set.

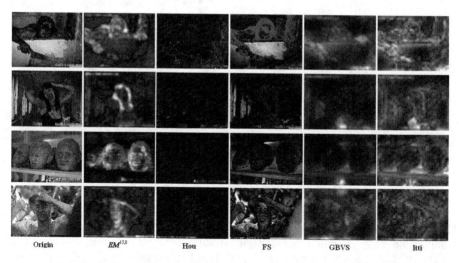

Origin $EM^{15,8}$ Hou   FS   GBVS   Itti

Fig. 4. Saliency maps comparison.

3.3 Experiments on MS Image Set

In this experiment, we compare our method with some visual saliency methods on Microsoft image set. The comparison results are shown in Fig. 5, from which we can find that the proposed $EM^{1,8}$ gets best performance among the proposed methods with different parameter settings. Furthermore, $EM^{1,8}$ outperforms ITTI, HOU, and FS algorithms, and is comparable to GBVS algorithm. Although the motivation of proposed method is to model emotional attention mechanism, its performance is also

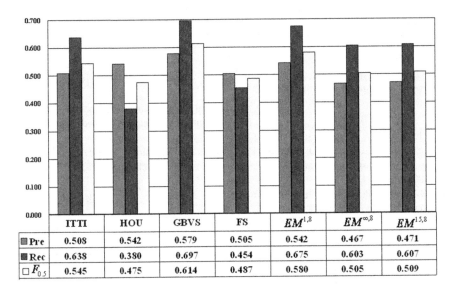

	ITTI	HOU	GBVS	FS	$EM^{1,8}$	$EM^{\infty,8}$	$EM^{15,8}$
■ Pre	0.508	0.542	0.579	0.505	0.542	0.467	0.471
■ Rec	0.638	0.380	0.697	0.454	0.675	0.603	0.607
□ $F_{0.5}$	0.545	0.475	0.614	0.487	0.580	0.505	0.509

Fig. 5. Comparison with existing visual saliency algorithm on MS image set.

comparable to visual saliency detection algorithms. This result proves that our method can also explain visual attention from emotional aspect.

4 Conclusion

A novel emotional saliency map computation method is proposed based on the emotional attention mechanism in this paper. Specially, we gives out the details of emotional saliency map computation by integrating the single pixel's saliency attribute and multi-scale contrast information in emotional color space. This strategy is in line with many psychological researchers' experiments that presented how affective context affects salient regions selection. Experiments on different image data set show that our proposed model outperforms other existing visual saliency map in its ability to recognize the emotional salient regions. In addition, the proposed framework can also be applied to other visual saliency maps generation.

Acknowlegment. This work is supported by the National Nature Science Foundation of China (No. 61303086, 61370038, 61571045, 61472227, 61503219, 61572296), National Nature Science Foundation of Shandong province (No. ZR2013FM015, ZR2015FL020) and Domestic Visiting Schoars Project of Shandong Provincial Education Department.

References

1. Fecteaua, J.H., Munoz, D.P.: Salience, relevance, and firing: a priority map for target selection. Trends Cogn. Sci. **10**(8), 382–390 (2006)

2. Itti, L., Koch, C., Niebur, E.: A model of saliency-based visual attention for rapid scene analysis. IEEE TPAMI **20**(11), 1254–1259 (1998)
3. Hou, X., Zhang, L.: Saliency detection: a spectral residual approach. In: CVPR, pp. 1–8 (2007)
4. Valenti, R., Sebe, N., Gevers, T.: Image saliency by isocentric curvedness and color. In: ICCV, pp. 2185–2192 (2009)
5. Harel, J., Koch, C., Perona, P.: Graph-based visual saliency. In: NIPS, pp. 545–552 (2006)
6. Achanta, R., Hemami, S., Estrada, F., Susstrunk, S.: Frequency-tuned salient region detection. In: CVPR, pp. 1597–1604 (2009)
7. Eastwood, J.D., Smilek, D., Merikle, P.M.: Differential attentional guidance by unattended faces expressing positive and negative emotion. Percept. Psychophysics **63**(6), 1004–1013 (2001)
8. Vuilleumier, P.: How brains beware: neural mechanisms of emotional attention. TRENDS Cogn. Sci. **9**(12), 585–594 (2005)
9. Anderson, A.K.: Affective influences on the attentional dynamics supporting awareness. J. Exp. Psychol. **134**(2), 258–281 (2005)
10. Carretie, L., Hinojosa, J.A., Martin-Loeches, M.: Automatic attention to emotional stimuli: neural correlates. Hum. Brain Mapp. **22**(4), 290–299 (2004)
11. Ou, L., Luo, M., Woodcock, A., Wright, A.: A study of colour emotion and colour preference. Part I. Col. Res. App. **29**(3), 232–240 (2004)
12. Andersen, S.K.: Color-selective attention need not be mediated by spatial attention. J. Vis. **9**(6), 1–7 (2009)
13. Liu, T., Sun, J., Zheng, N.N., Tang, X.: Learning to detect a salient object. In: CVPR, pp. 1–8 (2007)
14. Li, B., Xiong, W., Wu, O., et al.: Horror image recognition based on context-aware multi-instance learning. IEEE Trans. Image Process. **24**(12), 5193–5205 (2015)
15. Peng, H., Li, B., Ji, R., Hu, W.: Salient object detection via low-rank and structured sparse matrix decomposition. In: The 27th AAAI Conference on Artificial Intelligence (2013)
16. Li, B., Xiong, W., Hu, W.: Visual saliency map from tensor analysis. In: The 26th AAAI Conference on Artificial Intelligence (2012)

Automatic Endmember Extraction Using Pixel Purity Index for Hyperspectral Imagery

Qianlan Zhou[1], Jing Zhang[1,2(✉)], Qi Tian[2],
Li Zhuo[1,3], and Wenhao Geng[1]

[1] Signal and Information Processing Laboratory,
Beijing University of Technology,
100 Ping Le Yuan, Chao Yang District, Beijing 100124, China
{orchidzqll, gengwh}@emails.bjut.edu.cn,
{zhj, zhuoli}@bjut.edu.cn
[2] Department of Computer Science, University of Texas at San Antonio,
One UTSA Circle, San Antonio, TX 78249, USA
wywqtian@gmail.com
[3] Collaborative Innovation Center of Electric Vehicles in Beijing,
100 Ping Le Yuan, Chao Yang District, Beijing 100124, China

Abstract. Pixel Purity Index (PPI) is one of effective endmember extraction algorithms, which is a processing technique designed to determine which pixels are the most spectrally unique or pure. This paper proposes an automatic endmember extraction using pixel purity index for hyperspectral imagery. In computing the PPI, projection vectors are generated by applying the Givens rotation firstly. Then, pixels are projected onto the projection vectors. Next, the pixels located at the extreme positions are recorded. At last, the PPI score can be obtained. In endmember extraction, the number of endmembers is determined by using the Noise Subspace Projection (NSP) method. Hyperspectral image dimension is reduced by improving the Noise Covariance Matrix (NCM) estimation for Minimum Noise Fraction (MNF) transformation. The endmembers can be extracted with the improved pixel purity index. Compared with traditional APPI algorithm, the experimental results show that the proposed algorithm can obtain more endmembers as well as improve the accuracy of endmembers.

Keywords: Hyperspectral image · Endmember extraction · Automatic pixel purity index · Projection vector

1 Introduction

Many endmember extraction algorithms have been explored for finding endmembers assumed to be pure signatures in the hyperspectral imagery. Because of the limitation of spatial resolution, the pixels in hyperspectral image acquired by the hyperspectral imaging instruments usually contain more than one feature spectrum [1]. An endmember can be defined as an idealized pure signature for a class, which is a spectral signature that is completely specified by the spectrum of a single material substance [2, 3].

In existing endmember extraction algorithms, the pixel spectral signature is usually utilized to find endmember as a similarity measurement. For example, Reference [4]

Q. Tian et al. (Eds.): MMM 2016, Part II, LNCS 9517, pp. 207–217, 2016.
DOI: 10.1007/978-3-319-27674-8_19

found endmembers using the Vertex Component Analysis (VCA) to search the vertices for a simplex with dimensionality reduction, which may cause instability after running repeatedly. Reference [5] selected endmembers using the Iterative Error Analysis (IEA) to obtain pixel spectrum with the largest residual error, which will lead to the result affected by so-called "nesting effect" [6]. Some algorithms are based on the geometric description of linear spectral mixture model, such as Reference [7] proposed the N-FINDR algorithm by identifying the endmembers as the vertices of the simplex with the largest volume. However, it will result in inconsistent for final endmember selection. Besides that, an effective endmember extraction algorithm is the Pixel Purity Index (PPI), which is a processing technique designed to determine which pixels are the most spectrally unique or pure [8, 9]. PPI has good stability and can produce a final set of required endmembers without manual intervention. However, because the HFC algorithm does not consider the process of white noise, the weak signals in the image may be lost. And mixing various features also will influence on the classification accuracy of MNF transformation. Due to the projection vectors are generated in random directions, it may affect the comprehensiveness and integrity of the extracted endmembers.

Therefore, an automatic endmember extraction using pixel purity index for hyperspectral imagery is proposed in this paper. In computing the PPI, the projection vectors are generated in a series of certain directions in order to guarantee the comprehensiveness and integrity of the endmember set. When it comes to the endmember extraction, the weak signals are kept by using the Noise Subspace Projection (NSP) method [10]. Besides that, to alleviate the effects of mixed features, the classic MNF transformation is replaced with the improved Noise Covariance Matrix (NCM) estimation of MNF transformation [11].

The remainder of this paper is organized as follows. Section 2 explains how the Pixel Purity Index works. Section 3 gives the proposed automatic endmember extraction algorithm in detail. Experimental results and analysis are presented in Sects. 3.3 and 4 is conclusions.

2 Pixel Purity Index

In this paper, the improved Pixel Purity Index can be summarized with 4 steps: (1) generating projection vectors using Gives rotation; (2) projecting pixels onto projection vectors; (3) finding the pixels located at the extreme positions; (4) recording the PPI score. At last, the PPI score of the pixels can be obtained. Figure 1 shows the flowchart of the improved Pixel Purity Index.

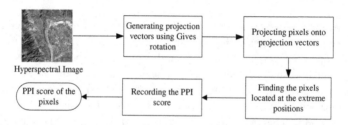

Fig. 1. Flowchart of the improved Pixel Purity Index

In traditional PPI, it computed by continually projecting n-dimensional scatter lots onto a random vector. By projecting pixels onto the projection vectors, a large number of unit vectors are generated in random directions, called as "skewers", and then the dot product of each skewer is computed with each data point. For each skewer, two extreme points are identified as well as their pixel purity index is incremented. The comparisons of projection principles before and after improvement are shown in Fig. 2.

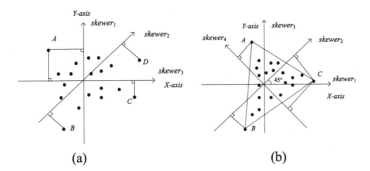

Fig. 2. The improved projection principle: (a) The projection principle of PPI (b) The improved projection principle

As can be seen, in traditional PPI, the skewers may be generated in a few directions so that some pixels which located at the extreme positions may be ignored. After improvement, the skewers are generated in different angles of the projection vector in certain directions. Assume that the pixel set is surrounded by triangle ABC after dimension reduction. Then we construct four vectors named $skewer_1$, $skewer_2$, $skewer_3$ and $skewer_4$, in which $skewer_1$ is the unit vector of the X axis. $skewer_2$, $skewer_3$ and $skewer_4$ can be obtained by anticlockwise rotating $skewer_1$ for 45°, 90°, and 135°. The rotated vector can be obtained by counter-clockwise rotation according to the angle difference in turn though Givens rotation [12]. Let $\mathbf{U}_1$ denotes $skewer_1$ and $\mathbf{U}_2$ denotes $skewer_2$, the unit vectors $\mathbf{U}_2$ and transform matrix $\mathbf{T}$ are defined as follows:

$$\mathbf{U}_2 = \mathbf{T}\mathbf{U}_1 \tag{1}$$

$$\mathbf{T} = \begin{bmatrix} \cos(-45°) & \sin(-45°) \\ -\sin(-45°) & \cos(-45°) \end{bmatrix} \tag{2}$$

After generating projection vectors, the endmembers are extracted by projecting pixels onto the projection vectors to find the pixels located at the extreme positions. Finally, the PPI score of the pixels can be obtained.

3   Automatic Endmember Extraction Using Pixel Purity Index

Next, we will apply PPI to automatically extract endmember. First, the number of endmembers is determined by using the NSP method. Then image dimension is reduced by improving the NCM estimation for MNF transformation. At last, the endmembers can be obtained according to improved pixel purity index.

3.1   Determining the Number of Endmembers Based on Noise Subspace Projection

Before extracting endmembers from the hyperspectral image, determining the number of endmembers is the first step [13]. The NSP method is utilized to replace HFC algorithm, which can be reviewed as follows:

1. Calculating the sample correlation matrix $R_{L \times L}$ and covariance matrix $K_{L \times L}$, which's corresponding eigenvalues are $\left\{ \overline{\lambda}_1 \geq \overline{\lambda}_2 \geq \ldots \geq \overline{\lambda}_L \right\}$ and $\{\lambda_1 \geq \lambda_2 \geq \ldots \geq \lambda_L\}$. Let $\{r_i\}_{i=1}^N$ be data sample vectors, μ is the sample mean vector given by $\mu = \sum\limits_{i=1}^N r_i$ and L is the number of spectral channels, so $R_{L \times L}$ and $K_{L \times L}$ can be expressed as:

$$R_{L \times L} = \sum\nolimits_{i=1}^N r_i r_i^T \tag{3}$$

$$K_{L \times L} = \sum\nolimits_{i=1}^N (r_i - \mu)(r_i - \mu)^T \tag{4}$$

2. Let $\overline{K}$ be the whitened result of $K_{L \times L}$, $\{u_l\}_{l=1}^L$ be a set of eigenvalues generated by $\overline{K}$, so the variances of the noise components can be whitened and normalized to unity by

$$\overline{K} = \sum\nolimits_{l=1}^{VD} \overline{\lambda}_L u_l u_l^T + \sum\nolimits_{l=VD+1}^L \overline{\lambda}_L u_l u_l^T \tag{5}$$

where $\{u_l\}_{l=1}^{VD}$ and $\{u_l\}_{l=VD+1}^L$ represent span signal subspace and noise subspace, respectively.

3. Determining the number of endmembers can be transformed into the following binary hypothesis testing problem H_0 and H_1:

$$H_0 : y_l = \overline{\lambda}_l = 1, (l = 1, 2, \ldots, L) \tag{6}$$

$$H_1 : y_l = \overline{\lambda}_l > 1, (l = 1, 2, \ldots, L) \tag{7}$$

where the null hypothesis H_0 represents the case that the correlation eigenvalue is greater than 1, and the alternative hypotheses is H_1 represents the case that the correlation eigenvalue is equal to 1. In other words, if H_1 is true (i.e., H_0 fails), it implies that there is an endmember contributing to the correlation eigenvalue in addition to noise. The random variable y_l under hypotheses H_0 and H_1 needs to satisfy the following conditional probability densities p_0 and p_1:

$$p_0(y_l) = p(y_l|H_0) \cong N\left(1, \sigma_{y_l}^2\right), (l = 1, 2, \ldots, L) \tag{8}$$

$$p_1(y_l) = p(y_l|H_1) \cong N\left(\mu_l, \sigma_{y_l}^2\right), (l = 1, 2, \ldots, L) \tag{9}$$

where u_l is an unknown constant and the variance $\sigma_{y_l}^2$ given by

$$\sigma_{y_l}^2 = Var\left[\overline{\lambda_l}\right] \cong \frac{2\overline{\lambda_l}^2}{N} \tag{10}$$

Under the hypothesis is H_0, Eq. (10) can be further simplified to

$$\sigma_{y_l}^2 \cong \frac{2}{N} \tag{11}$$

4. Determining the number of endmembers though the false-alarm probability P_F and detection probability P_D as follows:

$$P_F = \int_r^{\infty} p_0(z)dz \tag{12}$$

$$P_D = \int_r^{\infty} p_1(z)dz \tag{13}$$

By maximizing the detection probability P_D, the false alarm probability P_F can be obtained. In the binary hypothesis testing, if $\overline{\lambda_l} > 1$, it means the test fails. In other words, there is signal energy assumed to contribute to the eigenvalue in the l-th data dimension. If $\overline{\lambda_l} = 1$, it means that there is only noise in this component.

3.2   Data Dimensionality Reduction by Improving Noise Covariance Matrix (NCM) Estimation for MNF Transformation

Due to the large amount of bands in the hyperspectral image, dimension reduction is important and necessary [14]. The NCM estimation is improved for traditional MNF transform in order to alleviate the effects of mixed features [15]. At First, the noise of pixels in each band should be estimated. Let $x_{i,j,k}$ be the signal part of one pixel (i,j) in k-th band in hyperspectral image. The estimation value of $\widehat{x}_{i,j,k}$ is the best unbiased estimator of $x_{i,j,k}$, which can be described as

$$\widehat{x}_{i,j,k} = \begin{cases} ax_{i,j,k+1} + bx_{i,j,k+1} + cx_{p,k} + d, k = 1 \\ ax_{i,j,k-1} + bx_{i,j,k+1} + cx_{p,k} + d, 1 < k < N \\ ax_{i,j,k-1} + bx_{i,j,k-1} + cx_{p,k} + d, k = N \end{cases} \tag{14}$$

$$x_{p,k} = \begin{cases} x_{i-1,j,k}, i > 1 \\ x_{i+1,j,k}, i = 1 \end{cases}, (1 \le i \le W, 1 \le j \le H) \tag{15}$$

where W, H are the height and width of image, N is the number of bands. Suppose $r = x - \widehat{x}$ is the noise estimation value of each point. Expect that $\sum r^2 = 0$, so $\widehat{x}_{i,j,k}$ can be got by linear regression. At the same time, the coefficient $[a, b, c, d]$ can be obtained by linear regression, which involves M equations in k-th band. The number of equations M and the coefficient $[a, b, c, d]$ are denoted as follows:

$$M = W \times H - 1 \tag{16}$$

$$\mathbf{W} \bullet [a, b, c, d]' = x \Rightarrow [a, b, c, d]' = \mathbf{W}^T (\mathbf{W}\mathbf{W}^T)^{-1} \bullet x \tag{17}$$

where x is a vector composed by all pixels, $\mathbf{W}$ is a matrix composed by $[x_{i,j,k-1}, x_{i,j,k+1}, x_{p,k}, 1]$ for linear regression.

Then, the noise covariance matrix should be determined. According to the coefficient of each band, the variance σ_k^2 of k-th band and covariance $\mathbf{C}_{kl}$ between k-th and l-th band are denoted as follows:

$$\sigma_k^2 = \frac{\sum_{i=1}^W \sum_{j=1}^H r_{i,j,k}^2}{M - 4}, (1 \le k, l \le N, (i,j) \ne (1,1)) \tag{18}$$

$$\mathbf{C}_{kl} = \frac{\sum_{i=1}^W \sum_{j=1}^H r_{i,j,k} \bullet r_{i,j,l}}{M - 4}, (1 \le k, l \le N, (i,j) \ne (1,1)) \tag{19}$$

Finally, after a series of transformations given in [16], the results of MNF transformation can be obtained. Therefore, the proposed automatic endmember extraction using pixel purity index for hyperspectral imagery can be summarized as the following steps.

(1) Using the above NSP method to determine the retained dimension number, denoted as p.

(2) Appling improved NCM estimation for MNF transformation into dimensionality reduction and setting a counter to $n = 1$.

(3) According to the Givens rotation, generating K skewers denoted as $\left\{ skewer_j^{(n)} \right\}_{j=1}^K$. The rotation angle between two skewers is $15°$.

(4) Let $\left\{ r_i^{(n)} \right\}_{i=1}^N$ be the i-th sample vector where the total number of sample vectors is N.

(5) For each $skewer_j^{(n)}$, all the data sample vectors are projected onto $skewer_j^{(n)}$ to find sample vectors at its extreme positions to form an extreme set for this particular skewer $skewer_j^{(n)}$, denoted as $S_{extrema}\left(skewer_j^{(n)}\right)$. Despite the fact that a different $skewer_j^{(n)}$ generates a different extreme set $S_{extrema}\left(skewer_j^{(n)}\right)$. Define an indicator function of a set S, the indicator function $I_S\left(r_i^{(n)}\right)$ is defined by

$$I_S\left(r_i^{(n)}\right) = \begin{cases} 1; & if\ r \in S \\ 0; & if\ r \in S \end{cases} and\ N_{PPI}\left(r_i^{(n)}\right) = \sum_j I_{S_{extrema}(skewer_j^{(n)})}\left(r_i^{(n)}\right) \qquad (20)$$

where $N_{PPI}\left(r_i^{(n)}\right)$ is defined as the PPI score of the sample vector $r_i^{(n)}$, which is the times of each pixel located at the extreme positions by projection. Let $\left\{score_{PPI}^{(n)}(j)\right\}_{j=1}^p$ denotes the highest PPI scores set where $score_{PPI}^{(n)}(j)$ is the j-th highest PPI score.

(6) Finding a set of sample vectors $E_j^{(n)}$ corresponding to $score_{PPI}^{(n)}(j)$. Let $\Omega^{(n)} = \bigcap_{j=1}^p E_j^{(n)}$ be set of generated endmember pixels at the n-th run.

(7) Finding $E^{(n)} = \bigcap_{m=1}^n \Omega^{(m)} = \bigcap_{m=1}^n \left(\bigcup_{j=1}^p E_{j=1}^{(m)}\right)$, the set of endmember pixels up to n, i.e., $1 \leq m \leq n$..

(8) If $E^{(n)} \neq E^{(n-1)}$, it means some endmember pixels have not be recorded yet, then go to Step 3, otherwise, the algorithm is terminated. Here, the sample pixel vectors in $E^{(n)}$ are the extracted endmembers.

3.3   Experimental Results and Analysis

In this Section, experiments on hyperspectral image dataset of AVIRIS data which contains 224 spectral bands between 0.4 and 2.5 micrometers are conducted. The spatial resolution is 20 m, and the spectral resolution is 10 nm. The sample image is shown in Fig. 3. The corresponding spectrums of the image are shown in Fig. 4. The numbers of endmembers extracted by three algorithms are 7, 9, 12 respectively. That is to say, our algorithm can extract more endmembers and has a better performance than the others.

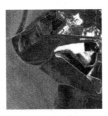

Fig. 3. The sample AVIRIS image

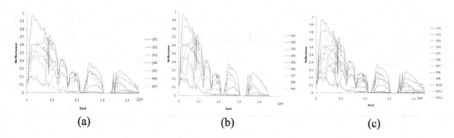

Fig. 4. The corresponding spectral curves of extracted endmembers: (a) PPI algorithm; (b) APPI Algorithm; (c) our algorithm

To prove the accuracy of endmembers set, the traditional APPI algorithm and our algorithm are then compared. The experiment scene was a part of the Cuprite AVIRIS image. We compared the spectral similarity between extracted endmembers spectrum and the actual spectrum in the spectral library to distinguish the extracted endmember types and measure the degree of similarity between them. In this paper, we used the Spectral Angel Mapper (SAM) and Spectral Feature Fitting (SFF) as the judgment. The weight of SAM and SFF were 0.4 and 0.6. The similarity Z is defined by

$$Z = 0.4 \frac{\sum_{i=1}^{n} t_i r_i}{\sqrt{\sum_{i=1}^{n} t_i^2 \sum_{i=1}^{n} r_i^2}} + 0.6 \sqrt{\frac{n \sum p_i \rho_i - \sum p_i \rho_i}{n \sum \rho_i^2 - (\sum \rho_i)^2} \bullet \frac{n \sum p_i \rho_i - \sum p_i \rho_i}{n \sum p_i^2 - (\sum p_i)^2}} \quad (21)$$

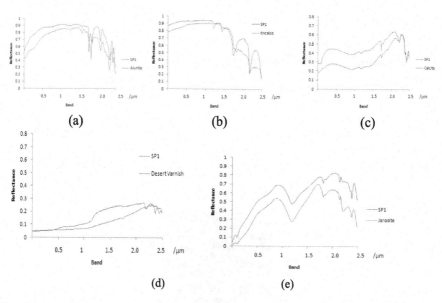

Fig. 5. The comparison of extracted endmember by APPI algorithm and USGS mineral spectrum: (a) Alunite; (b) Tincalconite; (c) Calcite; (d) Desert_Varnish; (e) Jarosite

where n is the number of bands, t_i and r_i are the pixel spectrum and reference spectrum respectively, p_i and ρ_i are the pixel spectral reflectance and reference spectral reflectance. The comparison of extracted endmember by traditional APPI algorithm and USGS mineral spectrum is shown in Fig. 5 and the comparison for our algorithm is shown in Fig. 6.

As can be seen from Figs. 5 and 6, five kinds of endmembers are extracted in Fig. 5 which are Alunite, Tincalconite, Calcite, Desert_Varnish and Jarosite respectively. In Fig. 6, Muscovite and Nontronite can also be extracted by our algorithm. In addition, the spectral curves of extracted endmember and actual spectrum are closerin Fig. 6, especially for the Fig. 6(a), (c). The similarity calculation result is given in Table 1.

As can be seen from Table 1, the similarity between endmembers spectrum obtained by our algorithm and the actual spectrum are increased than the traditional APPI algorithm. For further demonstration, the Root Mean Square Error (RMSE) is utilized to evaluate the extracted endmembers by our algorithm, which is denoted by

$$RMSE = \sqrt{\frac{1}{m}\sum_{i=1}^{m}(A(i)-B(i))^2} \qquad (22)$$

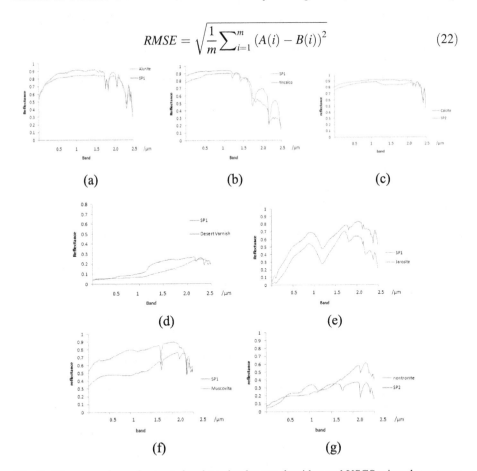

Fig. 6. The comparison of extracted endmember by our algorithm and USGS mineral spectrum: (a) Alunite; (b) Tincalconite; (c) Calcite; (d) Desert_Varnish; (e) Jarosite; (f) Muscovite; (g)Nontronite

Table 1. The similarity between endmembers spectrum extracted by two methods and the actual spectrum

Methods	APPI	ImprovedAPPI
Alunite	0.885	0.894
Tincal-conite	0.821	0.834
Calcite	0.874	0.906
Desert_Varnish	0.849	0.858
Jarosite	0.875	0.896
Muscovite	–	0.873
Nontronite	–	0.862

where m is the total number of pixels in the image, $A(i)$ is the vector-valued of the i-th pixel of original image, $B(i)$ is the vector-valued of the i-th pixel of the reconstructed image obtained by the abundance values of endmembers in the mixed pixels.

After calculation, the RMSE of the image by using three algorithms are 0.5437, 0.5205 and 0.5029 respectively. The RMSE of the image by using our algorithm is the lowest, which means that our algorithm performs better than the other two algorithms. That is to say, the accuracy of endmembers set extracted by our algorithm has improved.

4   Conclusions

In this paper, an automatic endmember extraction using pixel purity index is proposed to find endmembers from hyperspectral image. Compared with traditional APPI algorithm, determining the number of endmembers, data dimensionality reduction and the projection vector generation are improved respectively. The experimental results show that the proposed algorithm can obtain more endmembers as well as improve the accuracy of endmembers. In the next work, we will focus on the categories of extracted endmembers and further improve the performance of endmembers classification, such as the automatic classification of extracted endmembers.

Acknowledgment. The work in this paper is supported by the National Natural Science Foundation of China (No. 61370189, No. 61372149, No. 61429201, and No. 61471013), the Importation and Development of High-Caliber Talents Project of Beijing Municipal Institutions (No. CIT&TCD 201304036, No. CIT&TCD20150311, No. CIT&TCD 201404043), the Science and Technology Development Program of Beijing Education Committee (No. KM20 1410005002), the part to Dr. Qi Tian by ARO grant W911NF-12-10057 and Faculty Research Awards by NEC Laboratories of America, the Specialized Research Fund for the Doctoral Program of Higher Education (No. 20121103110017), the Natural Science Foundation of Beijing (No. 4142009), Funding Project for Academic Human Resources Development in Institutions of Higher Learning Under the Jurisdiction of Beijing Municipality.

References

1. Ngadi, M.O., Liu, L.: Hyperspectral Image Processing Technology: Hyperspectral Imaging for Food Quality Analysis and Control. Elsevier (2010)
2. Xu, M., Du, B., Zhang, L.: Spatial-spectral information based abundance-constrained endmember extraction methods. IEEE Trans. Geosci. Remote Sens. **7**, 2004–2015 (2014)
3. Kowkabi, F., Ghassemian, H., Keshavarz, A.: Using spatial and spectral information for improving endmember extraction algorithms in hyperspectral remotely sensed images. In: ICCKE Conference, Mashhad, pp. 548–553 (2014)
4. Nascimento, J.M.P., Bioucas-Dias, J.M.: Vertex component analysis: a fast algorithm tounmix hyperspectral data. IEEE Trans. Geosci. Remote Sens. **43**, 898–910 (2005)
5. Wang, L., Shen, C., Hartley, R.: On the optimality of sequential forward feature selection using class separability measure. In: DICTA Conference, Adelaide, pp. 203–208 (2011)
6. Chang, C., Xiong, W., Wen, C.: A theory of high-order statistics-based virtual dimensionality for hyperspectral imagery. IEEE Trans. Geosci. Remote Sens. **52**, 188–208 (2014)
7. Sun, Q., An, J., Song, M., Lin, B., Zhang, Y.: The improved N-FINDR endmember extraction algorithm and its application in the oil analysis. In: ICACI Conference, Hangzhou, pp. 601–604 (2012)
8. Heylen, R., Scheunders, P.: Multidimensional pixel purity index for convex hull estimation and endmember extraction. IEEE Trans. Geosci. Remote Sens. **51**, 4059–4069 (2013)
9. Chang, C., Wu, C.: Iterative pixel purity index. In: WHISPERS Conference, Lausanne, pp. 1–4 (2012)
10. Hasanlou, M., Samadzadegan, F.: Comparative study of intrinsic dimensionality estimation and dimension reduction techniques on hyperspectral images using K-NN classifier. IEEE Trans. Geosci. Remote Sens. **9**, 1046–1050 (2012)
11. Acito, N., Corsini, G., Diani, M.: A novel technique for hyperspectral signal subspace estimation in target detection applications. In: IGARSS Conference, Boston, pp. 95–98 (2008)
12. Zhang, D., Yu, L.: Support vector machine based classification for hyperspectral remote sensing images after minimum noise fraction rotation transformation. In: ICICIS Conference, pp. 132–135 (2010)
13. Li, Z., Chang, C.: Adaptive noise variance identification for data fusion using subspace-based technique. In: ICACC Conference, Beijing, pp. 764–770 (2010)
14. Zhang, J., Cao, Y., Zhuo, L., Wang, C., Zhou, Q.: Improved band similarity-based hyperspectral imagery band selection for target detection. J. Appl. Remote Sens. **9**, 095091.1–095091.12 (2015)
15. Gao, L., Zhang, B., Zhang, X., Zhang, W., Tong, Q.: A new operational method for estimating noise in hyperspectral images. IEEE Trans. Geosci. Remote Sens. **5**, 83–87 (2008)
16. Cheng, B.: Hyperspectral data unmixing algorithm comparative analysis based on linear spectral mixture model. In: IHMSC Conference, Hangzhou, pp. 11–14 (2013)

A Fast 3D Indoor-Localization Approach Based on Video Queries

Guoyu Lu[1]([✉]), Yan Yan[2], Abhishek Kolagunda[1], and Chandra Kambhamettu[1]

[1] Department of Computer and Information Sciences,
University of Delaware, Newark, USA
luguoyu@udel.edu
[2] Department of Information Engineering and Computer Science,
University of Trento, Trento, Italy

Abstract. Image-based localization systems are typically used in outdoor environments, as high localization accuracy can be achieved particularly near tall buildings where the GPS signal is weak. Weak GPS signal is also a critical issue for indoor environments. In this paper, we introduce a novel fast 3D Structure-from-Motion (SfM) model based indoor localization approach using videos as queries. Different from an outdoor environment, the captured images in an indoor environment usually contain people, which often leads to an incorrect camera pose estimation. In our approach, we segment out people in the videos by means of an optical flow technique. This way, we filter people in video and complete the captured video for localization. A graph matching based verification is adopted to enhance both the number of images that are registered and the camera pose estimation. Furthermore, we propose an initial correspondence selection method based on local feature ratio test instead of traditional RANSAC which leads to a much faster image registration speed. The extensive experimental results show that our proposed approach has multiple advantages over existing indoor localization systems in terms of accuracy and speed.

1 Introduction

Indoor localization methods are widely used to assist visitors to navigate large indoor environments, such as museums, hospitals or shopping malls, which aim to provide positioning information promptly. Most current indoor localization methods are based on pre-deployed beacons and GPS, whose satellite signals can be attenuated and scattered by obstructions such as roofs and walls, resulting in inaccurate estimation of position. Other methods relying on beacon and cellular base stations require a proliferation of equipment to achieve the localization, like increasing density of beacon distribution. A further disadvantage of such radio signal-based methods is that these methods usually fail to detect the requisite orientation while locating a person, in cases when it is necessary to know which direction a person or object is facing.

Image-based localization is mainly developed for outdoor tasks with an aim towards overcoming the inherent deficiency of weak GPS signals in areas with

© Springer International Publishing Switzerland 2016
Q. Tian et al. (Eds.): MMM 2016, Part II, LNCS 9517, pp. 218–230, 2016.
DOI: 10.1007/978-3-319-27674-8_20

large buildings. Recent research applies this technique to indoor localization tasks as well, by means of using a query image to match an image database in order to return the best matched image associated with the location information. Users, however, are still unable to obtain direction and orientation information based on the 2D image returned from the database. Moreover, location information, depending on the database size, may be inaccurate.

SfM reconstruction techniques are increasingly being used in outdoor localization tasks. Using 2D images captured by the one or more cameras, the 3D coordinates of an object's feature points can be calculated. Along with the 3D position information, the camera pose of each image can also be computed. Compared with stereo reconstruction which must use at least two calibrated cameras aimed towards the same object, SfM does not require cameras to be calibrated. Under this lesser restriction, images used for SfM are easier to obtain, allowing application of the technique to large-scale reconstruction tasks. Additionally, due to the ease of image availability the updating of indoor environment changes and the cost of building the model are both lessened, compared to other indoor localization techniques. During the localization process, a user only needs to capture an image and match it with the reconstructed 3D model. Based on local feature matches, the camera pose can be calculated, obtaining precise orientation information. To make the location more meaningful for navigation, a map showing the position should be indicated as well. For this purpose, the reconstructed 3D model would be the ideal choice to illustrate position in a 3D map.

The whole process of the system is illustrated in Fig. 1. Unlike outdoor localization tasks, indoor environments usually involve many people walking around. This heightens the difficulty of locating a captured image, as features extracted

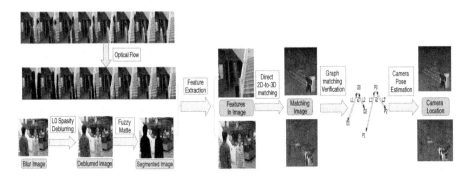

Fig. 1. Localization approach based on graph matching verification: the queries are either images or videos based on the camera capability. For images, de-blurring and interactive foreground segmentation processes are involved. For videos, optical flow is utilized to complete the image frames. After feature extraction from the images, we search query descriptors' approximate nearest neighbor for building correspondences. After the direct 2D-to-3D feature correspondence searching, graph matching is performed to verify the inliers. The final camera pose is estimated based on the verified inliers.

from people in the image will mis-match the features in the reconstructed 3D model. In dealing with this problem, we filter out the people in the image and use the features in the background to match those from the SfM model, thus avoiding wrong pose estimation because of incorrect 2D-to-3D feature correspondences. Most contemporary mobile phones are equipped with video capturing capability. Taking advantage of this, we record a short video, use the optical-flow method to detect and filter out moving people from the video, and complete the background image using the filtered video frames. The completed image is then applied as the query image for localization purpose. We next perform a graph match to verify inlier correspondences. Through graph matching based verification, both image registration number and camera-pose estimation accuracy are increased. Meanwhile, by substituting RANSAC with our initial correspondence selection method based on a SIFT ratio test, random selection is avoided, leading to a faster image registration speed.

To summarize, we address the indoor-localization problem using graph matching verification; we make use of videos to filter out people to perform localization; we consider practical issues involved in an indoor image-capturing process (such as occlusion and image blur) and propose a solution corresponding to each of these related issues, unlike previous localization methods. Experiments demonstrate that when our system is integrated with certain state-of-the-art techniques, localization accuracy and efficiency are improved.

2   Related Work

Image-based localization, first introduced by Robertson and Cipolla [19], is widely applied in localization problems especially in areas with weak GPS signal. Schindler et al. [21] selected vocabularies [17] with the most informative features to improve image retrieval performance on a large street-side image database. Xiao et al. [22] further improved object localization accuracy by using bag-of-words method together with geometric verification. Similar idea is realized for feature matching based on symmetric geometry constraints [1]. Lu et al. [14] proposed a multi-view image localization framework for predicting the image location and orientation under a multi-task learning framework, which is commonly used for daily activity recognition [24] and head pose estimation [25]. Thermal images are also applied for localization task in a dark environment [13] under an active transfer learning framework.

With the improvement of SfM techniques for reconstructing 3D point clouds [2], a 3D SfM model is used on image-based localization problems in enhancing localization accuracy. Irschara et al. [5] proposed retrieving images containing most descriptors matching the 3D points, and Li et al. [9] realized the 3D-to-2D matching through mutual visibility information. Sattler et al. [20] proposed a framework to directly match descriptors extracted from 2D images to the descriptors of the 3D model in order to improve localization accuracy. The proposed localization framework accelerates the matching process and achieves a high image registration rate, which is further improved by mapping local high dimensional descriptors to a low dimensional Hamming space [11]. Zhang et al. [26] used a graph mining algorithm to localize discriminative object parts in unannotated images, which the foreground images are segmented out. Nie et al. [16] presented a web image retrieval re-ranking scheme by complex queries through exploring visual information. Lu et al. [12] proposed a 3D-to-3D framework for localization.

For indoor image-based localization, Kosecka et al. [7] conducted edge detection of room images, generating edge histograms for each image. The matched histogram of the query image is returned together with position information. [18] extracted color histograms and shapes of the images and by matching these with the image database and returned the image position. Liu et al. [10] introduced an approach of indoor mobile 3D location estimation based on a Transfer Regression Model. [6] matched the query image against the image database using local features after hashing in order to save space for storing of local features. The system in [8] involved RANSAC inside matching to filter out outliers and apply ASIFT (Affine-SIFT) features to perform affine invariant image matching.

3   Fast 3D Indoor-Localization

3.1   Pipeline

Image-based localization is designed to provide users with navigation information using a query image provided by users. With just a mobile phone, a user can take a picture of his or her surroundings and transmit it to the image-based localization system. Then, a feedback with navigation information is received from the image-based localization system, as shown in Fig. 2.

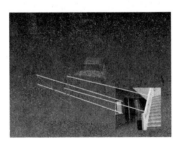

Fig. 2. 2D-to-3D image matching

Originally, image-based localization method was designed to search for the best candidate within an image database, i.e., a response image having the most correspondences to the query image. To achieve higher accuracy, we used a reconstructed 3D model that allows orientation estimation, as descriptors from a 3D reconstruction model are much denser than descriptors extracted from an image database. Compared with other state-of-the-art methods [5,9] which make use of 3D model for outdoor scene localization, we make use of 3D reconstructed scene for indoor environments.

Our image-based localization method uses the direct 2D-to-3D matching similar to that of Sattler et al. [20], the basic idea being to find correspondences between 2D features and 3D points. These correspondences are built by searching for the 2D descriptor's nearest neighbors from all descriptors in a 3D space. We use a kd-tree based approach, supported by FLANN library [15], to detect the approximate nearest neighbor descriptors. Each 3D point is represented by the mean value of all descriptors belonging to this point. The 2D-to-3D correspondence is accepted if it passes the SIFT ratio test. When more than one 2D feature match the same 3D point, the 2D feature with the smallest distance in Euclidean space is accepted as the matching feature. Finally, only those images consisting of at least 12 inliers are registered. Inliers are found by the Random Sample Consensus (RANSAC) algorithm [3]. A 6-point-direct-linear-transformation (6-point DLT) [4] is used in RANSAC loop to estimate the camera pose.

To achieve better image registration performance, we prune the 3D points to reduce wrong feature correspondences. During the Structure-from-Motion reconstruction step, the points may be reconstructed into several reference coordinates. We subsequently select those points within the dominant reference coordinate as the reconstruction model to be used for image-based localization. In indoor environments, light conditions usually affect feature retrieval and matching accuracy to a large extent. Large illumination changes create unreliable matches. Although the SIFT feature has the property of invariance to transformation and small illumination changes, the large-bin value of the SIFT feature is emphasized by computing Euclidean distance between features.

Occasionally, images with few distinct features will be rejected by the 2D-to-3D localization pipeline due to the small number of correct correspondences. In this case, we search through the images used for SfM reconstruction, accepting the one with most correspondences as the matching image. As camera pose information of images used for reconstruction has already been estimated during the Bundler reconstruction process, we can then acquire the camera pose information from the returned image, assigning this information to the query image.

3.2   Deblurring Query Images

When using a slow shutter speed while capturing weak light indoor scenes, even the slightest camera shake can result in a motion blur. Blurred images make extracting features difficult to localize. To de-blur images, we use the single image de-blurring technique described in [23], which uses L0 approximation function as the regularization term in a MAP-based optimization framework to iteratively estimate the kernel for deconvolution. The use of the implicit regularization term (which closely approximates the L0 sparsity) to regularize high frequencies makes this technique rapid when compared to other MAP-variant frameworks. The effect of deblurred images is shown in Fig. 3. Here, we can notice that the images

(a)   (b)   (c)   (d)

Fig. 3. Examples for showing deblurring effect on images: (a) Original blur image without people. (b) Deblurred image without people. (c) Original blur image with people. (d) Deblurred image example with people in the image.

are much more clear than the original image. Thus, the key points' descriptors extracted are more reliable.

3.3 Interactive Foreground Segmentation

In contrast to an outdoor environment, an indoor environment usually involves people walking around. As a result, features detected from the people in the images generate wrong correspondences in the matching process. Foreground objects that occlude the background might cause errors in localization/camera-pose estimation due to false matches. To solve this, we perform interactive foreground segmentation to remove a given object, and then we perform localization using the remainder of the image region. For segmentation, we use the fuzzy-matte technique described in [27]. The user interactively selects a few known foreground and background pixels. The algorithm then computes a fuzzy connectedness from the unknown pixels to the known regions using adjacency and similarity metrics. The connectedness is used to compute the alpha value for matting which we threshold to separate foreground from background. The segmented foreground contains holes and discontinuities which are filled with morphological operations (image closing and image opening). The whole process is described in Fig. 4. We also provide segmentation result for multiple people inside the images in Fig. 5.

We can notice from Fig. 4, after the fuzzy-matte interactive segmentation and morphological operations, the foreground of the image is accurately filtered out, leaving the features extracted only from the background. This helps us filter a large number of inaccurate correspondences. By reducing the unnecessary descriptors, we can accelerate the correspondence building process as well.

3.4 Dynamic Scene Query for Localization

Obstruction removal can help remove the foreground confusion. However, if the background area is too small and the number of features remaining is too small, a localization system will be unable to generate sufficient correspondences. In this circumstance, we can use video as the query to perform localization. Indoor

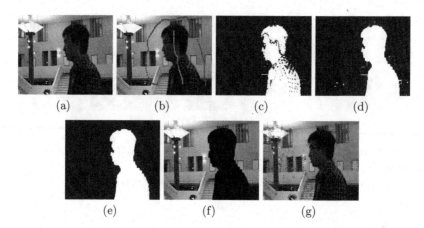

Fig. 4. Examples for segmenting the images: (a) Original image. (b) User specified foreground and background region. (c) Mask generated by Fuzzy matte. (d) Mask processed by image closing. (e) Mask further processed by image opening. (f) Image after masking the people. (g) Features after being filtered by the mask.

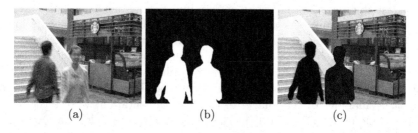

Fig. 5. Examples for segmenting the images with multi people inside. (a) Original image, (b) Mask, (c) Image after masking

localization relies on matching stationary landmark points, which lie on the background in an image. Indoor scenes usually have people walking around, a situation posing a challenge to capture query images with clear background. As a workaround to this problem, we propose to capture a short query video instead of an image. By segmenting out the moving foreground objects using optical flow measurements we extract the background from the video. Optical flow measurements between each adjacent set of frames are thresholded for obtaining moving foreground objects. As the foreground objects are moving, every segmented frame might expose a region of background not visible in previous frames. These regions are then added together to obtain the entire background. Figure 6 shows the background reconstructed using 30 consecutive frames.

From Fig. 6, optical flow can find out the movement of the foreground pixels and fill with the background from the adjacent frames, leading to a completed image. A video can provide more reliable discriminant descriptors than images.

Fig. 6. Optical flow based detection of people walking. The first row shows the original video frames. The second row is the mask based on optical flow. The bottom row displays the real-time image completion.

4   Graph Matching Verification

RANSAC can be used to estimate pose from the correspondences. RANSAC is an iterative method. In each iteration, 6 candidates from the correspondences are randomly selected to estimate the pose and the pose is determined to be correct if there are at least 12 inliers. However, RANSAC is a time consuming operation in the localization pipeline due to its random selection and iterative process. While improving speed, running RNASAC for fewer iterations does not guarantee a solution due to the random process of selection. We propose to use a graph matching based method to search for the candidates for estimating pose. Using the graph-matching method, we select 6 correspondences to estimate the camera pose.

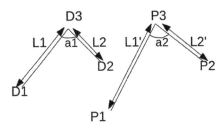

Fig. 7. Inlier verification based on graph matching.

Among all the correspondences, we select 100 best correspondences (same candidate number setting as [20]) having the smallest SIFT ratios as the seed candidate set for estimating pose. All the following steps use points from the seed candidate set. We choose the 6 point correspondences from the set of candidates through graph matching verification method to obtain an accurate 2D-to-3D transformation matrix. For projection of each point in the candidate set on a 2D image, we pick 2 farthest projected points (according to pixel coordinate)

Table 1. The data used for evaluation. Size describes the binary .info file size with all descriptors and 3D points' information.

Dataset	Number of 3D points	Number of images for reconstruction	Size (MB)	Number of query images
Indoor data	189,787	774	27	120

in the image to form a triangle. Then we find the corresponding points in the 3D space to form a 3D triangle, as shown in Fig. 7. In Fig. 7, D1, D2, and D3 are on a 2D image. D1 and D2 are the farthest neighbours of D3 in Euclidean distance. P1, P2 and P3 in the 3D model are the corresponding points to D1, D2 and D3 respectively. We calculate the angle and the ratio of the 2 lengths inside the triangle for both 2D image feature and 3D point. In Fig. 7, $a1$, $a2$, $L2/L1$ and $L2'/L1'$ are calculated and used as features for graph matching. Thus, each feature in the 2D image has an associated angle and edge ratio as a 2 dimensional feature. The same for the corresponding point in 3D space. After normalizing for both angles and edge ratios, we calculate the distance of graph matching feature between 2D feature and corresponding 3D point. The 6 correspondences with the smallest graph matching distance are selected to estimate the transformation model.

Based on the transformation model, we find the inliers to fit this model. If there are at least 12 inliers found to fit the model, we terminate the inlier searching process and re-estimate the camera pose based on all the inliers. If not enough inliers are found to fit in the transformation matrix, we add another 6 correspondences with the smallest graph matching feature distance to estimate the transformation model and find the inliers fitting this model. Once enough inliers are found or all the 100 seed correspondences are added to estimate transformation matrix, the whole process stops. The final transformation matrix estimated by all the inliers is then output as the camera pose. The edge length scale instead of the absolute length provides the property of invariance to scaling in our global graph matching feature.

5   Experiments

To evaluate our proposed methods, we conducted a series of experiments on various query scenes, capturing 774 images using 3 uncalibrated cameras for reconstructing our university departments' buildings. Meanwhile, we captured 120 images to do localization, among which 60 have people inside; another 60 images are taken without people in the images. The dataset we use is shown in Table 1.

In [20], Sattler et al. specify that clustering descriptors into fine visual words can improve localization accuracy by reducing the number of confusing descriptors within the same visual words. We observe that this situation is the case for

Table 2. The image registration number, average registration and average rejection time per image based on the decrease of visual word number.

Visual word number	Number of registered images	Average registration time per image (s)	Average rejection time per image (s)
100k	58	0.703	0.683
10k	83	0.773	0.797
1k	93	0.967	1.038
500	94	1.315	1.503
100	95	3.964	4.264
1	99	10.684	12.783

Table 3. The registration number for images without people inside.

Without people	Number of image registration (1k words)	Number of image registration (500 words)
Our method	48	48
P2F [9]	41	41
Direct 2D-to-3D [20]	45	46

city-scale localization tasks. For indoor localization, separating descriptors to an excess of visual words potentially reduces localization accuracy, as the potential correspondence to the query descriptor may not be found within the visual word. By increasing the size of each visual word, the probability of detecting correct correspondences is increased, thus facilitating camera pose estimation in the later process. We show the image registration result in Table 2.

From Table 2, we can observe that with a decrease from 100k visual words to 1k visual words, the image registration number increases significantly, with only small loss in the time performance. After 1k visual words, however, there is little improvement in the number of registered images; the average registration and rejection time for each image rises substantially. Grouping all the descriptors into 1 visual word amounts to the same thing as linearly searching through the whole dataset, resulting in very bad time performance. As a trade-off, clustering features into 1k visual words gives positive results in both accuracy and time performance. In our dataset, we have an equal number of query images with and without people showing in the image. The number of registered images based on 1K and 500 visual words is reported separately in Tables 3 and 4.

From Tables 3 and 4, we can observe that with and without people, the registered image number does not differentiate much, demonstrating that our technique is effective in localizing query images. By substituting RANSAC with

Table 4. The registration number for images with people inside.

With people	Number of image registration (1k words)	Number of image registration (500 words)
Our method	45	46
P2F [9]	38	38
Direct 2D-to-3D [20]	41	42

Table 5. Time performance for both images with and without people inside.

Methodology	Average registration time per image (s)	Average rejection time per image (s)
Our method	0.967	1.038
P2F [9]	6.327	9.435
Direct 2D-to-3D [20]	1.324	1.538

our SIFT ratio-based, initial-correspondence selection method, random selection is avoided, leading to a much faster image registration speed. We show the image registration and rejection time in Table 5.

From Table 5, we can notice after avoiding random correspondences selection, the time performance is largely increased. The initial selection of correspondence with SIFT feature ratio increases the time performance largely. We captured 20 videos with the frame numbers ranging from 30 to 60 frames, which are approximately 1- to 3- s long videos. After employing our image completion method based on optical flow, all the videos have been successfully localized. The background images generated from our optical flow image completion method achieves accurate estimation of camera pose.

6   Conclusion

In summary, we have developed a new indoor localization system based on a 3D Structure-from-Motion model. The utilization of such a model helps to find correspondences between 2D and 3D features. We have modified the original localization pipeline to make the localization process faster and more accurate. To tackle people in the indoor environments, we have developed a system for de-blurring images and segmenting people out of the images, leaving filtered images to be used for localization purposes. These operations help to reduce irrelevant descriptors detected for localization. Also, video is used in our localization pipeline to complete the background in the video based on optical flow techniques, which can be applied to crowded indoor areas. Graph matching verification based method largely enhance the inlier selection efficiency and the inlier rate. Experiments show that our indoor localization pipeline is suitable for real life applications.

References

1. Cheng, Z.Q., Chen, Y., Martin, R.R., Lai, Y.K., Wang, A.: Supermatching: feature matching using supersymmetric geometric constraints. TVCG **19**(11), 1885–1894 (2013)
2. Crandall, D., Owens, A., Snavely, N., Huttenlocher, D.: Discrete-continuous optimization for large-scale structure from motion. In: CVPR, pp. 3001–3008 (2011)
3. Fischler, M.A., Bolles, R.C.: Random sample consensus: a paradigm for model fitting with applications to image analysis and automated cartography. Commun. ACM **24**(6), 381–395 (1981)
4. Hartley, R.I., Zisserman, A.: Multiple View Geometry in Computer Vision. Cambridge University Press, Cambridge (2004). ISBN: 0521540518
5. Irschara, A., Zach, C., Frahm, J.M., Bischof, H.: From structure-from-motion point clouds to fast location recognition. In: CVPR, pp. 2599–2606 (2009)
6. Kawaji, H., Hatada, K., Yamasaki, T., Aizawa, K.: Image-based indoor positioning system: fast image matching using omnidirectional panoramic images. In: ACM International Workshop on Multimodal Pervasive Video Analysis, pp. 1–4 (2010)
7. Kosecka, J., Zhou, L., Barber, P., Duric, Z.: Qualitative image based localization in indoors environments. In: CVPR, vol. 2, pp. II-3 (2003)
8. Li, X., Wang, J.: Image matching techniques for vision-based indoor navigation systems: performance analysis for 3D map based approach. In: IPIN, pp. 1–8 (2012)
9. Li, Y., Snavely, N., Huttenlocher, D.P.: Location recognition using prioritized feature matching. In: Daniilidis, K., Maragos, P., Paragios, N. (eds.) ECCV 2010, Part II. LNCS, vol. 6312, pp. 791–804. Springer, Heidelberg (2010)
10. Liu, J., Chen, Y., Zhang, Y.: Transfer regression model for indoor 3D location estimation. In: Boll, S., Tian, Q., Zhang, L., Zhang, Z., Chen, Y.-P.P. (eds.) MMM 2010. LNCS, vol. 5916, pp. 603–613. Springer, Heidelberg (2010)
11. Lu, G., Sebe, N., Xu, C., Kambhamettu, C.: Memory efficient large-scale image-based localization. Multimedia Tools Appl. **74**(2), 479–503 (2015)
12. Lu, G., Yan, Y., Li, R., Song, J., Sebe, N., Kambhamettu, C.: Localize me anywhere, anytime: a multi-task point-retrieval approach. In: ICCV (2015)
13. Lu, G., Yan, Y., Ren, L., Saponaro, P., Sebe, N., Kambhamettu, C.: Where am I in the dark: exploring active transfer learning on the use of indoor localization based on thermal imaging. Neurocomputing (2015)
14. Lu, G., Yan, Y., Sebe, N., Kambhamettu, C.: Knowing where i am: exploiting multi-task learning for multi-view indoor image-based localization. In: BMVC (2014)
15. Muja, M., Lowe, D.G.: Fast approximate nearest neighbors with automatic algorithm configuration. In: VISAPP, pp. 331–340 (2009)
16. Nie, L., Yan, S., Wang, M., Hong, R., Chua, T.S.: Harvesting visual concepts for image search with complex queries. In: ACM MM, pp. 59–68 (2012)
17. Nister, D., Stewenius, H.: Scalable recognition with a vocabulary tree. In: CVPR, pp. 2161–2168 (2006)

18. Ravi, N., Shankar, P., Frankel, A., Elgammal, A., Iftode, L.: Indoor localization using camera phones. In: IEEE WMCSA, pp. 49–49 (2006)
19. Robertson, D., Cipolla, R.: An image-based system for urban navigation. In: BMVC, pp. 819–828 (2004)
20. Sattler, T., Leibe, B., Kobbelt, L.: Fast image-based localization using direct 2d-to-3d matching. In: ICCV, pp. 667–674 (2011)
21. Schindler, G., Brown, M., Szeliski, R.: City-scale location recognition. In: CVPR, pp. 1–7, June 2007
22. Xiao, J., Chen, J., Yeung, D.-Y., Quan, L.: Structuring visual words in 3D for arbitrary-view object localization. In: Forsyth, D., Torr, P., Zisserman, A. (eds.) ECCV 2008, Part III. LNCS, vol. 5304, pp. 725–737. Springer, Heidelberg (2008)
23. Xu, L., Zheng, S., Jia, J.: Unnatural l0 sparse representation for natural image deblurring. In: CVPR, pp. 1107–1114 (2013)
24. Yan, Y., Ricci, E., Liu, G., Sebe, N.: Egocentric daily activity recognition via multitask clustering. TIP **24**(10), 2984–2995 (2015)
25. Yan, Y., Ricci, E., Subramanian, R., Lanz, O., Sebe, N.: No matter where you are: flexible graph-guided multi-task learning for multi-view head pose classification under target motion. In: ICCV, pp. 1177–1184 (2013)
26. Zhang, L., Yang, Y., Zimmermann, R.: Fine-grained image categorization by localizing tinyobject parts from unannotated images. In: ICMR, pp. 107–114 (2015)
27. Zheng, Y., Kambhamettu, C., Yu, J., Bauer, T., Steiner, K.: Fuzzymatte: a computationally efficient scheme for interactive matting. In: CVPR, pp. 1–8 (2008)

Smart Ambient Sound Analysis via Structured Statistical Modeling

Jialie Shen[1]([✉]), Liqiang Nie[2], and Tat-Seng Chua[2]

[1] School of Information Systems,
Singapore Management University, Singapore, Singapore
jlshen@smu.edu.sg
[2] School of Computing, National University of Singapore, Singapore, Singapore
nieliqiang@gmail.com, chuats@comp.nus.edu.sg

Abstract. In this paper, we introduce a novel framework called SASA (Smart Ambient Sound Analyser) to support different ambient audio mining tasks (e.g., audio classification and location estimation). To gain comprehensive ambient sound modelling, SASA extracts a variety of acoustic features from different sound components (e.g., music, voice and background), and translates them into structured information. This significantly enhances quality of audio content representation. Further, distinguished from existing approaches, SASA's multilayered architecture seamlessly integrates mixture models and aPEGASOS (adaptive PEGASOS) SVM algorithm into a unified classification framework. The approach can leverage complimentary strengths of both models. Experimental results based on three large test collections demonstrate the SASA's advantages over existing methods on various analysis tasks.

Keywords: Ambient intelligence · Environmental sound analysis

1 Introduction

Rapid advances in mobile computing and multimedia technologies have led to an explosive growth of various kinds of audio information related with our daily life. In particular, ambient audio (environmental sound) contains rich information (semantic concepts) about activity, event, emotion and venue. As a consequence, smart techniques for ambient sound understanding have become more and more important due to potential applications such as home care, health monitoring, intelligent personal assistant and security protection. Essentially, ambient audio understanding can be modelled as an S-class categorization. The performance of technical solutions is largely dependent on their capabilities to model and capture discriminative features to identify one category of sound from others. Although traditional audio analysis schemes or algorithms designed for speech or music recognition could be applied to solve the problem, it is difficult for them to achieve promising performance in terms of accuracy and robustness. This is because most of ambient sounds contain rich sets of basic audio components

© Springer International Publishing Switzerland 2016
Q. Tian et al. (Eds.): MMM 2016, Part II, LNCS 9517, pp. 231–243, 2016.
DOI: 10.1007/978-3-319-27674-8_21

coming from different sources (e.g., human voice, animal sound, music, background events or activities). The acoustic structure and interplay between the elements could be highly complex and dynamic. For example, from the sound track recorded in open market or restaurant, we can easily find that voice or music is often intertwined with the non-stationary background signals from different events or activities (e.g., car engine start or music from shop or party). To develop robust and effective modelling of ambient sounds, it is essential to identify those basic audio elements and design advanced approach to model the highly unstructured information.

In recent years, several approaches have been proposed to apply statistical models or machine learning techniques for ambient sound analysis [4, 6, 13]. These methods commonly consist of two main steps: audio modelling and label identification via machine learning algorithms. In audio modelling, low level feature is extracted and used as content representation of raw ambient sound. Based on the features extracted, specific statistical models or machine learning algorithms (e.g., SVM, KNN or artificial neural network) can be constructed to identify label of the ambient sound. However, the schemes based on this paradigm suffer from low accuracy and poor robustness. The main reasons are:

- many of them only use single type of acoustic feature, which is not able to characterize complex ambient audio comprehensively.
- as mentioned before, ambient sound's structure and content could be very complex and this requires combination multiple types of acoustic features as effective content signature. However, existing studies simply ignore the effects of multiple acoustic features.
- they are mainly based on simple machine learning algorithms instead of advanced scheme, which could lead to more accurate and robust performance.

Motivated by the above discussion, a novel system called SASA (Smart Ambient Sound Analyser) is proposed to facilitate ambient sound characterization and analysis. Distinguished from the previous approaches only considering very limited amount of features and directly applying classic machine learning algorithms, our main research contributions include:

- In order to achieve effective audio modelling, we propose a novel structural analysis framework using the multiple features extracted from various kind of components to improve the system's performance. To the best of our knowledge, this is the first attempt to characterize unstructured ambient sound using a structured way.
- A probabilistic sound characteristic modelling method is designed based on mixture models and aPEGASOS (adaptive PEGASOS) SVM classifier to bridge the "semantic gap" between low level features and high level audio concept.

2   Multilayer Based Ambient Sound Understanding

SASA applies multilayered architecture consisting of three major functionality modules: sound preprocessing, structured sound modelling and effective understanding with advanced SVMs. Figure 1 illustrates detail architecture of SASA.

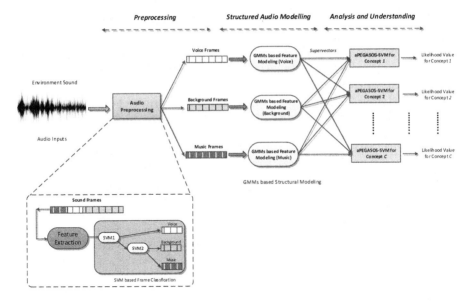

Fig. 1. Multilayer based ambient sound understanding framework

Audio preprocessing module aims to separate an incoming environmental sound into voice, music and background segments, and to extract related audio features from those segments. In the second layer of SASA, there are a collection of statistical models based on GMMs, one for each frame category. To analyse input environmental sound, different feature vectors are firstly extracted from the various segments. The feature vectors are then fed into the statistical models, generating a set of GMM supervectors, which serves as the input to the third module - effective understanding with advanced SVMs. The main functionality of the third module is to estimate which is the most relevant concept for input audio. The core of the third module is a set of SVM classifiers trained using aPEGASOS algorithm. Based on GMM supervectors, overall likelihood score for each concept is computed and the query audio is assigned to the k concepts (labels) with the top k likelihood scores. In the following subsections, we give a comprehensive introduction of the various modules and core algorithms used in the system.

2.1 Audio Preprocessing

In the first stage of the understanding process, our system classifies and labels the music, voice and background segments via the preprocessing module. We use the approach similar to the one presented in [9]. This process can be modelled as a problem of audio frame classification. A learning framework based on SVM can be applied and inputs are acoustic feature vectors combining MFCC and Wavelet. The algorithm demonstrates a promising performance because audio segments containing human voices, music and other background events have very significant differences inside the spectral features.

The preprocessing phase consists of two sub-processes: acoustic feature extraction and frame classification with SVMs. After raw environment sound is received, it will be divided into multiple fixed length time-frames without overlapping. In our implementation, length of frame is set to be 0.75 sec because based on empirical study, it leads to the best performance in our experiments in terms of identification accuracy. The acoustic features extracted from each frame include: MFCC features and Wavelet. The acoustic features serve as input to a set of SVMs, which classify each frame into three possible categories: voices, music or background. We select SVM as classifier because it has demonstrated excellent performance on a range of categorization problems. In our framework, the LIBSVM [3] library is used for implementation and the kernel type is linear kernel. After the process, environment sound input au is segmented into three major components: voice frames au_v, music frames au_u and background frames au_b.

2.2   Structured Environmental Sound Modelling

Given that environmental sound can include multiple concepts and come from multiple sources, The second layer of SASA system consists of multiple modality models, one for each basic audio element (e.g., voice, music or background). Each model is made up of two parts: feature extractors and modelling via Gaussianization. In below, we will provide details about each component.

Acoustic Feature Extraction. To effectively represent and model the complex contents of ambient audio, our system extracts various features from different types of segments. In total, four different features are extracted to model ambient sound from various perspectives: timbre feature (TF), pitch feature (PF), instrument-based feature (IF) and wavelet-based feature (WF) [1]. In particular, the TF and PF capture information from the voice segments generated by human in ambient audio. The wavelet-based feature (WF) characterizes the music style and background acoustic events. The instrument feature (IF) is used to model the characteristics of typical instrument(s) in music frames. The details of the features used by the our framework:

- **Timbre Feature (TF):** Voice is a special instrument for human being and each person's timbre texture is unique due to the physical structure of voice fold. In SASA, LPCCs (Linear Prediction-based Cepstral Coefficients) from vocal segments are extracted to characterize this information (LPCCs are Linear Prediction Coefficients (LPCs) represented in the cepstrum domain).
- **Pitch Feature (PF):** To gain comprehensive modelling on human voice, it is crucial to take the harmonic and structural information about each person's voice into account. Since the algorithm proposed by Tolonen and Karjalainen

---

[1] Note that our method can be easily extended to consider more acoustic features.

is computational efficient and has superior capability in capturing human auditory perception, we apply it to extract pitch feature from human voice segments.

- **Instrument Based Feature (IF):** Instrument appearing in music segments can provide a lot of details about ambience. The main goal of IF is to characterize instrument configuration information of music frames. Our framework applies MFCC features as signature of instrument configuration. This is because MFCCs have been widely used to model timbre for purpose of instrument identification.

- **Wavelet based Feature (WF):** In SASA, WF is used as content representation to capture local and global dynamics of ambient sounds [1]. Wavelet analysis has been widely applied to model and characterize a wide range of audio information (e.g., background events [5] and music genre [7]). In our system, WF feature is based on the Daubechies Wavelet Coefficient Histograms (DWCHs) and mainly characterizes music genre and background events.

Multi-modality Based Modeling. The second layer of SASA system is a multi-modality based environmental sound characterization model. We apply the Gaussian Mixture Models (GMMs) for statistically modelling from different audio component perspectives since it has strong flexibility and effectiveness to represent complex data distribution. In SASA, one frame category corresponds to one GMMs and thus there are three GMMs. However, construction of GMMs based scheme on highly diverse distribution associated with environmental sound is not easy task. Since the number of sound frames for robust training is limited, parameter estimation of a GMMs robustly and accurately becomes very time-consuming. To solve this problem, a two-step adaptation approach is applied to develop GMMs including generative adaptation and sound segment based adaptation [2]. The key advantage of the approach include: (1) efficiency - a complex model can be developed using a small set of data and (2) simplicity - the model's output space is the Euclidean space, which can support fast search.

In generative adaptation, the GMMs is constructed with all training audio and then generate Universal Background Model (UBM). The UBM can be denoted as,

$$G = P(X|\theta) = \sum_{k=1}^{K} w_k N(X; \mu_k, \Sigma_k), \tag{1}$$

where w_k, μ_k and Σ_k are the weight, mean and covariance matrix of the kth Gaussian component, respectively. $X = \{x_1, x_2, ..., x_T\}$ is a set of input feature vectors extracted from audio segments. K is the total number of Gaussian components and the probabilistic density can be calculated using a weighted combination of K Gaussian densities,

$$N(X; \mu_k, \Sigma_k) = \frac{1}{(2\pi)^{d/2} |\Sigma_k|^{1/2}} e^{-\frac{1}{2}(z-\mu_k)^T \Sigma_k^{-1}(z-\mu_k)}. \tag{2}$$

The covariance matrix Σ_k is set to a diagonal matrix for reducing computational cost. To estimate the optimal values of key parameters $\{w_k, \mu_k, \Sigma_k\}$ of

UBM, we apply expectationmaximization (EM) algorithm, which is an iterative scheme to identify maximum a posteriori (MAP) estimates of model parameters.

2.3   Segment Based Adaptation

The goal of segment based adaptation is to modify the parameters of UBM to fit into the data distribution of audio segments. In SASA, each audio segment is represented as an collection of feature vectors, extracted from 30 ms window with step size of 5 ms. For different types of audio segments, we calculate different acoustics features and their details can be found in Sect. 2.2. The adaptation is carried out using maximum a posteriori (MAP). For Gaussian component k in the mixture model, we firstly compute,

$$pr(k, x_i) = \frac{w_k P_k(x_i|\theta_k)}{\sum_{j=1}^{K} w_j P_j(x_i|\theta_j)}, \tag{3}$$

$$\eta_k = \sum_{t=1}^{T} pr(k|x_i) \tag{4}$$

$$E_k(X) = \frac{1}{\eta_k} \sum_{t=1}^{T} pr(k|x_i)x_i, \tag{5}$$

The statistical values shown above can be then applied to adapt mean vector μ_k of each Gaussian component via the iteration process, in which $\hat{\mu}_k$ value at iteration l - $\hat{\mu}_k^l$ can be estimated by using:

$$\hat{\mu}_k^l = \alpha_k E_k(X) + (1 - \alpha_k)\hat{\mu}_k^{l-1} \tag{6}$$

where $\alpha_k = \eta_k/(\eta_k + r)$, $l = 1, ..., L$ is iteration number. r is smoothing factor, which can be fine-tuned empirically based on the total number of feature vectors extracted from each audio segment. It ranges from 5 to 20. We follow the approach introduced in [2] to gain an approximation of KL divergence of two models by:

$$d(\mu^\alpha, \mu^\beta) = \frac{1}{2} \sum_{k=1}^{K} w_k(\mu_k^\alpha - \mu_k^\beta) \Sigma_k^{-1}(\mu_k^\alpha - \mu_k^\beta) \tag{7}$$

where μ^α and μ^β denote supervector for model α and β. After audio segment based adaptation, the audio segment can be represented by super-vector,

$$SV = [v_1, v_2, \ldots, v_K] \tag{8}$$

where $v_k = \sqrt{\frac{w_k}{2}}\Sigma_k^{-\frac{1}{2}}\mu_k$. The supuervector serves as input to audio concept estimation module in the third layer of SASA.

2.4   Audio Concept Estimation Using SVM

The third layer of SASA system consists of a set of Support Vector Machines (SVMs) for the purpose of probabilistic estimation over different audio concepts. In order to support fast and effective SVM training, we develop an advanced extension of PEGASOS [12] algorithm called aPEGASOS (adaptive PEGASOS), which enhances SVM based classification from two perspectives: (1) probabilistic based audio concept estimation and (2) adaptive sampling to effectively select discriminative audio segments as training examples instead of random projection.

With a given training set $\mathcal{S} = \{(\mathbf{x}_i, y_i)\}_{i=1}^m$, where $\mathbf{x}_i \in R^n$ and $y_i \in \{+1, -1\}$, the PEGASOS algorithm is an efficient scheme aiming to effectively solve primal form of SVM $\mathbf{w}$ in an iterative fashion. At each iteration of the training algorithm, there are two key substeps: a stochastic gradient descent step and a projection step. The optimization goal of PEGASOS is to minimize the training error defined as:

$$f(\mathbf{w}; \mathcal{SA}_t) = \frac{\lambda}{2} \| \mathbf{w} \|^2 + \frac{1}{k} \sum_{(\mathbf{x}, y) \in \mathcal{SA}_t} \max\{0, 1 - y\langle \mathbf{w}, \mathbf{x}\rangle\} \tag{9}$$

The process requires T iterations and k the training samples for computing sub-gradients at each iteration. $\mathcal{SA}_t \subset \mathcal{S}$ consists of k samples selected using adaptive sampling scheme introduced late in the section from $\mathcal{S}$ at each iteration t. In the initial, we set the values of $\mathbf{w}$ to be zero. With learning rate $\eta_t = 1/(\lambda t)$ and a set of training samples $\mathcal{SA}_t^+$, parameter updating process at each iteration t has two steps,

$$\mathbf{w}_{t+\frac{1}{2}} = (1 - \eta_t \lambda)\mathbf{w}_t + \frac{\eta_t}{k} \sum_{(\mathbf{x}, y) \in \mathcal{A}_t^+} y\mathbf{x} \tag{10}$$

$$\mathbf{w}_{t+1} = \min\{1, \frac{1/\sqrt{\lambda}}{\| \mathbf{w}_{t+\frac{1}{2}} \|}\}\mathbf{w}_{t+\frac{1}{2}} \tag{11}$$

$\mathbf{w}$ has non-zero training error when using $\mathcal{SA}_t^+$. Similar to the approach introduced in [15], once SVM training is completed, the learning method proposed in [10] is used to infer posterior probability $p^+(c = 1|x)$ of a given input belonging to certain class $c = 1$ as:

$$p^+(c = 1|x) = \frac{1}{1 + \exp(A\langle \mathbf{w}, \mathbf{x}\rangle + B)} \tag{12}$$

where scalar A and B can be estimated via seeking minimization of the error function by using the training data.

The quality of learning examples plays an important role in SVM training. Our system is trained using the acoustic features extracted from audio segments and thus notion of discriminative frame is very important because some of the segments enjoy more informative or distinctive cues about basic audio events or concepts (e.g., gun shot, party, dance and happy). However, how to select high quality training example is very challenging task because they generally need to satisfy two main requirements: (1) excellent representativeness and (2) high distinctiveness. To achieve this goal, we proposed a simple but effective algorithm based on semi-supervised principle. It involves two main steps:

- Seed selection: In our approach, human subjects are invited to provide a few good quality samples as seeds - a set of representative examples for three frame categories. In our current implementation, totally 5 human subjects are invited to select examples.
- Propagation: Based on seeds, we apply a common and robust neighborhood learning method proposed in [16] to construct n nearest neighborhood graph G_{ij} and identify training examples from unselected samples.

3   Experimental Configuration

In this section, we present the experimental settings for the performance evaluation, including competitive systems, testing datasets, evaluation task and performance metrics. All methods evaluated in this study have been fully implemented and tested on a server with 2.2 GHz Intel Xeon processor and 8 GB RAM.

3.1   Data Collections

Test collections play very important role in empirical study. The size of data sets used in existing study is quite small. To ensure accuracy, robustness and fairness of our empirical results, three large benchmark datasets are selected as testbeds in our evaluation. For three datasets, the sound files were converted to 16 kHz, 16 bit, mono audio files. More details about the three datasets can be found as below,

- Dataset I (DSI) is UrbanSound8K [11], which consists of 8,732 audio clips up to 4 sec in duration. The sound files are extracted from field recording crawled from the Freesound online archive [2]. Each clip contains one of 10 possible sound sources including: air conditioning, car horn, children playing, dog bark, drilling, idling engines, gun shot, jackhammer, siren, street music. Those sources are carefully selected from the Urban Sound Taxonomy [11] based on the frequency with which they appear in noise complaints provided by New York City's 311 service. All the sound clips have been manually annotated with human subject and a subjective judgment about whether the sound is in the foreground or background has been given.
- Dataset II (DSII): It consists of 1,873 audio clips and covers 25 concepts (e.g., dancing, singing, beach, playground, graduation, group of 3+,......) belonging to 6 main categories including: activities, locations, occasions, objects, scenes and sounds [6]. The 25 concepts was defined by starting from a full ontology of over 100 concepts generated via user study done by the Eastman Kodak company [8]. To develop this dataset, totally 4,539 video was downloaded from YouTube by using most related keywords associated with the definition of these 25 concepts. To remove irrelevant commercial contents, raw dataset was manually checked before used for extracting accompanying sound tracks.

---

[2] http://www.freesound.org.

- Dataset III (DSIII) consists of 10,000 sound clip and covers 10 different recording locations including: Library, Office, Bathroom, Cafe, Restaurant, Kitchen, Living-room, Classroom, Subway, and Open mark. All of sound clips in the collection were recorded using Sony PCM-D100 high resolution audio-recorder. The duration of the audio files ranges from 5 sec to 30 sec. It covers 35 different concepts including high music, walking, chatting, cheering, typing, phone tone, door opening, door closing, crying, baby, male's voice, female's voice, TV sound, engine starting, crowd and others. Total duration of the whole collection is 20 h. All the concepts have been manually verified by human subjects and each sound in test collection could be associated with 1 - 5 concepts.

3.2 Methodology and Evaluation Metrics

Environmental sound analysis is one of the most fundamental components in various kinds of ambient intelligence applications. In order to conduct a comprehensive performance comparison of different schemes, our proposed system and the competitors are tested on the following two application driven tasks. They are,

- Task I - Sound understanding: The goal of the test is to evaluate and compare what are the accuracies achieved by different approaches in classifying the input environmental sounds. The datasets used for this task include DSI and DSII.
- Task II - Location estimation: Based on the input environmental sounds, we would like to test and compare how accurate different approaches can infer the venue. The dataset used for this test is DSIII.

As discussed above, the main goal of the system is to identify the suitable concepts related to input sound. Thus, our evaluation method focuses on how accurate the identification process is with different approaches for a particular database. We use the *accuracy* as the metric for evaluation: $Accuracy = \frac{NA}{NT}$, where NA is the number of sound correctly identified and NT is the total number of sounds used in the evaluation.

3.3 Competitors for Performance Comparison

We introduce several state-of-the-art methods on environmental sound recognition and location estimation based environmental sound analysis for comparison. For Task I (sound classification), we compare the performance of our system SASA against three state-of-the-art approaches including LEE [6], MP [4] and ESCLH [13]. For our, we consider three modality configurations (voice modality denoted by VM, music modality denoted by MM, background modality denoted by BM). SASA(VM), SASA(MM), SASA(BM), SASA(ALL) denote our proposed model built based on voice modality, music modality, background modality and the combination of all three modalities. To demonstrate advantages of our approach in location detection (Task II), we examine a wide range of possible methods, including ABS [14], LEE [6], and MP [4]. For both Task I and Task II, mixture component number k for GMM in LEE and SASA is set to be 4, which is optimal value.

4 Experiment Results

This section presents a set of empirical studies to test and compare the performance of different systems on two tasks including environmental sound classification and location estimation.

On Environmental Sound Classification Table 1 shows the results of our experiments to test the accuracy of environmental sound classification using different schemes. The test was carried out on two different data sets - DSI and DSII. For each of the classifiers, fivefold cross validation is applied to ensure robustness of classification results.

Table 1. Environmental Sound Classification Accuracy Comparison.

Model	DSI	DSII
SASA(ALL)	84.3 %	89.3 %
SASA(BM)	73.5 %	77.5 %
SASA(MM)	50.2 %	56.9 %
SASA(VM)	50.6 %	55.1 %
ESCLH	73.2 %	80.2 %
LEE	74.2 %	81.2 %
MP	69.6 %	77.3 %

The first four rows of Table 1 indicate how the proposed SASA system performed using DSI and DSII. We find that the accuracies achieved by SASA(VM) and SASA(MM) (between 50 % and 60 % for all two datasets) are quite low. By taking background modality into consideration, SASA(BM) improves classification accuracy around 20 %. This result verifies the claim that background contains rich information about sound category and plays very crucial role in effective ambient audio classification and modelling. Further, SASA(ALL), which considers all three different sound elements, achieve a significant performance gain in classification accuracy, 84.3 % for DSI and 89.3 % for DSII. The results demonstrate that combining various acoustic cues from different sources can enhance classification effectiveness greatly. Meanwhile, the results provide strong empirical evidence that accurate classification cannot be achieved by considering only a single sound components.

In comparison with ESCLH, LEE and MP, sound classification with SASA(ALL) results in a great improvement in accuracy for all of the different datasets. For example, in case of DSI, SASA(ALL) improves accuracy by 16.1 % against MP. For DSI, the improvement is 15.5 % . Among all classification methods, SVMs give the best results, whatever kind of music descriptor is used. On the other hand, good scalability is particularly important for large audio information systems, because the size of modern sound collection can be huge and changed frequently. Thus, it is important for analysis schemes to maintain stable accuracy against dataset size change. Based on Table 1, all the methods suffer from accuracy loss at some level when size of testbed becomes larger. However, SASA yields the lowest accuracy drop rates than do all other approaches. For example, when tested on DSI, SASA(ALL)'s accuracy is decreased by only 5 % comparing to SASA(ALL) on DSII. However, under same testing configuration, performance drops of other methods range from 8.6 % to 10 % , which is significantly higher.

On Location Estimation, Table 2 summaries location estimation effectiveness of the SASA, ESCLH, LEE and MP techniques. It is shown in the first four columns that comparing to other SASA variants, SASA(MM) built based on music components is the worst in terms of estimation accuracy rates. Furthermore, although the SASA(VM) demonstrates better performance than SASA(VM), gain is very marginal. This is because music and voice segments only capture very limited amount information about one location. In fact, the results clearly demonstrate that SASA(BM) outperforms the SASA(VM) and SASA(BM) greatly. Once again, this results provide strong support on the claim that background elements contains more information about one location than other two basic elements. More importantly, SASA(ALL) achieves the best estimation accuracy comparing to six other methods. In addition, it is worth noticing that integrating effects of additional sound elements bring SASA nice lift in accuracy improvement. For example, by considering two additional sound elements, accuracy improvement over SASA(VM), SASA(MM), and SASA(BM) is 54.3 %, 60.1 %, and 12.0 %, respectively.

Table 2. Location Estimation Accuracy Comparison.

Schemes	DSIII
SASA(ALL)	81.2 %
SASA(BM)	72.5 %
SASA(MM)	50.7 %
SASA(VM)	52.6 %
ABS	70.2 %
LEE	71.2 %
MP	56.6 %

5   Conclusions

In this paper, we present an intelligent framework, called SASA, to facilitate effective environmental sound analysis. The system has been fully implemented and tested using different datasets. As shown in our experimental evaluation, the SASA system not only demonstrates significantly better effectiveness on audio classification and location estimation over the state-of-the-art systems, but also achieves good robustness against acoustic distortion. The research opens up several promising directions for future study. Firstly, we plan to test the framework over larger dataset with higher complexity. Further, it is interesting to investigate how to develop advanced acoustic modelling scheme to support accuracy and robustness improvement.

Acknowledgments. This work was partly supported by Singapore Ministry of Education Academic Research Fund Tier 2 (MOE2013-T2-2-156), Singapore.

References

1. Bailey, T., Sapatinas, T., Powell, K.J., Krzanowski, W.J.: Signal detection in underwater sound using wavelets. J. Am. Statist. Ass **93**, 73–83 (1998)
2. Campbell, W.M., Sturim, D.E., Reynolds, D.A.: Support vector machines using GMM supervectors for speaker verification. IEEE Signal Process. Lett. **13**(5), 308–311 (2006)
3. Chang, C., Lin, C.: LIBSVM: A library for support vector machines. ACM TIST **2**(3), 27 (2011)
4. Chu, S., Narayanan, S.S., Kuo, C.C.J.: Environmental sound recognition with time-frequency audio features. IEEE Trans. Audio Speech Lang. Process. **17**(6), 1142–1158 (2009)
5. Feng, Z.R., Zhou, Q., Zhang, J., Jiang, P., Yang, X.W.: A target guided subband filter for acoustic event detection in noisy environments using wavelet packets. IEEE/ACM Trans. Audio Speech Lang. Proc. **23**(2), 361–372 (2015)
6. Lee, K., Ellis, D.P.W.: Audio-based semantic concept classification for consumer video. IEEE Trans. Audio Speech Lang. Proc. **18**(6), 1406–1416 (2010)
7. Li, T., Ogihara, M., Li, Q.: A comparative study on content-based music genre classification. In: Proceedings of ACM SIGIR Conference, pp. 282–289 (2003)
8. Loui, A.C., Luo, J., Chang, S., Ellis, D., Jiang, W., Kennedy, L.S., Lee, K., Yanagawa, A.: Kodak's consumer video benchmark data set: concept definition and annotation. In: Proceedings of ACM MIR 2007, pp. 245–254 (2007)
9. Lu, L., Zhang, H., Li, S.Z.: Content-based audio classification and segmentation by using support vector machines. Multimedia Syst. **8**(6), 482–492 (2003)
10. Platt, J.C.: Probabilistic outputs for support vector machines and comparisons to regularized likelihood methods. In: Advances In Large Margin Classifiers, pp. 61–74. MIT Press, Cambridge (1999)
11. Salamon, J., Jacoby, C., Bello, J.P.: A dataset and taxonomy for urban sound research. In: 22st ACM International Conference on Multimedia (ACM-MM 2014) (2014)
12. Shalev-Shwartz, S., Srebro, N.: SVM optimization: inverse dependence on training set size. In: Proceedings of ICML 2008, pp. 928–935 (2008)
13. Su, F., Yang, L., Lu, T., Wang, G.: Environmental sound classification for scene recognition using local discriminant bases and HMM. In: Proceedings of ACM MM 2011, pp. 1389–1392 (2011)

14. Tarzia, S.P., Dinda, P.A., Dick, R.P., Memik, G.: Indoor localization without infrastructure using the acoustic background spectrum. In: Proceedings of ACM MobiSys 2011, pp. 155–168. ACM (2011)
15. Zhang, B., Shen, J., Xiang, Q., Wang, Y.: Compositemap: a novel framework for music similarity measure. In: Proceedings of ACM SIGIR Conference, pp. 403–410 (2009)
16. Zhou, D., Schölkopf, B., Hofmann, T.: Semi-supervised learning on directed graphs. In: NIPS 2004, pp. 1633–1640 (2004)

Discriminant Manifold Learning
via Sparse Coding for Image Analysis

Meng Pang[1], Binghui Wang[1], Xin Fan[1], and Chuang Lin[2]([✉])

[1] School of Software,
Dalian University of Technology, Dalian City, People's Republic of China
[2] Department of Neurorehabilitation Engineering,
Bernstein Focus Neurotechnology (BFNT) Göttingen,
Bernstein Center for Computational Neuroscience (BCCN),
University Medical Center Göttingen, Georg-August University, Göttingen, Germany
`lin.chuang@bccn.uni-goettingen.de`

Abstract. Traditional subspace learning methods directly calculate the statistical properties of the original input images, while ignoring different contributions of different image components. In fact, the noise (e.g., illumination, shadow) in the image often has a negative influence on learning the desired subspace and should have little contribution to image recognition. To tackle this problem, we propose a novel subspace learning method named Discriminant Manifold Learning via Sparse Coding (DML_SC). In our method, we first decompose the input image into several components via dictionary learning, and then regroup the components into a More Important Part (MIP) and a Less Important Part (LIP). The MIP can be regarded as the clean part of the original image residing on a nonlinear submanifold, while LIP as noise in the image. Finally, the MIP and LIP are incorporated into manifold learning to learn a desired discriminative subspace. The proposed method is able to deal with data with and without labels, yielding supervised and unsupervised DML SCs. Experimental results show that DML_SC achieves best performance on image recognition and clustering tasks compared with well-known subspace learning and sparse representation methods.

Keywords: Subspace learning · Dictionary learning · Sparse coding · Image decomposition · Image analysis

1 Introduction

In image processing filed, the input image is often of very high dimensionality, which makes it difficult for applying statistical techniques to conduct visual analysis. Hence, there is the need to detect a low-dimensional hidden subspace to represent the original image. To achieve this goal, subspace learning based techniques, which lower down the input dimensionality, have received considerable attention in recent years [1–5].

One reasonable assumption is that naturally generated high-dimensional data probably lies on or near a relatively lower dimensional manifold [6]. Based on this assumption, manifold learning has been studied as one of the most successful subspace learning techniques in the past decades [7–13]. Typical manifold

© Springer International Publishing Switzerland 2016
Q. Tian et al. (Eds.): MMM 2016, Part II, LNCS 9517, pp. 244–255, 2016.
DOI: 10.1007/978-3-319-27674-8_22

learning algorithms include locality preserving projection (LPP) [7], locality sensitive discriminant analysis (LSDA) [9] and manifold discriminant analysis (MDA) [11] etc. These manifold learning algorithms do perform well in face recognition. Nevertheless, almost all of them calculate directly the statistical properties of the original input data as a whole to learn the subspace, without eliminating the noise in image in advance. Former research has demonstrated that natural images can be represented by a small number of basis functions chosen from an over-complete code set [14], which provides the theoretical feasibility for image decomposition. With the development of norm minimization techniques [15,16], many sparse coding (SC) approaches [17–21], e.g., sparse representation classifier (SRC) [17], graph regularized sparse coding (GraphSC) [18], metaface learning (MFL) [19], have been proposed to achieve the sparse representations of original images under some pre-given or learnt basis. Furthermore, researchers [22] began to consider the importance of each image component, and utilized the criterion, e.g., PCA, to regroup these components after dictionary learning. The work in [22] is encouraging, however the image decomposition is incomplete and the regrouped representations cannot provide an explicit concept explanation.

To handle the aforementioned drawbacks existed in recent manifold learning and sparse coding techniques, we propose a novel subspace learning method named Discriminant Manifold Learning via Sparse Coding (DML_SC). Our DML_SC explicitly exploits the contributions of different image components as well as the discriminative structure of the images. DML_SC is a two-stage algorithm, i.e., dictionary learning and feature regrouping followed by graph embedding. In the first stage, an effective dictionary learning strategy and a regrouping criterion are utilized to decompose the input image into a More Important Part (MIP) and a Less Important Part (LIP). Subsequently, the MIP and LIP are embedded into the spectral graph [10] to learn a desired discriminative subspace, where the MIP is preserved while LIP is suppressed. It is worth noting that both MLP and LIP have explicit concept explanation, the MIP can be regarded as the clean part of the original image residing on a nonlinear submanifold, and LIP as noise or trivial structure. Since the regrouping criterion can be designed in either a supervised or unsupervised way, our DML_SC is thus performed in supervised (SDML_SC) or unsupervised (UDML_SC) versions for image recognition and clustering, respectively. The rest paper is organized as follows: in Sect. 2, we introduce Discriminant Manifold Learning via Sparse Coding (DML_SC) in details. Experimental results on image recognition and clustering are presented in Sect. 3. Finally, some conclusion remarks are given in Sect. 4.

2   Discriminant Manifold Learning via Sparse Coding (DML_SC)

2.1   Motivation

Suppose $\mathbf{X} = [\mathbf{x}_1, \cdots, \mathbf{x}_N] \in \Re^{m \times N}$ is a set with N training samples. We aim to decompose each training sample $\mathbf{x}_i$ into two parts, i.e., a more important

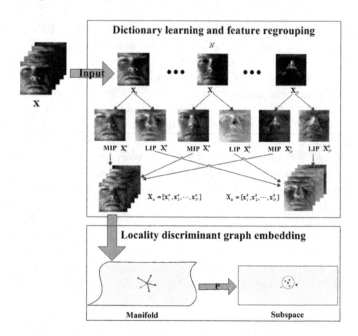

Fig. 1. The pipeline of the proposed supervised DML_SC (SDML_SC). The points with same color belong to the same class.

part (MIP) $\mathbf{x}_i^a$ and a less important part (LIP) $\mathbf{x}_i^b$. MIP reserves main features of the images and contains vital discriminant information, while LIP is noise containing interference information. Intuitively, the training samples $\mathbf{X}$ can be split into two parts, i.e., $\mathbf{X}_a$, $\mathbf{X}_b$, and rewritten as $\mathbf{X} = \mathbf{X}_a + \mathbf{X}_b$. Next, we try to design specific graphs to characterize the geometry property for MIP and LIP. Subsequently, we try to learn a projection $\mathbf{P}$ to project the $\mathbf{X}$ into a lower dimensional subspace, where the topology structure of $\mathbf{X}_a$ on the manifold is preserved while the energy of $\mathbf{X}_b$ is suppressed. The pipeline of the proposed supervised DML_SC (SDML_SC) is shown in Fig. 1. The detailed structure of our DML_SC can be interpreted into following steps: dictionary learning, feature regrouping and graph embedding.

2.2 Dictionary Learning and Feature Regrouping

The dictionary learning and feature regrouping process are as the follows.

Dictionary Learning. In this phase, the important issue is how to decompose the training sample images $\mathbf{X}$ into representation components. Inspired by the success of dictionary learning and sparse coding in image processing [17,19], we also choose to learn an adaptive over-complete dictionary $\mathbf{\Phi}$ to represent $\mathbf{X}$ under sparse coding constraints. Since the dimensionality of input image $\mathbf{x}_i$ is often very high, it is difficult to learn a over-complete dictionary directly for $\mathbf{X}$. Different from [22], we prefer to generate Randomfaces (or Eigenfaces) $\boldsymbol{\Omega} \in \Re^{d \times m}$ to perform dimension reduction rather than learn a patch based dictionary. Now, our aim is to learn a k-atom dictionary $\boldsymbol{\Phi} = [\mathbf{d}_1, \mathbf{d}_2, \cdots, \mathbf{d}_k] \in \Re^{m \times k}$, where $\mathbf{d}_i^T \mathbf{d}_j = 1$, so as to minimize:

$$\min \|\boldsymbol{\Lambda}\|_1, \quad s.t. \ \boldsymbol{\Omega}\mathbf{X} = \boldsymbol{\Omega}\boldsymbol{\Phi}\boldsymbol{\Lambda} \tag{1}$$

or

$$J(\boldsymbol{\Phi}, \boldsymbol{\Lambda}) = \arg\min_{\boldsymbol{\Phi}, \boldsymbol{\Lambda}}(\|\boldsymbol{\Omega}\mathbf{X} - \boldsymbol{\Omega}\boldsymbol{\Phi}\boldsymbol{\Lambda}\|_F^2 + \gamma\|\boldsymbol{\Lambda}\|_1) \tag{2}$$

where $\boldsymbol{\Lambda} = [\alpha_1, \alpha_2, \cdots, \alpha_N]$ is the sparse matrix of $\mathbf{X}$ over $\boldsymbol{\Phi}$, and γ is the regularization parameter. The optimization problem of (2) is not jointly convex to $\boldsymbol{\Phi}$ and $\boldsymbol{\Lambda}$, but it is convex to $\boldsymbol{\Phi}$ or $\boldsymbol{\Lambda}$ when the other is fixed. Thus, we solve (2) by optimizing $\boldsymbol{\Phi}$ and $\boldsymbol{\Lambda}$, respectively. In this paper, we use metaface learning (MFL) in [19] to work it out.

After obtaining $\boldsymbol{\Phi}$ and $\boldsymbol{\Lambda}$, for each face image $\{\mathbf{x}_i\}_{i=1}^N$, we have $\mathbf{x}_i = \boldsymbol{\Phi}\alpha_i = \sum_{z=1}^k \mathbf{d}_z\alpha_{i,z}$, and $\alpha_{i,z}$ is the zth element of α_i. Let $\mathbf{x}_i^z = \mathbf{d}_z\alpha_{i,z}$, then $\mathbf{x}_i = \sum_{z=1}^k \mathbf{x}_i^z$. Thus each $\mathbf{x}_i$ can be rewritten as the summation of k parts and each part is denoted by $\mathbf{x}_i^z$.

Feature Regrouping. Considering whether or not the label information of face images is known, we utilize two reasonable criteria, i.e., *unsupervised* presentation regrouping and *supervised* presentation regrouping, to group the parts into a MIP and LIP for more effective graph embedding.

For *unsupervised* learning, the label information is unknown. If one representation has a relative larger variance, then we believe that it is more informative to separate the data. The variance of the zth presentation is defined as:

$$var(z) = \sum_{i=1}^N (\mathbf{x}_i^z - \mu^z)^T(\mathbf{x}_i^z - \mu^z) \tag{3}$$

where μ^z is the mean of zth presentation of all images, i.e. $\mu^z = \frac{1}{N}\sum_{i=1}^N \mathbf{x}_i^z$.

To implement presentation regrouping, we first reorder $\mathbf{x}_i^z$ according to its $var(z)$ into a descending order, and then put those presentations having relative larger variance into the MIP $\mathbf{x}_i^a$, i.e. $\tau = 80\%$ of total variance $\sum_{z=1}^k var(z)$ and the remaining into the LIP $\mathbf{x}_i^b$.

For *supervised* learning, as the label information is given, if one presentation owns relatively better discriminant ability, then it is more discriminative for data separation. Considering that the Maximum Margin Criterion (MMC) [23] is often exploited to evaluate the discriminant ability of labeled samples, thus the supervised representation regrouping can be designed according to MMC. We define the discriminative factor d_M^z of $\mathbf{x}_i^z$ as:

$$d_M^z = \sum_{c=1}^C (m_z^c - m_z)^2 - \sum_{c=1}^C \frac{1}{N_c} \sum_{\mathbf{x}_j \in \mathbf{X}_c} (\mathbf{x}_j^z - m_z^c)^2 \tag{4}$$

where $\mathbf{X}_c$ is image set of the cth class, m_z is the mean vector of $\mathbf{x}_i^z$, m_z^c is the mean vector of $\mathbf{x}_i^z$ that belong to the cth class, and N_c denotes the number of the cth class.

Similar to unsupervised presentation regrouping, the presentations having larger d_M^z are added up to form the MIP $\mathbf{x}_i^a$ and the remaining to form the LIP $\mathbf{x}_i^b$ of face image $\mathbf{x}_i$. Finally, we achieve the MIP $\mathbf{X}_a = [\mathbf{x}_1^a, \cdots, \mathbf{x}_N^a]$, and the LIP $\mathbf{X}_b = [\mathbf{x}_1^b, \cdots, \mathbf{x}_N^b]$.

2.3   Graph Embedding

After presentation regrouping, the main problem is how to utilize the MIP and LIP of the input images to learn a desired subspace for classification or clustering. Obviously, the MIP $\mathbf{X}_a$ plays a more important role than the LIP $\mathbf{X}_b$ on image recognition. However, we cannot ignore the contribution of LIP $\mathbf{X}_b$ because it is also useful to determine the projection directions. *What we need to do is to learn a subspace where the energy and manifold structure of MIP $\mathbf{X}_a$ are both preserved, while the energy of LIP $\mathbf{X}_b$ is suppressed simultaneously.*

Inspired by the graph embedding (GE) framework [10], we propose locality preserving embedding and locality discriminant embedding, to keep consistency with above *unsupervised* regrouping and *supervised* regrouping, respectively.

Locality Preserving Embedding. As mentioned before, naturally generated data lies on or close to a sub-manifold. Similar to LPP, we first construct two graphs G^a and G^b to represent the MIP $\mathbf{X}_a$ and LIP $\mathbf{X}_b$ of the face images respectively. Accordingly, the weight matrices $\mathbf{W}^a$ and $\mathbf{W}^b$, the Laplacian matrices $\mathbf{L}^a$ and $\mathbf{L}^b$, the diagonal matrices $\mathbf{D}^a$ and $\mathbf{D}^b$ of the MIP $\mathbf{X}_a$ and LIP $\mathbf{X}_b$ are defined, where $\mathbf{L}^a = \mathbf{D}^a - \mathbf{W}^a$ and $\mathbf{L}^b = \mathbf{D}^b - \mathbf{W}^b$.

Suppose the linear projection vector is $\mathbf{p}$, we consider the problem of mapping the MIP $\mathbf{X}_a$ on the graph G^a to a line so that connected points stay as close as possible. In accordance with LPP, we have:

$$\min \mathbf{p}^T \mathbf{X}_a \mathbf{L}^a \mathbf{X}_a^T \mathbf{p} \tag{5}$$

In addition to preserving the graph structure, we also aim at preserving the global variance of the MIP $\mathbf{X}_a$ on the manifold, while suppressing the variance of LIP $\mathbf{X}_b$. Suppose the MIP $\mathbf{X}_a$ and LIP $\mathbf{X}_b$ has a zero mean, then the variance of $\mathbf{X}_a$ and $\mathbf{X}_b$ can be estimated as follows:

$$\sum_i ||\mathbf{p}^T \mathbf{x}_i^a||^2 \mathbf{D}_{ii}^a = \mathbf{p}^T \mathbf{X}_a \mathbf{D}^a \mathbf{X}_a^T \mathbf{p} \tag{6}$$

$$\sum_i ||\mathbf{p}^T \mathbf{x}_i^b||^2 \mathbf{D}_{ii}^b = \mathbf{p}^T \mathbf{X}_b \mathbf{D}^b \mathbf{X}_b^T \mathbf{p} \tag{7}$$

we seek for the projection $\mathbf{p}$ to maximize the variance of $\mathbf{X}_a$ while minimizing the variance of $\mathbf{X}_b$, and at the same time to preserving the manifold structure. Therefore, we obtain the following objective function:

$$\min_{\mathbf{p}} \frac{\mathbf{p}^T \mathbf{X}_a \mathbf{L}^a \mathbf{X}_a^T \mathbf{p} + \rho \mathbf{p}^T \mathbf{X}_b \mathbf{D}^a \mathbf{X}_b^T \mathbf{p}}{(1-\rho)\mathbf{p}^T \mathbf{X}_a \mathbf{D}^a \mathbf{X}_a^T \mathbf{p}} \tag{8}$$

where ρ is a suitable balanced scalar controlling the variances of the LIP. When $\rho = 0$, the Locality preserving embedding reduced to LPP.

Locality Discriminant Embedding. Similar to LSDA [9] and NSPE [13], we first construct two graphs, i.e., within-class graph G_w^a and between-class graph G_b^a to represent the MIP $\mathbf{X}_a$ of all face images. Meanwhile, we construct a graph G^b to represent the LIP $\mathbf{X}_b$ of all face images. Let $l(\mathbf{x}_i^a)$ be the class label of $\mathbf{x}_i^a$ on MIP, For each $\mathbf{x}_i^a$, its k nearest neighborhood set $N(\mathbf{x}_i^a) = \{\mathbf{x}_{i,1}^a, \mathbf{x}_{i,2}^a, \cdots, \mathbf{x}_{i,k}^a\}$ is split into $N_w(\mathbf{x}_i^a)$ and $N_b(\mathbf{x}_i^a)$. $N_w(\mathbf{x}_i^a)$ contains the neighbors sharing the same class label with $\mathbf{x}_i^a$, while $N_b(\mathbf{x}_i^a)$ contains the neighbors having different class labels with $\mathbf{x}_i^a$.

Let $\mathbf{W}_w^a$ and $\mathbf{W}_b^a$ be the weight matrices of G_w^a and G_b^a, respectively. $\mathbf{W}_{w\,ij}^a = 1$ if $\mathbf{x}_i^a \in N_w(\mathbf{x}_j^a)$ or $\mathbf{x}_j^a \in N_w(\mathbf{x}_i^a)$ and $\mathbf{W}_{b\,ij}^a = 1$ if $\mathbf{x}_i^a \in N_b(\mathbf{x}_j^a)$ or $\mathbf{x}_j^a \in N_b(\mathbf{x}_i^a)$. Let $\mathbf{W}^b$ be the weight matrix of G^b. Then, the Laplacian matrices $\mathbf{L}_w^a$, $\mathbf{L}_b^a$ and $\mathbf{L}^b$ are defined, where $\mathbf{L}_w^a = \mathbf{D}_w^a - \mathbf{W}_w^a$, $\mathbf{L}_b^a = \mathbf{D}_b^a - \mathbf{W}_b^a$ and $\mathbf{L}^b = \mathbf{D}^b - \mathbf{W}^b$.

Now, we consider the problem of mapping G_w^a and G_b^a to a line so that connected points of G_w^a stay as close as possible while connected points of G_b^a stay as distant as possible. Thus, for MIP, we have following objective functions:

$$\min \sum_{i,j} (\mathbf{p}^T \mathbf{x}_i^a - \mathbf{p}^T \mathbf{x}_j^a)^2 \mathbf{W}_{w\,ij}^a \iff \max \mathbf{p}^T \mathbf{X}_a \mathbf{W}_w^a \mathbf{X}_a^T \mathbf{p} \tag{9}$$

$$\max \sum_{i,j} (\mathbf{p}^T \mathbf{x}_i^a - \mathbf{p}^T \mathbf{x}_j^a)^2 \mathbf{W}_{b\,ij}^a \iff \max \mathbf{p}^T \mathbf{X}_a \mathbf{L}_b^a \mathbf{X}_a^T \mathbf{p} \tag{10}$$

with constraint $\mathbf{p}^T \mathbf{X}_a \mathbf{D}_w^a \mathbf{X}_a^T \mathbf{p} = 1$.

To suppress the LIP $\mathbf{X}_b$, we seek to minimize its variance on G^b. Finally, the optimization problem reduces to finding:

$$\arg_{\mathbf{p}} \max \mathbf{p}^T \{\mathbf{X}_a(\eta \mathbf{W}_w^a + \beta \mathbf{L}_b^a)\mathbf{X}_a^T - \varepsilon \mathbf{X}_b \mathbf{D}^b \mathbf{X}_b^T\}\mathbf{p}^T$$
$$s.t. \mathbf{p}^T \mathbf{X}_a \mathbf{D}_w^a \mathbf{X}_a^T \mathbf{p} = 1 \tag{11}$$

where η, β, ε are balanced factors with $\eta + \beta + \varepsilon = 1$. When $\varepsilon = 0$, the Locality discriminant embedding reduced to LSDA. It is worthwhile to noting that, in the following experiments the optimal hyperparameters, i.e., α, β, and ε, are chosen via grid search based cross validation.

3   Experiment Results

In this section, several experiments are performed to show the effectiveness of our proposed algorithm. We denote the proposed unsupervised DML_SC as UDML_SC, and supervised DML_SC as SDML_SC. We conduct face recognition experiments on two real-world face databases (Extended YaleB and CMU PIE) to test the classification performance of SDML_SC, and clustering experiment on COIL20 image library to evaluate the clustering ability of UDML_SC. The data preparation and presentation is described in Sect. 3.1.

3.1 Data Preparation and Representation

Two representative facial image databases: Extended YaleB [24] and CMU PIE [25] are adopted to evaluate the performance of the proposed SDML_SC algorithm. Subsequently, we conduct clustering experiment on COIL20 image library to test the clustering ability of UDML_SC. Figure 2 shows some sample images in the E-YaleB, PIE and COIL20 databases respectively.

The Extended YaleB (E-YaleB) dataset [24] consists of 2414 frontal face images of 38 subjects under 9 poses and 64 illumination conditions. They are captured under various lighting conditions and cropped and normalized to pixels. For our experiment, we randomly take 32 images of 20 subjects thus in total 640 images as the training set and the corresponding 622 images (nearly 32 images per person) as the testing set.

The PIE dataset [25] consists of 41,368 images of 68 people, each person under 13 different poses, 43 different illumination conditions, and with 4 different expressions. We selected a subset of 1700 images of 10 subjects that only contains five near frontal poses (C05, C07, C09, C27, C29) and all the images under different illuminations and expressions. So, there are nearly 170 images for each individual. In the experiment, we randomly selected 85 images of each class thus 850 images in total for training and the remaining 850 images for testing.

COIL 20 database [28] contains gray scale images of 20 objects viewed from varying angles and each object has 72 images. Four popular clustering algorithms: K-means, NMF [26], LapSC [20], GNMF [27] and GRNMF_SC [21] were used as comparing algorithms. After dimensionality reduction, we have a lower dimensional representation of each image. The clustering is then performed in this lower dimensional space.

In dictionary learning (DL) process, to learn an over-complete dictionary, Randomface (Eigenface) [14] method is first applied to reduce the dimensionality of face (object) images to 300. In each face database, in order to provide an intuition of MIP and LIP, we conduct a face representation experiment using the SDML_SC method to show the MIP and LIP of a given face image. The MIP and LIP of given face images in the two databases using supervised presentation regrouping criteria are shown in Fig. 3. It is clear to see that, the dictionary learning and feature regrouping process of our proposed DML_SC algorithm can separate the noise or trivial structure from the original image successfully. The learned MIP and LIP have explicit concept explanation.

3.2 Face Recognition Results

In dictionary learning phase, the atom numbers of the over-complete dictionary on E-YaleB and PIE databases are 500 and 680, respectively. To demonstrate how

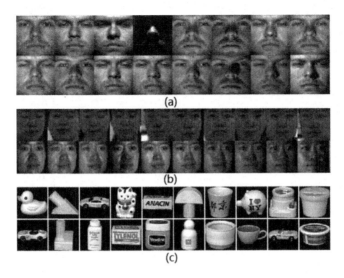

Fig. 2. Randomly selected training samples for one subject of three face databases: (a) E-YaleB; (b) PIE; (c) COIL20.

(a) (b)

Fig. 3. (a) and (b) are the results on the E-YaleB and PIE databases. From left to right are: a face sample; The MIP and the LIP after supervised feature grouping

the face recognition performance can be improved by our method, some popular subspace learning methods, i.e., PCA [1], graph-based LDA [7], supervised LPP (sLPP) [7], supervised NPE (sNPE) [8], LSDA [9], NSPE [13] and recent dictionary learning methods, i.e., original Sparse Representation Classifier (SRC) [17], metaface Learning (MFL) [19] are used for comparison, respectively. In this experiment, to conduct a fair comparison with SRC and MFL, our DML_SC, SRC and MFL all generate Eigenface to perform DR, the reduced dimension of each feature is 300.

Tables 1, 2 summarize the top average recognition rates of competing subspace learning methods and dictionary learning methods on E-YaleB and PIE database, respectively. From Table 1, it is clear to see that the top average recognition rates of SDML_SC are higher than other competing subspace learning methods. Especially on E-YaleB database, SDML_SC even boosts nearly 5% recognition rates compared with the second best method. Though NSPE and LSDA are comparable to SDML_SC on PIE database, they perform much worse than SDML_SC on the E-YaleB database. In addition, as the representative

Table 1. Top Average Recognition Rates (%) and the associated Dimensionality of different subspace learning methods.

Methods	YaleB	PIE
PCA	53.0(130)	83.9(140)
LDA	75.2(180)	91.6(30)
sLPP	76.5(150)	90.3(40)
sNPE	83.2(140)	92.1(20)
LSDA	78.0(200)	91.4(20)
NSPE	76.2(190)	92.1(190)
SDML_SC	88.0(150)	94.9(80)

Table 2. Top Average Recognition Rates (%) and corresponding number of dictionary atoms of different dictionary learning methods.

Methods	YaleB	PIE
SRC	70.1(640)	92.5(850)
MFL	82.9(440)	93.3(590)
SDML_SC	88.0(500)	94.9(680)

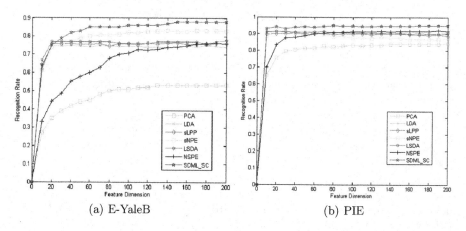

(a) E-YaleB   (b) PIE

Fig. 4. (a) and (b) are the recognition rates of all competing methods versus dimensions on the E-YaleB and PIE databases.

one-graph based subspace learning methods, sNPE and sLPP obtain comparative recognition results. PCA performs the worst among all the involved methods.

As for the Table 2, we observe that SDML_SC still achieves the best recognition results among the involved dictionary learning methods. For the reason that, compared to SRC and MFL methods, SDML_SC not only considers the contributions of different image components, but also introduces spectral graph theory to detect intrinsic topology structure and discriminant information embedded in original data. Additionally, MFL works stably across all the databases and obtains the second best results. SRC gets comparative recognition results with MFL and SDML_SC on PIE database, while is much worse than MFL and SDML_SC on E-YaleB database. It is because SRC simply utilizes the original training samples to construct dictionaries, the severe variations of illumination or pose on E-YaleB database would lead dictionary atoms (training samples) of each category hard to represent the probe image from the same category.

Figure 4 (a) and (b) presents the top average recognition rates of involved subspace learning methods versus the variation of feature dimensions on the E-YaleB and PIE database, respectively. It is clear to see that SDML_SC outperforms other competing methods almost across all the dimensions, which confirms the rationality and effectiveness to explicit exploit the contributions of different image components for image recognition.

3.3   Clustering Experiment on COIL20 Database

In this subsection, we conducted clustering experiment on COIL20 image library. In order to randomize the experiments, we evaluate the clustering performance with different number of clusters ($k = 8, \cdots, 18, 20$). For each given cluster number, 5 tests were conducted on different randomly chosen classes. For UDML_SC, the dimension of images on COIL20 reduced to 300 in DR process. Consequently, to construct over-complete dictionary for each test ($atom\ number \geq 300$), we selected 50 % number of samples when $k = 10, \cdots, 18, 20$ and 60 % number of samples when $k = 8$ to be the atom numbers of the over-complete dictionaries, respectively. To be specific, the atom numbers of over-complete dictionary for each $k = 8, \cdots, 18, 20$ were defined to be 344, 360, 432, 504, 576 and 648. Subsequently the average performance was computed over these 5 tests. About the parameter configuration, there are three key parameters in UDML_SC algorithm: the number of nearest neighbors k_a in Graph G^a, the regularization parameter ρ and the threshold τ to divide the MIP and LIP parts. We empirically set $k_a = 5$, $\rho = 0.3$ and $\tau = 0.8$.

The clustering result is evaluated by comparing the obtained label of each sample with the label provided by the dataset. Two metrics, the accuracy (AC) and the normalized mutual information metric (NMI) are used to measure the clustering performance of involved methods. Please see [21,27] for detailed definitions of these two metrics. Table 3 shows the clustering results on the COIL20, the average clustering performance of different number of clusters are reported in the table.

As shown in Table 3, our UDML_SC always result in the best performances in all the cases. UDML_SC aims to seek a desired subspace representation, which considers the intrinsic manifold structure as well as the importance of different image components. Consequently, UDML_SC possess superior discriminant capacity, by simply using K-means on the low-dimensional sparse representation, UDML_SC achieves very impressive clustering performance. In addition, GRNMF_SC gets better clustering results compared to GNMF and LapSC, while NMF and K-means perform the worst among the involved methods.

Table 3. Clustering performance on COIL20.

	AC(%)						NMI(%)					
k	Kmeans	NMF	LapSC	GNMF	GRNMF_SC	UDML_SC	Kmeans	NMF	LapSC	GNMF	GRNMF_SC	UDML_SC
8	66.4	65.4	72.0	83.0	85.4	90.1	68.0	72.8	76.3	86.7	86.4	89.9
10	64.3	63.5	76.8	82.5	84.3	87.1	68.5	73.5	81.2	85.5	86.2	89.2
12	63.8	62.7	72.2	80.8	83.0	90.9	68.7	72.1	79.5	84.7	87.3	92.9
14	61.2	61.5	72.7	78.9	82.4	89.8	69.2	71.8	80.3	83.6	87.5	93.4
16	62.8	62.2	71.8	79.0	81.9	83.5	70.0	70.3	80.5	83.2	86.3	89.7
18	62.6	60.5	69.4	77.6	80.4	82.3	71.2	71.1	79.4	82.5	86.1	91.3
20	61.5	60.4	71.1	77.2	77.9	88.9	73.9	70.3	79.8	85.4	87.2	93.4
Avg	63.2	62.3	72.3	79.8	82.2	87.5	69.9	71.7	79.6	84.5	86.7	91.4

4   Conclusion

In this paper, we have proposed a subspace learning method named Discriminant Manifold Learning via Sparse Coding (DML_SC). DML_SC aims to decompose the original image into More Important Part (MIP) and Less Important Part (LIP), then learn a desired intrinsic subspace where the energy and topology structure of MIP $\mathbf{X}_a$ are both preserved, while the energy of LIP $\mathbf{X}_b$ is suppressed simultaneously. As a result, DML_SC can both explicitly exploits the contributions of image components as well as the manifold structure information. The experiments corroborate that DML_SC delivers consistently promising results compared with recent popular subspace learning and sparse representation methods, with respect to image recognition and clustering tasks.

References

1. Turk, M., Pentland, A. P.: Face recognition using eigenfaces. In: IEEE International Conference on Computer Vision and Pattern Recognition (CVPR), pp. 586–591 (1991)
2. Nikitidis, S., et al.: Maximum margin projection subspace learning for visual data analysis. IEEE Trans. Image Process. (TIP) **23**(10), 4413–4425 (2014)
3. Jin, W., Liu, R., et al.: Robust visual tracking using latent subspace projection pursuit. In: IEEE International Conference on Multimedia and Expo (ICME) (2014)
4. Jiang, X.: Linear subspace learning-based dimensionality reduction. IEEE Signal Process. Mag. **28**(2), 16–26 (2011)
5. Huang, Z., Wang, R., Shan, S., Chen X.: Projection metric learning on grassmann manifold with application to video based face recognition. In: International Conference on Computer Vision and Pattern Recognition (CVPR), pp. 140–149 (2015)
6. Roweis, S.T., Saul, L.K.: Nonlinear dimensionality reduction by locally linear embedding. Science **290**(5500), 2323–2326 (2000)
7. He, X., Yan, S., et al.: Face recognition using laplacianfaces. IEEE Trans. Pattern Anal. Mach. Intell. (TPAMI) **27**, 328–340 (2005)
8. Zeng, X., Luo, S.-W.: A supervised subspace learning algorithm: supervised neighborhood preserving embedding. In: Alhajj, R., Gao, H., Li, X., Li, J., Zaïane, O.R. (eds.) ADMA 2007. LNCS (LNAI), vol. 4632, pp. 81–88. Springer, Heidelberg (2007)

9. Cai, D., He, X., Zhou, K., Han, J., Bao, H.: Locality sensitive discriminant analysis. In: IJCAI, pp. 708–713 (2007)

10. Yan, S., Xu, D., Zhang, B., Zhang, H.J., Yang, Q., Lin, S.: Graph embedding and extensions: a general framework for dimensionality reduction. IEEE Trans. Pattern Anal. Mach. Intell. (TPAMI) **29**(1), 40–51 (2007)

11. Wang, R., Chen, X.: Manifold discriminant analysis. In: International Conference on Computer Vision and Pattern Recognition (CVPR), pp. 429–436 (2009)

12. Wang, R., Shan, S., Chen, X., Chen, J., Gao, W.: Maximal linear embedding for dimensionality reduction. IEEE Trans. Pattern Anal. Mach. Intell. (TPAMI) **33**(9), 1776–1792 (2011)

13. Wang, B.H., Lin, C., et al.: Neighbourhood sensitive preserving embedding for pattern classification. IET Image Proc. **8**(8), 489–497 (2014)

14. Olshausen, B.A., Field, D.J.: Sparse coding with an over-complete basis set: a strategy employed by V1? Vision. Res. **37**(23), 3311–3325 (1997)

15. Kim, S.J., Koh, K., Lustig, M., Boyd, S., Gorinevsky, D.: An interior-point method for large-scale l1-regularized least squares. IEEE J. Sel. Top. Signal Process. **1**(4), 606–617 (2007)

16. Aharon, M., Elad, M., Bruckstein, A.: The K-SVD: an algorithm for designing of overcomplete dictionaries for sparse representation. IEEE Trans. Image Process. (TIP) **54**(11), 4311–4322 (2006)

17. Wright, J., Yang, A.Y., Ganesh, A., Sastry, S.S., Ma, Y.: Robust face recognition via sparse representation. IEEE Trans. Pattern Anal. Mach. Intell. (TPAMI) **31**(2), 210–227 (2009)

18. Zheng, M., Bu, J., Chen, C., Wang, C., Zhang, L., Qiu, G., Cai, D.: Graph regularized sparse coding for image representation. IEEE Trans. Image Process. (TIP) **20**(5), 1327–1336 (2011)

19. Yang, M., Zhang, L., et al.: Metaface learning for sparse representation. In: International Conference on Image Processing (ICIP), pp. 1601–1604 (2010)

20. Gao, S., Tsang, I. W. H., Chia, L. T., Zhao, P.: Local features are not lonely-laplacian sparse coding for image classification. In: Computer Vision and Pattern Recognition (CVPR), pp. 3555–3561 (2010)

21. Wang, B., Pang, M., Lin, C., Fan, X.: Graph regularized non-negative matrix factorization with sparse coding. In: IEEE China Summit & International Conference On Signal and Information Processing (ChinaSIP), pp. 476–480 (2013)

22. Zhang, L., Zhu, P., et al.: A linear subspace learning approach via sparse coding. In: IEEE International Conference on Computer Vision (ICCV), pp. 755–761 (2011)

23. Li, H., Jiang, T., Zhang, K.: Efficient and robust feature extraction by maximum margin criterion. IEEE Trans. Neural Netw. **17**(1), 157–165 (2006)

24. Georghiades, A.S., et al.: From few to many: illumination cone models for face recognition under variable lighting and pose. IEEE Trans. Pattern Anal. Mach. Intell. (TPAMI) **23**(6), 643–660 (2001)

25. Sim, T., Baker, S., Bsat, M.: The CMU Pose Illumination, and Expression (PIE) database. In: FG, pp. 46–51 (2002)

26. Lee, D.D., Seung, H.S.: Learning the parts of objects by non-negative matrix factorization. Nature **401**(6755), 788–791 (1999)

27. Cai, D., He, X., Han, J., Huang, T.S.: Graph regularized non-negative matrix factorization for data representation. IEEE Trans. Pattern Anal. Mach. Intell. (TPAMI) **33**(8), 1548–1560 (2011)

28. Nene, S. A., Nayar, S. K., Murase, H.: Columbia object image library (coil-20). Technical Report CUCS-005-96 (1996)

A Very Deep Sequences Learning Approach for Human Action Recognition

Zhihui Lin[1(✉)] and Chun Yuan[1,2]

[1] Graduate School at Shenzhen, Tsinghua University, Shenzhen 518055, China
lin-zhl4@mails.tsinghua.edu.cn,
yuanc@sz.tsinghua.edu.cn
[2] Departmnt of Computer Science and Technology,
Tsinghua University, Beijing 100084, China

Abstract. Human action recognition is a popular study in computer vision. The most difficult challenge is capturing the movement features of image sequences or videos. In recent years, deep convolutional networks have achieved great success in many image classification and recognition tasks. But in videos interpretation tasks, the deep-learning has not done well. There were [18, 19] earlier models which were built on convolutional networks for human action recognition tasks. We propose an approach based on CNNs and RNN-like models which have abilities to extract spatial and temporal features both, a CNN model can get static scores, a LSTM or GRU layer which gets dynamic class scores of human action. In another side, compared to a two-stream ConvNet [18, 24], we do not need an optical-flow CNN stream that saves us considerable time, RNN-like models just need a few hours to get convergence. And, we have achieved a quite remarkable performance.

1 Introduction

Recognition and classification of images are traditional topics in the field of computer vision. In recent years, more and more researchers have concentrated on these tasks of videos. Human action recognition [1, 2] is part of these tasks that get more interest in the research community. This is the fact that it's widely application in the real world, as human computer interaction, public safety, and video retrieval and so on. Another reason is the rapid development of deep-learning, in the image field, models based on the deep-learning have got unprecedented success [3–5], also in the natural language processing field. Researchers believe that deep-learning can also achieve significant success in video analysis or other tasks on videos. Compared to traditional state-of-the-art methods based on hand-craft features [6, 7], models based on the deep-learning extract meaningful hierarchical features automatically. However, there are a lot of difficult problems in this challenging task. Realistic videos from the internet or movies always be more complicated than many previous benchmark datasets [8, 9], they have numerous components of movement, like objects in movement, background movement and camera movement. These factors cause a lot of difficulties to learn meaningful visual representation. How to extract efficient features from videos is a core work of numerous researchers.

Q. Tian et al. (Eds.): MMM 2016, Part II, LNCS 9517, pp. 256–267, 2016.
DOI: 10.1007/978-3-319-27674-8_23

Most approaches are mainly founded on two types of features. One of them is the hand-craft local feature, include STIP (space time interest points) [10], trajectories [11, 12], and exemplar-based and state-based features. These features always need many complicated descriptors of HOG, HOF. They have got excellent performance on various challenging datasets as HMDB51 [13], UCF101 [14] datasets. Nevertheless, these approaches based on a hand-craft structure cannot get dense dynamic features and static image features both.

Approaches based on deep-learning, especially the convolutional structure, naturally accept the raw image data, which can extract dense hierarchical features automatically. It just needs several initialization parameters. Krizhevsky and Hinton's work [3] has proven their convolutional structure has strong ability in image features extraction. In recent works [4, 5, 15], deeper convolutional neural networks have got higher performances. Although there is not a theoretical proof of these results. Various tasks which have used VGGNet [5] and SPPNet [15] have achieved state-of-the-art performance. With the same framework, approaches based on VGGNet always can get higher performance than those which have used AlexNet [3].

Typical convolutional neural networks can receive raw images. [16] has extended the traditional CNN to 3D-CNN that can process sequences of images, so it can also process video data, and it gets competitive performance on various simple tasks [8]. In recent research works, [17] has extracted dense features use both a CNN model and improved trajectories to obtain multi-features. Therefore, it has spatial features which were extracted by a CNN model and temporal features that were represented by improved trajectories. It had the most competitive results on the UCF101 dataset by 91.5 %. And there are two competitive models only based on the deep-learning methods, Two-stream ConvNet [18] and LRCN [19], they can also get much higher performance than similar methods in a recent period of time.

We propose a similar architecture as LRCNs, but we did not train our recognition system in an end-to-end manner. It has abandoned some useful spatial features of raw images. We combine the static visual features and the sequences' temporal features by stacking them that have been processed by a LSTM-like [20, 21] structure afterwards. We have done experiments on the UCF101 dataset that have been adopted by most researchers and got the competitive performance.

The remainder of this article is organized by follows. In Sect. 2, we will introduce related works. In Sect. 3, there is a thorough introduction of our approach and related knowledge. We will introduce experiment details of our work in Sects. 4 and 5 is the evaluation, finally, in Sect. 6, there is a conclusion of our work and the future's work plan.

2   Related Work

2.1   Convolutional Neural Networks

Typical CNNs were most used on image based tasks [3, 22], and they have been developed very fast with the improvement of GPUs' computational power. Compared to CPUs, we can save tens of times by using GPUs, so we can design more complicate and deeper CNNs to extract dense hierarchical features of images. In 2012, Hinton [1] got a champion of ImageNet [23] challenge using a CNN structure with five convolution layers and three fully connected layers. Two years later, several models based on CNNs have got the top3 results in ImageNet 2014. GoogleNet has got 6.67 % top-5 error in 1000 classes. VGGNet has got a result of 7.32 % top-5 error. These results are close to human level top-5 5.1 % error rate on imagenet dataset. We can trust that these very deep convolutional architectures are capable of extracting dense hierarchical features in the vision field.

In the video processing field, [16] has designed a 3D-CNN that can receive 3-dimensions data based on typical CNNs. A video can be split into several sequences of images, as the input of a 3D-CNN. Features extracted through the 3D-CNN have been stacked together as a n-dimensions vector. Finally, author has utilized a SVM classifier to categorize the human activity.

2.2 Long Short-Term Memory Block

LSTM is a kind of RNNs, it has proved more effective than the traditional RNN. It is designed to process sequences, it has been usually used on speech, text recognition or generation tasks [26, 27].

2.3 Recent Researches

In recent years, there is quite a lot [17–19, 24] works based on winner CNNs models of ImageNet challenge, especially classical AlexNet and VGGNet. [17] has used a AlexNet-like structure to extract spatial feature maps, furthermore, extracting trajectories, combine these two type of feature maps by spatiotemporal normalization and channel normalization, finally, through a trajectory-constrained pooling step, those two type feature maps have been transformed as TDD features (trajectory-pooled deep convolutional descriptors). And in experiments, Wang et al. [17] has used a PCA dimension reducer to reduce the dimension of TDDs and GMM training then Fisher vector for encoding, and a linear SVM as classifier. [18, 24] covert videos as images and obtain optical flows, row images for training spatial CNN stream, stacked flows data have been trained to extract dynamic features. Also a linear SVM has been used to fuse softmax scores of two stream ConvNets. [19] propose an end-to-end manner by connecting CNNs and a LSTM layer.

Above works all use pre-trained models as initialization of their methods. [17] has achieved the highest performance. [24] has achieved a quite close result using two-steam ConvNet by instead the AlexNet with VGGNet. But VGGNet is a very deep network which needs more time to train. They have used four Titan X GPUs to train the model. We propose a single-stream ConvNet using VGGNet, abandon the flow stream used by [18, 19, 24] that needs more time to train. Instead, we have used a GRU [25] layer with a simpler structure than LSTM to capture dynamic features. Unlike [19]'s end-to-end manner, we train the model separately, therefore we can fuse the two scores of a VGGNet-like model and a GRU layer to achieve outperform performance.

3   Essential Knowledge

3.1   Convolutional Neural Networks

AlexNet is the winner of imagenet 2012 challenge. It consists of five convolution layers and three fully connect layers. The kernel sizes are 11×11, 5×5, 3×3, 3×3, 3×3, the full6 layer and the full7 layer's dimension is 4096, the full8 layer output the vector with the same dimension as the number of categories. It has around 50 million parameters that are required to adjust.

   VGG-Net is a type of development on AlexNet. VGG-Net has smaller kernel size, it is 3×3 in all convolution layers and smaller convolutional stride (2 vs 1), smaller pooling window (3×3 vs 2×2). Hence, it has more layers, one 16 layers and one 19 layers with 13 and 16 convolutional layers. It has made use of architecture that has been called Network in Network, which can transform a shallower structure to a deeper one with above techniques. VGG-Net has benefited from this way to obtain a big performance improvement compared to an original AlexNet model. However, VGGNet has more parameters. It is about three times than AlexNet which means it has higher time complexity. We use the same structure as [24] based on VGGNet of the spatial features extracting stream, but with different learning strategy.

3.2   Sequence Features Extraction

LSTM (Long Short-Term Memory). Traditional RNNs can accept a sequence with an arbitrary length, therefore, RNNs have achieved higher performance than state-of-the-art models on many tasks such as speech recognition [26] and natural language processing [27]. But RNNs keep the risk of the vanishing and exploding gradients problem. LSTM based models have resolved the problem by introducing memory units. This architecture can avoid the gradient vanishing and keep the long-term memory, so it is a kind of very powerful neural network.

1. The cell block like a notebook to remember things with parameter state;
2. The input gate and the output gate are compute units;
3. The forget gate choses to forget the context information through two time-steps.

It has additional gates and its' architecture is below:

$$o_i^t = \sigma\left(W_{ix}x^t + W_{ih}h^{t-1} + W_{ic}s^{t-1}\right) \tag{1}$$

$$o_f^t = \sigma\left(W_{fx}x^t + W_{fh}h^{t-1} + W_{fc}s^{t-1}\right) \tag{2}$$

$$s_c^t = o_f^t s_c^{t-1} + o_i^t g\left(W_{cx}x^t + W_{ch}h^{t-1}\right) \tag{3}$$

$$o_o^t = \sigma\left(W_{ox}x^t + W_{oh}h^{t-1} + W_{oc}s^{t-1}\right) \tag{4}$$

$$o_c^t = o_o^t \phi\left(s_c^t\right) \tag{5}$$

c represents the cell, *h* represents hidden state and parameters through times. *i* denotes the index of the input gate. *W* denotes the weights matrix, and subscript means different gates. *t* denotes the time step, *x* means the input vector, *s* is the cell input activation vector, σ is the logistic sigmoid function and *o* denotes the output vector. *f* is the forget gate, other symbols have the same meanings as the above description. *g* denotes the tanh function. This cell is the black point below the cell circle in the above Fig. 1. And *s* means state. The formulation of the output gate is similar as the input gate and forget gate. o^t_c is the final output of a LSTM block, ϕ is the activation function which can be tanh or a logistic sigmoid function. Then it sets $h^t = o^t_c$ as the hidden vector at time *t*.

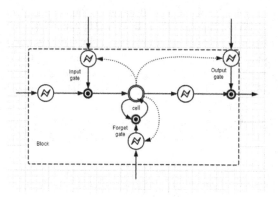

Fig. 1. A LSTM block (was described by Alex [21])

GRU (Gated Recurrent Unit). GRU [25] has simpler structure and fewer parameters than LSTM. A GRU block does not have the forget gate and cell block, but has added the reset and the update gates. Gates' computation follows bellow formulations:

$$h^t_j = \left(1 - z^t_j\right)h^{t-1}_j + z^t_j \tilde{h}^t_j \tag{6}$$

$$z^t_j = \sigma\left(W_{zx}x^t + W_{zh}h^{t-1}\right)^j \tag{7}$$

$$\tilde{h}^t_j = tanh\left(W_{\tilde{h}x}x^t + W_{\tilde{h}h}\left(r^t \odot h^{t-1}\right)\right)^j \tag{8}$$

$$r^t_j = \sigma\left(W_{rx}x^t + W_{rh}h^{t-1}\right)^j \tag{9}$$

h^t_j is the activation of the GRU at time *t*, $\tilde{h}^t_j$ denotes the candidate activation, z^t_j is an update gate, r^t_j is the reset gate, in our Fig. 2, it is computed in the input gate similarly to the update gate, *W*, *x* and σ have the same meanings as the LSTM. The $\odot$ is an element-wise multiplication. In our experiments, use GRUs can avoid the over fitting that LSTMs cannot do well.

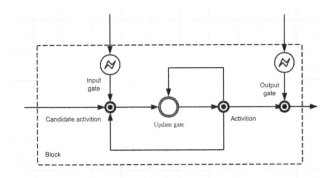

Fig. 2. A GRU block, a GRU block has a simpler structure, lesser parameters than a LSTM block.

4 Multi-Feature Framework

We use a VGG-Net pre-trained with the image classification dataset of the ImageNet challenge [28] as the initialization of the model based on CNNs. The input to the neural network are raw images by splitting original videos. Final a fully connected layer output softmax scores that can classify the human action. After fine-tuning a VGGNet model, we have extracted features of the full6 and the full8 layer, stacked them together in sequences, then we executive a sequences processing method, LSTMs or GRUs, also capture softmax scores. Last step, we fuse two scores with a weighted average (Fig. 3).

4.1 Experiments Details

Dataset Detail. We have utilized the UCF101 [14] dataset to conduct experiments. The UCF101 dataset contains 13, 320 video clips with 101 classes. We have followed the evaluation scheme of the THUMOS13 challenge [29], done experiments on three training/testing splits independent of each other.

Dataset Preprocessing. The caffe-toolbox [30] just accepts image data now, so we have cut videos into images for each frame. Then we have used these about 250 million images to fine-tune a 16-layers VGGNet model. After that we have extracted features of the full6 layer and the full8 layer through the fine-tuned model.

Multi-scale Data. *Multi-scale Approach on The 2D-image Data.* As the image classification's setting, we have resize images to different scales with multi-spatial-scale. In the experiments, we have made the same settings with [18, 24], there are four different scale factors which were 1, 0.875, 0.75 and 0.66, the original image size is 256×340, we randomly chose scale factors to sample the width and height of input images. Yet, we implement a cropping method to resize the image to 224×224 as the input of a

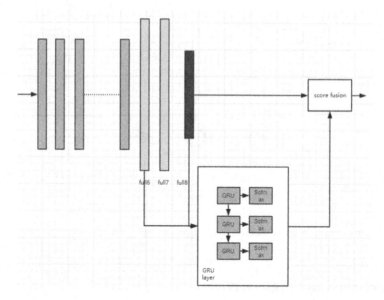

Fig. 3. Our pipeline for the human recognition task. First, we extract spatial features through training a single frame network, and then extract features of the full6 or the full8 layer, then features are received by a LSTM or GRU layer, last we combine the scores from two ways by weights averaging.

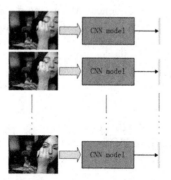

Fig. 4. Sequences extraction. We have extracted static features which were painted yellow by a trained VGGNet-like model, if these are extracted from the full6 layer, the dimension of vectors is 4096, and 101 for the full8 layer(Color figure online).

VGGNet model. Each training image has been transformed to 10 images, include five crops of four corners and the center region, for each cropped image, we have made a mirror. Multi-scale data can vary the inputs and reduce the risk of over fitting effectively.

Multi-scale Approach on The Time Dimension. After capturing the full6 and the full8 features, we have got the sequences of features by stacking these features of video sequences. These sequences can be trained by a sequences processor, as we used a GRU layer witch can extract high level temporal features, these features can be classified by a single softmax layer. Also, for robustness, we have set different time strides to sample sequences to different time scales, we call it multi-temporal-scale, and we set it to 1, 2, and 4 as time strides, it can be a sampling of the original video, we have speed up the video by 2 and 4 times to various the training data. There is an example in Fig. 5.

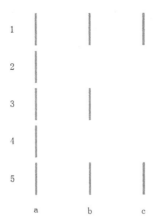

Fig. 5. This instance has explained the multi-time-scale approach, 1-5 means the index of features in Fig. 4. These five features make up a feature sequence with the length 5. a is the original sequence, then we sample it by strides 2 and 4 to get sequences b and c.

Training Settings. A VGG-like model has been trained with the pre-trained models [5] as the initialization, we have changed the last layer to 101 dimensions as the number of categories on UCF101 human action dataset. We have used the popular caffe-toolbox [30], and set the learning rates of the full8 and the full7 layer 10 times than other layers to keep the parameters with discriminations trained on ImageNet dataset, we set 0.9 and 0.9 dropout ratios for the full6 layer and the full7 layer. The mini-batch size is 64. For every 15000 iterations, we have reduced the learning rate to 1/10 of the current one, after 50000 iterations, we stop the training process.

A LSTM or GRU layer has been applied to feature sequences processing and recognition. In order to prevent over fitting, we finally use a single GRU layer with 128 blocks, the length of each one input sequence is 50. The dropout ratio is 0.85. After that, a softmax layer can help us to get a classify score of human action. Also, we have processed two type sequence features, the full6 layer features with a dimension of 4096, another and the full8 layer outputs with a dimension of 101 the same as the category number of UCF101 dataset. And we finally use a PCA dimension reducer to process the full6 layer features to improve the performance and reduce the cost of memory and storage, we have reduced the dimension from 4096 to 128. The implementation of LSTM or GRU models are based on Theano [31].

From above paragraphs, we have used a 16-layer VGGNet model to train every frame of videos. We have got a basic classifier of human action. And then we have extracted static features through this model, finally, we have got sequence features by stacking these static features by an order of the original video. We have processed these sequences by a GRU layer which is commonly used for various sequences classification tasks in natural language processing, speech recognition and synthesis fields. The invariable method can get a good performance which benefits from a deeper structure, a GRU layer can learn movement features or inter-frame information to improve the performance of human action recognition. Finally, we use the weights 1 and 2 of above two parts to average the scores to get a final recognition result, the combination tends to be able to get $5 \sim 6$ % accuracy improvement [18, 24].

5   Evaluation

5.1   Single Frame Models

Our one-way training approach can get a really high performance that benefits from a very deep structure and the big data information. Compared to setting 1 [24], we have fine-tuned both the full7 and the full8 layer instead of their single full8 layer, fully

Table 1. Single frame accuracy of UCF101 dataset split1. (1) single frame accuracy of Two-stream ConvNets;(2) VGG-like models accuracy on single frame RGB image stream, which has fine-tuned the full8 layer with the same learning rate as other layers; (3) The VGG-like model we used, we have fine-tuned the full7 and the full8 layer with the 10 times learning rate than others.

Method	Single frame
(1) Single Frame CNN Model (Images) [18]	72.6 %
(2) VGGNet-single frame setting 1[24]	79.8 %
(3) VGGNet-single frame setting 2	80.4 %

connected layers contain the most parameters of a CNN based model, so our setting can learn more unique features of the video dataset, it can help to improve the performance of the spatial stream (Table 1).

5.2   Sequence-Related Models

PCA Approach. Using original 4096 dimensions features of the full6 layer, we can get a result close to single frame approach 80.4 % on UCF101 dataset split1, the PCA dimension reduction can help us to achieve 82.3 % accuracy compared to LRCN's 71.12 % [19]. If we use features extracted from the full8 layer, we just get around 76 %, the full8 layer output the softmax scores of categories, it would confuse the GRU layer with a limitation of single frame stream accuracy.

RNN-like Block. Compared to LSTMs, GRUs have fewer parameters, can avoid the over fitting. A LSTM layer can achieve an accuracy around 80.1 %, with the same setting, a GRU layer has an improvement about 2.2 %.

Multi-Time-Scale. We have got an accuracy of 81.2 % when we just used original settings, if we use multi-time-scale data as description in Sect. 4.1, the accuracy has been improved to 82.3 % on UCF101 dataset split1. The multi-time-scale approach can extend the training dataset and reduce the risk of over fitting.

Table 2. The accuracy results on UCF101 dataset all three splits. Almost methods utilise the weight average. One kind of weight is (1/2,1/2). Another setting is (1/3, 2/3)

Method	Fusion method	
LRCN-fc6 [19]	(1/2 1/2) 80.62 %	(1/3 2/3) 82.66 %
Very deep two-stream ConvNets [24]	——	(1/3 2/3) 91.4 %
Two-stream ConvNets [18]	(averaging)86.9 %	(SVM) 88 %
Our method	(1/2 1/2) 85.8 %	(1/3 2/3) 87.2 %

5.3 Fusion Models

Although we did not get best results on UCF101 dataset, we get competitive performance with less computational complexity, optical-flow has been proved that it can help to improve the accuracy of the video classification task, without it, we can still get close results with a sequence model which just need several hours to be trained (Table 2).

LRCN [19] is an end-to-end approach of sequences processing, it is more convenient than our method, but we have at least two advantages:

1. We have used a deeper convolutional neural network model, VGGNet-like 16 layers model vs AlexNet-like model;
2. A LRCN model does not use static image features to get a classifier, we not only use dynamic features that were extracted by a GRU layer, but also use static features that were extracted by a VGGNet model to build a full classifier.

5.4 Discussion of the Time Cost

If we use the same method like [18], we have to train another model of the optical-flow data, it will cost around 3.5 s for every iteration on a single Titan X GPU and the full training process will need about 5 days. But our GRU based method just needs about $5 \sim 7$ h on 128 dimension features, and less than a day for full 4096 dimension features.

6 Conclusions

In this work, we have proposed a combined method of CNNs and sequences-related models like LSTMs or GRUs. It can process sequences of image, so we have used it on human action recognition with videos, and achieve competitive performance without optical-flow information, furthermore, it needs fewer time cost. In the future works, we will concentrate on finding more discriminating dynamic features and at the same time, we will use our methods on more tasks, like object detection in videos.

Acknowledgements. This work is supported by the National High Technology Development Plan (863), under Grant No. 2011AA01A205, by the National Significant Science and Technology Projects of China, under Grant No. 2013ZX01039001-002-003; by the NSFC project under Grant Nos. U1433112, and 61170253.

References

1. Aggarwal, J.K., Cai, Q.: Human motion analysis: a review. Comput. Vis. Image Underst. (CVIU) **73**(3), 428–440 (1999)
2. Allen, J.F.: Maintaining knowledge about temporal intervals. Commun. ACM **26**(11), 832–843 (1983)
3. Krizhevsky, A., Sutskever, I., Hinton, G.E.: ImageNet classification with deep convolutional neural networks. In: NIPS, pp. 1106–1114 (2012)
4. Szegedy, C., Liu, W., Jia, Y., Sermanet, P., Reed, S., Anguelov, D., Erhan, D., Vanhoucke, V., Rabinovich, A.: Going deeper with convolutions. CoRR, abs/1409.4842 (2014)
5. Simonyan, K., Zisserman, A.: Very deep convolutional networks for large-scale image recognition. CoRR, abs/1409.1556 (2014)
6. Wang, H., Ullah, M.M., Kläser, A., Laptev, I., Schmid, C.: Evaluation of local spatio-temporal features for action recognition. In: BMVC (2009)
7. Kläser, A., Marszalek, M., Schmid, C.: A spatio-temporal descriptor based on 3D-gradients. In: BMVC (2008)
8. Schuldt, C., Laptev, I., Caputo, B.: Recognizing human actions: a local SVM approach. In: International Conference on Pattern Recognition (ICPR), vol. 3, pp. 32–36 (2004)
9. Blank, M., Gorelick, L., Shechtman, E., Irani, M., Basri, R.: Actions as space-time shapes. In: IEEE International Conference on Computer Vision (ICCV), pp. 1395–1402 (2005)
10. Laptev, I.: On space-time interest points. IJCV **64**(2–3), 107–123 (2005)
11. Wang, H., Kläser, A., Schmid, C., Liu, C.-L.: Dense trajectories and motion boundary descriptors for action recognition. IJCV **103**(1), 60–79 (2005)
12. Wang, H., Schmid, C.: Action recognition with improved trajectories. In: ICCV (2013)
13. Kuehne, H., Jhuang, H., Garrote, E., Poggio, T., Serre, T.: HMDB: a large video database for human motion recognition. In: ICCV (2011)
14. Soomro, K., Zamir, A.R., Shah, M.: UCF101: a dataset of 101 human actions classes from videos in the wild. CoRR, abs/1212.0402 (2012)
15. He, K., Zhang, X., Ren, S., Sun, J.: Spatial pyramid pooling in deep convolutional networks for visual recognition. In: Fleet, D., Pajdla, T., Schiele, B., Tuytelaars, T. (eds.) ECCV 2014, Part III. LNCS, vol. 8691, pp. 346–361. Springer, Heidelberg (2014)
16. Ji, S., Xu, W., Yang, M., Yu, K.: 3D convolutional neural networks for human action recognition. TPAMI **35**(1), 221–231 (2013)

17. Wang, L., Qiao, Y., Tang, X.: Action recognition with trajectory-pooled deep-convolutional descriptors. In: Proceedings of the IEEE Conference on Computer Vision and Pattern Recognition, pp. 4305–4314 (2015)
18. Simonyan, K., Zisserman, A.: Two-stream convolutional networks for action recognition in videos. In: NIPS (2014)
19. Donahue, J., Anne Hendricks, L., Guadarrama, S., et al.: Long-term recurrent convolutional networks for visual recognition and description. In: Proceedings of the IEEE Conference on Computer Vision and Pattern Recognition, pp. 2625–2634 (2015)
20. Hochreiter, S., Schmidhuber, J.: Long short-term memory. Neural Comput. **9**(8), 1735–1780 (1997)
21. Graves, A.: Supervised sequence labelling. In: Graves, A. (ed.) Supervised Sequence Labelling with Recurrent Neural Networks. SCI, vol. 385, pp. 5–13. Springer, Heidelberg (2012)
22. Tan, C.C., Jiang, Y.-G., Ngo, C.-W.: Towards textually describing complex video contents with audio-visual concept classifiers. In: MM (2011)
23. Deng, J., Dong, W., Socher, R., Li, L.-J., Li, K., Fei-Fei, L.: ImageNet: a large-scale hierarchical image database. In: CVPR (2009)
24. Wang, L., Xiong, Y., Wang, Z., et al.: Towards good practices for very deep two-stream ConvNets. arXiv preprint (2015). arXiv:1507.02159
25. Chung, J., Gulcehre, C., Cho, K.H., et al.: Empirical evaluation of gated recurrent neural networks on sequence modeling (2014). arXiv preprint arXiv:1412.3555
26. Vinyals, O., Ravuri, S.V., Povey, D.: Revisiting recurrent neural networks for robust ASR. In: ICASSP (2012)
27. Sutskever, I., Martens, J., Hinton, G.E.: Generating text with recurrent neural networks. In: ICML (2011)
28. Russakovsky, O., Deng, J., Su, H., Krause, J., Satheesh, S., Ma, S., Huang, Z., Karpathy, A., Khosla, A., Bernstein, M.S., Berg, A.C., Li, F.-F.: ImageNet large scale visual recognition challenge. CoRR abs/1409.0575 (2014). http://arxiv.org/abs/1409.0575
29. Jiang, Y.-G., Liu, J., Roshan Zamir, A., Laptev, I., Piccardi, M., Shah, M., Sukthankar, R.: THUMOS challenge: action recognition with a large number of classes. In: ICCV 2013-Action-Workshop (2013)
30. Jia, Y., Shelhamer, E., Donahue, J., Karayev, S., Long, J., Girshick, R.B., Guadarrama, S., Darrell, T.: Caffe: convolutional architecture for fast feature embedding. CoRR, abs/1408.5093
31. Bastien, F., Lamblin, P., Pascanu, R., Bergstra, J., Goodfellow, I., Bergeron, A., Bouchard, N., Warde-Farley, D., Bengio, Y.: Theano: new features and speed improvements. In: Proceedings NIPS 2012 Deep Learning Workshop (2012)

Attribute Discovery for Person Re-Identification

Takayuki Umeda[✉], Yongqing Sun, Go Irie, Kyoko Sudo,
and Tetsuya Kinebuchi

NTT, 1-1 Hikari-no-oka, Yokosuka, Kanagawa 239-0847, Japan
{umeda.takayuki,yongqing.sun,irie.go,
sudo.kyoko,kinebuchi.t}@lab.ntt.co.jp

Abstract. An incremental attribute discovery method for person re-identification is proposed in this paper. Recent studies have shown the effectiveness of the attribute-based approach. Unfortunately, the approach has difficulty in discriminating people who are similar in terms of the pre-defined semantic attributes. To solve this problem, we automatically discover and learn new attributes that permit successful discrimination through a pair-wise learning process. We evaluate our method on two benchmark datasets and demonstrate that it significantly improves the performance of the person re-identification task.

Keywords: Person re-identification · Attribute discovery · Incremental leaning

1 Introduction

Person re-identification addresses the problem of automatically linking individuals across camera images at different times. Although known approaches show promising results in matching a target (a probe) person given a gallery of people based on the similarity of low-level features [1,2], they are too sensitive to changes in appearance (e.g. lighting conditions, occlusion, background complexity). Recent studies have proposed the attribute-based approach [3–6] which uses visual attributes (e.g. gender, hair color and cloth patterns) as features for calculating similarity instead of low-level features. Attributes are learned for different people and situations, so it is robust to changes in appearance.

Most traditional re-identification methods assume only static settings, i.e., the number of people to be identified and the attributes are fixed. However in real world re-identification services, new persons come continuously. In such situation, pre-defined attributes may identify the people in the beginning but not in the end, because the number of similar people increases with the increasing number of people to identify. To overcome this limitation of the attribute-based approach, we propose an 'incremental' attribute discovery method that suits continuous person re-identification.

Our approach is motivated by the fact that, observing the relationship between people and their associated attribute sets [3] (e.g., *Person199* and *Person221* have *Patterned Cloth* and *Has Backpack*; *Person583* and *Person593*

© Springer International Publishing Switzerland 2016
Q. Tian et al. (Eds.): MMM 2016, Part II, LNCS 9517, pp. 268–276, 2016.
DOI: 10.1007/978-3-319-27674-8_24

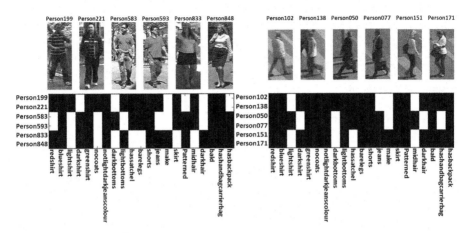

Fig. 1. Person-attribute matrix generated from the part of the annotation results in VIPeR [2] (left) and PRID [7] (right): the rows are the person ids and the columns are the attributes. The lighter the cell, the stronger the association between attribute and person.

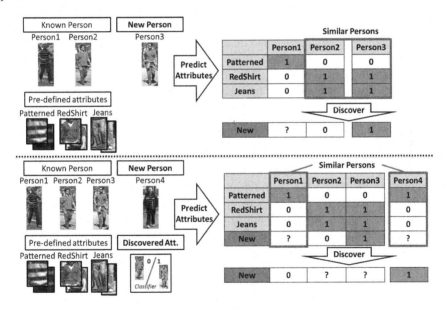

Fig. 2. Illustration of new attribute discovery in ideal case.

have *Red Shirt* and *Jeans*) as in left of Fig. 1, we can see that some people are visually dissimilar, but their attributes are exactly same. Inspired by the above observation, given known people and pre-defined attributes, our approach uses the existing attributes to find similar people, whose attributes are highly correlated with a new person to be identified, and learns a small number of new attributes by pair-wise max-margin training to separate these people (Fig. 2).

Note that our method does not discard pre-defined attributes but adds new attributes incrementally; it permits the concatenation of any attributes that can be discovered by other methods. We evaluate our method on PRID and VIPeR datasets and show that it significantly improves the performance of the person re-identification task.

2   Related Work

Re-Identification. While initial attribute-based methods have succeeded in the static problem setting [3–5], some recent studies tackled real world re-identification [6,8]. Our motivation is very similar to [6], i.e., how to discover new attributes that can discriminate new people who are similar to known people. Unlike their method, which demands manual attribute discovery, our method offers fully automatic discovery.

Attribute Discovery. The common task in applying the attribute-based app-roach to computer vision problems is to define and learn useful attributes, also known as attribute discovery. The traditional approach is to define 'seman-tic attributes', which are human understandable properties (nameability) pro-vided by experts in the area [9] or vision researchers [10,11]. After preparing vocabularies of semantic attributes, images with attributes labels are needed to learn attribute classifiers, which are typically collected by using crowdsourcing. Although semantic attributes lead to intuitive results due to their nameabil-ity, the necessity of manual labeling makes it difficult to get new attributes. In addition, nameability may not always be optimum with regard to discrimination which is of primarily importance for effective person re-identification.

Aiming to discover non-semantic attributes which have high discrimination power for classification task, some recent studies have proposed automatic dis-covery techniques [12–16]. Most existing methods are based on batch optimiza-tion. For example, given the training images, the methods in [12,13] try to dis-cover new discriminative attributes by splitting a low-level feature space so as to maximize the margin between the object categories. While these discriminative attributes are helpful for classification, they may not make sense for human. Another approach, [14], discovers attributes that are both human understand-able and discriminative by combining the max-margin framework and human interaction. However, adding new people to the training image may change the distribution of the data points in the low-level feature space. Therefore, such batch-based methods need to re-train all attributes from scratch every time a new person is added, which may be too computationally expensive. Our method is much more efficient as it incrementally discovers discriminative attributes.

3   The Proposed Method

3.1   Key Idea: Ideal Case

To demonstrate the key idea behind our approach, let us first start with dis-cussing the 'ideal' situation, i.e., given attributes, each person can be deterministically represented by a subset of these attributes, which is equivalently represented by a binary string, denoted here as $\mathbf{a}^k \in \mathcal{A}^d$, where $\mathcal{A}^d$ is d-dimensional binary space $\{0, 1\}^d$ and $\mathbf{a}^k := (a_1^k, \ldots, a_d^k)^\top$ is the binary attribute representation of the k-th person. $a_q^k = 1$ if the person has the q-th attribute and 0 otherwise. In this case the following observations can be made for person re-identification.

1. Two people are not distinguishable if all bits of their binary strings are identical, i.e., the hamming distance between the two strings is zero.
2. Two people are distinguishable if at least one bit is different.

This gives us the hint that attribute discovery and person re-identification can be realized by a pair-wise learning process, and that just one attribute is enough to discriminate 'similar' people (Fig. 2). In the top image, at the beginning, in the dataset there are two people (*person1* and *person2*) with three predefined attributes (*Patterned*, *RedShirt* and *Jeans*). When *person3* is added to the dataset, the binary string of *person3* is compared to those of *person1*, and *person2*. As a result, *person3* and *person2* are found be similar, i.e., $\mathbf{a}^3 = \mathbf{a}^2$, here $\mathbf{a}^3$ is attribute representation of *person3* and $\mathbf{a}^2$ is that of *person2*. Next, a new bit is learned, some attribute that distinguishes *person3* from *person2*. Last, the new attribute is used to update the dataset. When *person4* is added, the above learning and updating process is iterated as shown at the bottom of Fig. 2. We can see that the above pair-wise learning process is effective in discriminating a new person from an identical (in terms of attributes) person in the dataset. This is because our work grasps the nature of binary attributes mentioned above and applies it to person re-identification. Our method has the following advantages.

1. it is efficient due to the pair-wise learning process.
2. it is easy to add new attributes, and expand the dataset.
3. it is easy to apply in real applications because it does not need to gather a large number of images.

The third advantage is especially important for person re-identification, because the system usually does not have a lot of images at the beginning and the method is highly effective in updating the image dataset sequentially.

3.2   Method Detail: Real Case

Above we discussed the ideal case where each person is deterministically described by some specific set of attributes. However, in practice, attribute strings of a person are probabilistic values. Several images of the same person may have different attribute representations because of occlusion or lighting conditions. Assume that an attribute representation of the i-th image for the k-th person is denoted as $\mathbf{a}_i^k \sim \phi_k$ which is sampled i.i.d. from the unknown probability distribution $\phi_k : \mathcal{A}^d \rightarrow \mathbb{R}_+$. Hence, unlike the ideal situation, in order to know whether two people are distinguishable or not, we need to check the identity of their distributions ϕ_a and ϕ_b.

In this paper, we estimate the identity of two unknown probabilistic distributions by applying the t-test to given data $A^1 := [\mathbf{a}_1^1, \ldots, \mathbf{a}_n^1]$ and $A^2 := [\mathbf{a}_1^2, \ldots, \mathbf{a}_m^2]$, where n is the number of images of the first person and m is that of the second person. We count the number of attributes $l \leq d$, for which the null hypothesis of no relationship between two distributions is 'not' rejected at the significance level of 5 %, i.e., the attributes of the two person are the same. Next, we calculate the similarity of the pair of interest, $S_{1,2} = l/d$. When similarity $S_{1,2}$ exceeds threshold $\tau = [0, 1]$, we determine that the people are similar. Threshold τ is the parameter that controls similarity strictness, and is equivalent to controlling the number of attributes to discover. For example, $\tau = 1$ requires all attribute distributions to be consistent in order to determine that the pair are similar. On the other hand, when $\tau = 0$, all people are determined as similar regardless of their distributions.

After determining sets of similar people, we discover new attributes that best discriminate the couple by pair-wise learning. Our solution is to simply use a one-vs-one linear SVM to discover the attribute or attributes necessary. In particular, we train the one-vs-one linear SVM using the low-level features using images of the new person as positive samples and those of the similar person as negative samples. Finally, we get a new attribute classifier with known people similar to the new person, $\mathcal{A}^d \rightarrow \mathcal{A}^{d+D}$, here D is the number of discovered attributes. Our method uses only low-level features corresponding to similar people and so differs from previous work, which attempts to optimize the discrimination for all people. Therefore, our approach offers superior efficiency.

4   Experiments

4.1   Experiment Setup

We evaluate our method on PRID [7] and VIPeR [2] which are benchmark datasets for person re-identification. We use the attribute labels provided in [3] which include *redshirt, blueshirt, lightshirt, darkshirt, greenshirt, nocoats, notlightdarkjeanscolour, darkbottoms, lightbottoms, hassatchel, barelegs, shorts, jeans, male, skirt, patterned, midhair, darkhair, bald, hashandbagcarrierbag* and *hasbackpack*. We follow the setting given for low-level features in [2,5] to train the attribute classifier. Images are divided into six equal sized horizontal strips, then color features (RGB, HS, YCbCr) and texture features (Gabor, Schmid) are extracted from each strip. We train a linear Support Vector Machine (SVM) to predict the attributes whose cost parameter is set to 0.5 for all attributes (including newly discovered attributes). Person re-identification is performed using the standard L2 nearest-neighbor technique. Evaluation metrics used in the following experiments are Rank n, Expected Rank (ER), Cumulative Match Characteristic (CMC) curve, and normalized area under the CMC curve (nAUC). Rank n is the probability of that the correct match is found in the n-th result (higher values = better performance). CMC curve plots Rank n value for each n. ER is the mean rank of correct matches (lower values = better performance).

Note that our method requires a number of images for each person, however most person re-identification datasets (like VIPeR) contain only single image for each person. Therefore, we modified our method to handle single-shot datasets. Implementation details are shown in Sect. 4.3.

4.2   Result on PRID

PRID has multi-shot and single-shot data captured from two cameras at fixed viewpoints. Our method requires several images for each person, thus we use the multi-shot dataset. This dataset contained 100 to 150 images for each of 200 people captured by two cameras with different views. We split randomly split the 200 people in half. The first half was used to train attributes and the second half for re-identification. We use camera A for gallery data and camera B for probe data. In this experimens, the similarity strictness, τ, was set at 0.6.

Figure 3 shows the results in the form of CMC curves. Table 1 lists the performance values. These results show that the discovered attributes significantly enhance re-identification performance, in particular, five times more accurate on Rank 1. The number of discovered attributes is 66 in this case, which is slightly more than half the probe people. Even though the number of probe people was only 100, the pre-defined attributes could not discriminate all of them. In a real

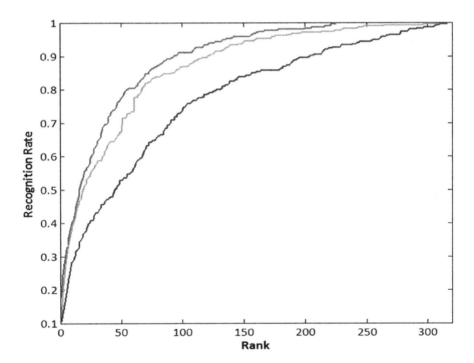

Fig. 3. CMC plots from PRID dataset. Blue: low-level features [2,5]. Green: pre-defined attributes. Red: discovered and pre-defined attributes (ours) (Color figure online)

Table 1. Re-identification performance from PRID dataset.

	ER	Rank 1	Rank 5	Rank 10	Rank 25	nAUC
Low-level features	37.5	0.04	0.18	0.24	0.45	0.64
Pre-defined attributes	27.2	0.04	0.19	0.31	0.65	0.74
Discovered (ours)	14.3	0.20	0.46	0.63	0.84	0.87

scene like an airport, at least ten thousand people will be captured by each camera, hence the incremental learning framework is effective.

4.3   Result on VIPeR

VIPeR, similar to PRID, contains 632 people captured from two cameras, and we split them in half. The first half was used for training attributes and the second half for re-identification. We use camera A for gallery data and camera B for probe data. However, unlike the PRID dataset, each person had only one image from each camera. Our method requires several images for each person, so we had to modify the proposed similarity S, and perform data augmentation for learning attributes.

In Sect. 3.2, we calculated the similarity between two people by applying the t-test to their attribute distributions. However in the single-shot case, due to insufficient number of samples, we can not get reasonable score. Thus we simply use the hamming distance as the similarity between the attribute vectors of the two people. When the distance is 0, i.e., the predicted attributes are perfectly matched, we learn new attributes which discriminate the people.

Our method mines similar two people in term of attributes based on above similarity. Then training new attributes classifier by linear SVM through images of the people. However in this experiment, we have only single image for each people, thus being unable to perform training classifier because at least three data points are required for linear SVM. Therefore we simply perform data augmentation (left-right flipping and four random cropping) for each image, which provides six images for each people.

Figure 4 shows the results in form of CMC curves. Table 2 lists the performance values. Despite the paucity of training samples, our discovered attributes showed better performance than the pre-defined attributes, so the incremental framework can enhance re-identification performance.

Finally, we compare our method to state-of-the-art attribute-based methods. Layne et al. [3] uses the same pre-defined attributes, but combines them with low-level features with optimized weights. Nguyen et.al. [4] uses the same attributes and information of the relationship between attributes (e.g. a *man* is not likely to wear a *skirt*). Both methods and our method use different information, thus they are complementary. Therefore, combining the static and incremental methods would yield more effective person re-identification in the real world.

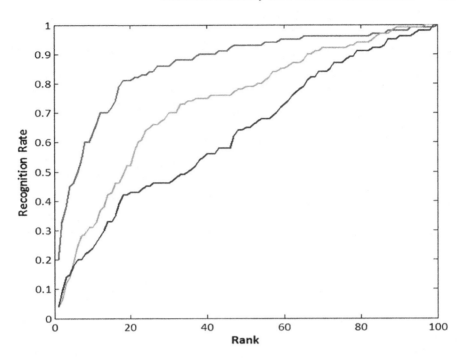

Fig. 4. CMC plots from VIPeR dataset. Blue: low-level features [2,5]. Green: Pre-defined attributes. Red: discovered and pre-defined attributes (ours) (Color figure online)

Table 2. Re-identification performance from VIPeR dataset.

	ER	Rank 1	Rank 5	Rank 10	Rank 25	nAUC
Low-level features	72.4	0.10	0.19	0.28	0.41	0.77
Pre-defined attributes	39.4	0.11	0.29	0.39	0.55	0.87
Discovered attributes (ours)	34.6	0.18	0.32	0.41	0.59	0.90
Layne et.al. [3]	29.0	0.19	0.40	0.54	0.71	0.91
Nguyen et.al. [4]	61.2	-	0.12	0.24	0.39	0.81

5   Conclusion

We proposed an incremental approach based on pair-wise learning to discover discriminative attributes for continuous person re-identification task. We discover attributes by focusing on just the discrimination of similar people. Experiments on two public datasets indicate that the discovered attributes enhance re-identification performance. Our future work includes an investigation of name-ability of discovered attributes.

References

1. Prosser, B., Zheng, W.S., Gong, S., Xiang, T., Mary, Q.: Person re-identification by support vector ranking. In: BMVC, vol. 2, p. 6 (2010)
2. Gray, D., Tao, H.: Viewpoint invariant pedestrian recognition with an ensemble of localized features. In: Forsyth, D., Torr, P., Zisserman, A. (eds.) ECCV 2008, Part I. LNCS, vol. 5302, pp. 262–275. Springer, Heidelberg (2008)
3. Hospedales, T.M., Layne, R., Gong, S.: Attributes-based re-identification. In: Person Re-Identification (2013)
4. Nguyen, N.-B., Nguyen, V.-H., Duc, T.N., Le, D.-D., Duong, D.A.: AttRel: an approach to person re-identification by exploiting attribute relationships. In: He, X., Luo, S., Tao, D., Xu, C., Yang, J., Hasan, M.A. (eds.) MMM 2015, Part II. LNCS, vol. 8936, pp. 50–60. Springer, Heidelberg (2015)
5. Layne, R., Hospedales, T.M., Gong, S., Mary, Q.: Person re-identification by attributes. In: BMVC (2012)
6. Das, A., Panda, R., Roy-Chowdhury, A.: Active image pair selection for continuous person re-identification. In: ICIP (2015)
7. Hirzer, M., Beleznai, C., Roth, P.M., Bischof, H.: Person re-identification by descriptive and discriminative classification. In: Heyden, A., Kahl, F. (eds.) SCIA 2011. LNCS, vol. 6688, pp. 91–102. Springer, Heidelberg (2011)
8. Cancela, B., Hospedales, T.M., Gong, S.: Open-world person re-identification by multi-label assignment inference. In: BMVC (2014)
9. Lampert, C.H., Nickisch, H., Harmeling, S.: Learning to detect unseen object classes by between-class attribute transfer. In: CVPR, pp. 951–958 (2009)
10. Farhadi, A., Endres, I., Hoiem, D., Forsyth, D.: Describing objects by their attributes. In: CVPR, pp. 1778–1785 (2009)
11. Patterson, G., Hays, J.: Sun attribute database: discovering, annotating, and recognizing scene attributes. In: CVPR, pp. 2751–2758 (2012)
12. Bergamo, A., Torresani, L., Fitzgibbon, A.W.: Picodes: learning a compact code for novel-category recognition. In: NIPS, pp. 2088–2096 (2011)
13. Rastegari, M., Farhadi, A., Forsyth, D.: Attribute discovery via predictable discriminative binary codes. In: Fitzgibbon, A., Lazebnik, S., Perona, P., Sato, Y., Schmid, C. (eds.) ECCV 2012, Part VI. LNCS, vol. 7577, pp. 876–889. Springer, Heidelberg (2012)
14. Parikh, D.: Grauman, K.: Interactively building a discriminiative vocabulary of nameable attributes. In: CVPR, pp. 1681–1688 (2011)
15. Berg, T.L., Berg, A.C., Shih, J.: Automatic attribute discovery and characterization from noisy Web data. In: Daniilidis, K., Maragos, P., Paragios, N. (eds.) ECCV 2010, Part I. LNCS, vol. 6311, pp. 663–676. Springer, Heidelberg (2010)
16. Yu, F.X., Cao, L, Feris, R.S., Smith, J.R., Chang, S.F.: Designing category-level attributes for discriminative visual recognition. In: CVPR, pp. 771–778 (2013)

What are the Limits to Time Series Based Recognition of Semantic Concepts?

Peng Wang[1]([✉]), Lifeng Sun[1], Shiqiang Yang[1], and Alan F. Smeaton[2]

[1] National Laboratory for Information Science and Technology,
Department of Computer Science and Technology,
Tsinghua University, Beijing, China
{pwang,sunlf,yangshq}@tsinghua.edu.cn
[2] Insight Centre for Data Analytics,
Dublin City University, Glasnevin, Dublin 9, Ireland
alan.smeaton@dcu.ie

Abstract. Most concept recognition in visual multimedia is based on relatively simple concepts, things which are present in the image or video. These usually correspond to objects which can be identified in images or individual frames. Yet there is also a need to recognise semantic concepts which have a temporal aspect corresponding to activities or complex events. These require some form of time series for recognition and also require some individual concepts to be detected so as to utilise their time-varying features, such as co-occurrence and re-occurrence patterns. While results are reported in the literature of using concept detections which are relatively specific and static, there are research questions which remain unanswered. What concept detection accuracies are satisfactory for time series recognition? Can recognition methods perform equally well across various concept detection performances? What affecting factors need to be taken into account when building concept-based high-level event/activity recognitions? In this paper, we conducted experiments to investigate these questions. Results show that though improving concept detection accuracies can enhance the recognition of time series based concepts, they do not need to be very accurate in order to characterize the dynamic evolution of time series if appropriate methods are used. Experimental results also point out the importance of concept selection for time series recognition, which is usually ignored in the current literature.

Keywords: Concept detection · Time series · Activity recognition · Attribute dynamics · Event classification

---

P. Wang—Work was part-funded by 973 Program under Grant No. 2011CB302206, National Natural Science Foundation of China under Grant No. 61272231, 61472204, 61502264, Beijing Key Laboratory of Networked Multimedia.

A.F. Smeaton—Supported by Science Foundation Ireland under grant number SFI/12/RC/2289.

Q. Tian et al. (Eds.): MMM 2016, Part II, LNCS 9517, pp. 277–289, 2016.
DOI: 10.1007/978-3-319-27674-8_25

1   Introduction and Background

The proliferation of online videos and personal media has created multimedia data which require effective indexing and recognition techniques to support flexible retrieval and management. The development of automatic concept based indexing of multimedia has shown the importance of concepts in supporting the understanding of such media. Such concept labels might include the occurrences of scenes, objects, persons, etc. Though various efforts have been tried such as providing large annotated corpora for training, improving the discriminative algorithms, utilising external ontological knowledge, post-processing the indexing results for enhancement, etc., the detection of concepts is still far from being perfect.

While low-level feature-based methods have been shown to be ill-suited for multimedia semantic indexing due to the lack of semantics for user interpretation, high dimensionality, etc., high-level concept attributes are widely employed in the analysis of complex semantics corresponding to things like events and activities. Since such semantic structures can be represented as typical time series, the recognition of events or activities usually involves two components, initially concept detection followed by dynamics-based recognition. This means that initial concept detection results are used as input for modeling the evolution of time-based semantics such as events or activities. This is usually carried out by representing the time series as a sequence of units such as video clips or image frames. After concatenating the results of concept detectors on each unit, time series can then be represented by a temporally-ordered sequence of vectors, as shown in Fig. 1.

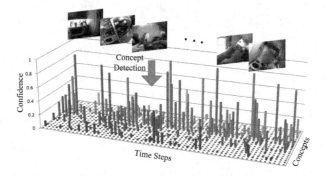

Fig. 1. The dynamics of concept attributes quantified by confidences returned by concept detections.

Attribute-based event and activity detection has attracted much research attention. For example, [1] presented an approach to learn a visually grounded storyline model of videos directly from weakly labeled data. In [3], a rule-based method is proposed to generate textual descriptions of video content based on concept classification results. [3] also found that although state-of-the-art concept detections are far from perfect, they still provide useful clues for event classification. In [4], a multimedia event recounting method is proposed based on detected concepts in order to build discriminative event models using a SVM. Similar work is carried out in [5] aiming at video event classification using semantic concept attributes of different categories like action, scene, object, etc. [6] employed an intermediate representation of semantic model vectors trained from SVMs as a basis for detecting complex events, and revealed that this representation outperforms – and is complementary to – other low-level visual descriptors for event modeling. [7] showed that concept-based temporal representations are promising in more complex event recognition. Other efforts as presented in the TRECVid event detection tasks [8,9] also demonstrated promising results for concept-based event detection. Similar work is also carried out using concept detections to characterize everyday activities as reported in [10,11] where activity recognition is built on the basis of concept detection. In [11], detection results are first binarized and then applied to learn temporal dynamics in order to train an activity model. In addition, the correlation of activity and concept detection performances is analyzed in [11].

Though effectiveness is shown using some of the above algorithms in recognizing segments of interest for multimedia time series analytics, there still exists research questions which remain unanswered, due to imperfect concept detection performances. Since these methods use *detection and assemble* schemes which aggregate concept detection results, how concept detection affects the final time series analysis is unclear. Because current research tends to report results based on their own concept detections, whether a proposed event or activity recognition method can adapt to other concept detections is not addressed, such as in cross-domain applications. To overcome these limitations, the following research questions need to be addressed:

- What levels of concept detection accuracy are needed for satisfactory time series analytics? In real-life applications, the pursuit of perfect concept detectors is non-trivial and only a manually annotated groundtruth can be regarded as an Oracle, which is time consuming to obtain. In most cases, however, certain levels of accuracy of concept detection are provided for time-series modeling and classification. In the work we report in this paper we imply that dynamic correlations among imperfect concept detections can still reflect patterns of time series which vary over context.
- How can different recognition methods adapt to varying concept detection accuracies? Most results are reported using individual researchers' own concept detections. It would be of help for researchers to choose methods if the correlations between these methods and the concept detection accuracies, are made clear. More importantly, whether downgraded concept detection accuracies will propagate over time and across models needs to be validated for choosing the right recognition method. As demonstrated later in this paper, the typical methods chosen in our experiments can adapt to varying concept

detection accuracies. This shows why state-of-the-art concept-based event/ activity recognitions are feasible and have satisfactory results.

- What factors affect concept-based time series analytics? While most research focuses on temporal modeling of concept occurrence dynamics, a systematic view of concept-based time series analytics involving both performances of concept detection and temporal modeling will provide guidance on this topic. In experiments reported in this paper, we quantify the factors affecting this and we point out that besides time series modeling methods, concept detection and concept selection also need to be considered to improve the performance of the final recognition.

2 Experimental Datasets

In our experiments, we take recognition of everyday activity as demonstration of the kind of time series based concept recognition we focus on and we explore the research questions on two datasets, namely lifelogged image streams (Dataset1) and egocentric video collections (Dataset2) respectively.

For lifelog activity recognition, the set of 85 everyday concepts investigated in [12] are used as semantic attributes. Dataset1 includes event samples of 16 activity types collected from 4 SenseCam wearers and consisting of 10,497 SenseCam images [11]. Meanwhile, for egocentric video analysis, we also evaluated various algorithms for recognising activities of daily living (ADL) [2] with 45 underlying semantic concepts. There are a total of 18 activity types and 23,588 frames used in this ADL corpus. To make full use of activity samples in both datasets, we decompose each set of positive samples into 50:50 ratios for training and testing respectively. The two datasets are summarized in Table 1.

Table 1. Summary of the lifelog and ADL datasets.

Datasets	#Types	#Concepts	#Samples	#Frames	Domain
Dataset1	16	85	500	10,497	Lifelog
Dataset2	18	45	624	23,588	ADL

For both datasets, concept detectors with different accuracy levels were applied by simulating the confidence score outputs from concept detectors as a probabilistic model of two Gaussians. In a state-of-the-art concept detector simulation by Aly *et al.* [13], concept detection performance is controlled by modifying the models' parameters based on manually annotated groundtruth of concept occurrences. These parameters are the mean μ_1 and standard deviation σ_1 for the positive class, as well as the mean μ_0 and the standard deviation σ_0 for the negative class. The performance of concept detection can be varied by controlling the intersection of the areas under the two probability density curves by changing the means or the standard deviations of the two classes for a single concept detector.

In addition to the simulation of the confidence distribution for the negative and positive classes as $N(\mu_0, \sigma_0)$ and $N(\mu_1, \sigma_1)$ respectively, we can also obtain the prior probability $P(c)$ for a concept c from the dataset by using manual annotations. Then the sigmoid posterior probability function is needed to fit which has the form of a sigmoid function as [11, 13]:

$$P(c|o) = \frac{1}{1 + exp(Ao + B)} \tag{1}$$

where a random confidence score o is drawn from the corresponding normal distribution after parameters A and B are fixed. The posterior probability of the concept is then returned using Eq. (1) for each image frame or video shot.

For each setting of parameters, we executed 20 repeated simulation runs and the averaged concept AP and MAP were both calculated. During the simulation procedure, we fixed the two standard deviations and the mean of the negative class and varied the mean of the positive class μ_1 in the range $[0.5...10.0]$. Figure 2 shows the improvement in concept MAP for both datasets with increased μ_1, and near-perfect detection performances are achieved when $\mu_1 \geq 4.0$, as segmented by the blue dashed line in Fig. 2.

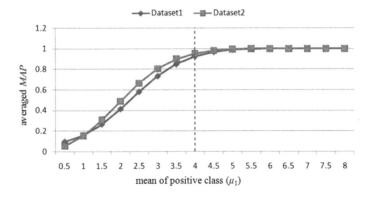

Fig. 2. Averaged concept MAP with different μ_1 values for two datasets. (Color figure online)

3   Methods

In this experiment, we provide discussions based on the investigation into various time series recognition methods on the above described datasets. To fully exploit the interacting correlations between concept detections and time series recognition, as well as providing investigation into affecting factors on the recognition performance, we validated a selection of methods including temporal and non-temporal features, generative and discriminative models, holistic and pyramid representations, signatures of dynamic systems, etc. The time series recognition methods investigated are summarized in Table 2. Whether they utilise

Table 2. Summary of time series recognition methods investigated.

Methods	Temp?	#Types	#Concepts
Max-Pooling	×	10 [14], 15 [19], 25 [7]	50 [14], 101 [19], 93 [7]
Bag-of-Features	×	10 [14], 18 [2], 3 [6]	50 [14], 42 [2], 280 [6]
Temporal Pyramids	✓	18 [2]	42 [2]
HMM	✓	16 [11]	85 [11]
Fisher Vector	✓	16 [11], 15 [17]	85 [11], 60 [17]
Dynamic System	✓	25 [7], 5 [18]	93([7], [18])

temporal features, the number of event/activity types and the number of concept attributes are all depicted in Table 2.

As we can see from Table 2, the number of categories and concepts used in our datasets (Sect. 2) are both within the prevalent range reported in recent literature. Therefore, the setup of our experiment is close to a realistic implementation and the conclusions should therefore be valid. The details of the different recognition methods we implemented are now outlined.

Max-Pooling (MP): As one of the fusion operations for concept detection results, Max-Pooling [14] has been demonstrated to give better performance compared to other fusions for most complex events. In Max-Pooling, the maximum confidence is chosen from all keyframe images (or video subclips) for each concept to generate a fixed-dimensional vector for an event or activity sample. Since by definition the maximum value cannot characterize a temporal evolution of concepts within a time series, this method can be regarded as non-temporal.

Bag-of-Features (BoF): Similar to Max-pooling, Bag of Features is a way of aggregating concept detection results by averaging the confidences over time window. Because Bag-of-Features and Max-Pooling reflect the statistical features within the holistic time series, they both ignore the temporal evolution of concept detection results.

Temporal Pyramids (TP): Motivated by the spacial pyramid method, the temporal pyramids proposed in [2] approximate temporal correspondence with a temporally-binned model [15]. In this method, the histogram over the whole time series represents the top level while the next level can be represented by concatenating two histograms of temporally segmented sequences. More fine-grained representations can be formalized in the same manner. By applying the multi-scale pyramid to approximate a coarse-to-fine temporal correspondence, this method generates fixed-length features with temporal embeddings.

Hidden Markov Model (HMM): A generative method based on HMMs as used in [11] is employed in this representation. HMMs are first trained for each activity class and we concatenate the log-likelihood representations of per-class posteriors into a vector. Assume there are l hidden states in the HMM and each pair of states have a transition probability $a_{ij} = P(s_i|s_j)$. The parameters of the

HMM can be denoted as $\lambda = (A, B, \pi)$, where $A = \{a_{ij}\}$, $\pi = \{\pi_i\}$ stands for the initial state distribution. $b_j(X_t)$ is the distribution of the concept observation X_t at time step t with respect to state j.

Fisher Vector (FV): The principle of the Fisher kernel is that similar samples should have similar dependence on the generative model, i.e. the gradients of the parameters [16]. Instead of directly using the output of generative models, such as in the HMM method, using a Fisher kernel tries to generate a feature vector which describes how the parameters of the activity model should be modified in order to adapt to different samples. Based on the above formalization of an HMM, X can be characterized as Fisher scores with regard to the parameters λ, $U_X = \nabla_\lambda log P(X|\lambda)$. Therefore, the Fisher kernel can be formalized as $K(X_i, X_j) = U_{X_i}^T I_F U_{X_j}^T$, where $I_F = E_X(U_X U_X^T)$ denotes the Fisher information matrix.

Liner Dynamic System (LDS): As a natural way of modeling temporal interaction within time series, Liner Dynamic Systems [7] can characterize temporal structure with attributes extracted from within a sliding window. The time series can be arranged in a block Hankel matrix H whose elements in a column have the length of sliding window (denoted as r) and successive columns are shifted with one time step. According to [7], singular value decomposition of $H \cdot H^T$ has achieved comparable performance to more complex representations. We constructed the feature using the k largest singular values along with their corresponding vectors.

4   Results

For recognition based on HMM log-likelihood and on Fisher Vector, generative models were obtained with two-state ergodic HMMs to model the sequence of concept occurrences. Because the confidence vector X_t has continuous values, we employed Gaussian emission distributions $b_j(X_t) = \mathcal{N}(X_t, \mu_i, \sigma_i)$ and $B = \{\mu_i, \sigma_i\}$. Parameters μ_i and σ_i are the mean and covariance matrix of the Gaussian distribution in state i respectively. This setting was applied both in HMM-based and Fisher Vector-based time series recognition. To alleviate the sub-optimal problem of Fisher kernels induced by (nearly) zero gradient representations of a generative model, we employed a model parameter learning as proposed in [20], to train the model so that samples of the same class will have more similar gradients than other classes. The Fisher kernel was then embedded in the SVMs for activity classification. To simplify the computation, we approximated I_F by the identity matrix in the implementation.

In implementing the Hankel matrix H for dynamic system characterization, the length of the sliding window r reflects the temporal influence range which can be regarded as one parameter. Besides r, the number k of largest singular values along with their corresponding vectors determines the final dynamic system signature from singular value decomposition of $H \cdot H^T$, and was accepted as the other parameter. Similar to work in [7], we examined performances in the assignment range $r \in \{2, 4, 8\}$ and $k \in \{1, 2, 4\}$. The final time series recognition accuracies were compared and we chose the best performances at $\{r = 2, k = 1\}$ for Dataset1 and $\{r = 4, k = 1\}$ for Dataset2 in the evaluation. As for the temporal pyramids method, two levels of feature histograms were extracted and concatenated to construct the final time series representation. This was chosen empirically without further optimization since two level pyramids have shown to be effective in our datasets which are also employed in [2].

After fixed-length features were extracted using the methods listed in Table 2, the same discriminative classifier SVM was employed to classify activities, with the same way of parameter optimization for fair comparison. The resulting accuracies on two datasets are shown in Figs. 3 and 4, across various concept detection accuracies controlled by the simulation parameter μ_1.

As shown by these two figures, the resulting accuracy curves for different recognition methods have very similar shapes on both datasets, implying similar correlation with concept detections. That is, while improving concept detection accuracies (increasing μ_1), the recognition accuracies of time series are enhanced for all methods, in both datasets. This enhancement is especially significant when the original concept detection accuracies are low, say, at $\mu_1 \leq 2$ in Fig. 3 on the left side of the blue dashed line. When concept detections are accurate enough, say on the right side of the blue dashed line in Fig. 3, time series recognitions converge with relatively stable performances.

In Figs. 3 and 4, the performances of time series recognitions differ across concept detection accuracies. For example, FV, BoF and TP achieve better recognition performances than the others. The advantage of FV is obvious when concept detections have very poor performances at $\mu_1 \leq 1$. This implies that

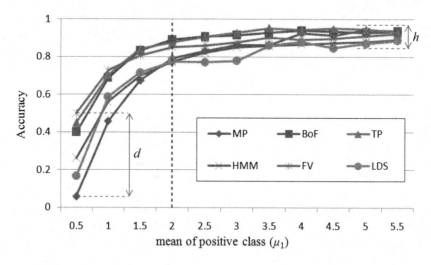

Fig. 3. Averaged recognition accuracy on lifelog dataset with different μ_1 values. Accuracy increases significantly when concept detections are low (left of blue dashed line) and converges when concept detections are high (right of blue dashed line) (Color figure online).

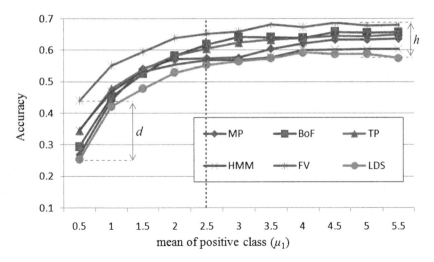

Fig. 4. Averaged recognition accuracy on ADL dataset with different μ_1 values. Accuracy increases significantly when concept detections are low (left of blue dashed line) and converges when concept detections are high enough (right of blue dashed line) (Color figure online).

FV can better adapt to less accurate concept annotations. However, when concept annotation burdens are relieved at high μ_1 values, FV is outperformed by BoF and TP. A similar tendency for performance variations at different concept detection levels is also common for other recognition methods in Figs. 3 and 4. From this we suggest that the evaluation of current concept-based time series recognitions on single concept detection performance is limited. More comprehensive assessments on various concept annotations are required to demonstrate the robustness and adaption capabilities of recognition methods.

5   Discussion

As previously presented in Sect. 2, concept detection accuracies converge around $\mu_1 = 4.0$ and near-perfect detection performances are achieved after $\mu_1 \geq 4.0$. It is interesting to note that this critical value is lower for recognition accuracy curves in Figs. 3 and 4, in which the best-performing value occurs around $\mu_1 = 2.0$ and $\mu_1 = 2.5$ respectively, as denoted by blue dashed lines. The earlier convergence in Figs. 3 and 4 is good news for time series recognition based on concept detections which are still not accurate. This also manifests why current concept-based event/activity recognition can outperform using low-level descriptors even though state-of-the-art concept detections are far from perfect. The deviation of the dashed line from $\mu_1 = 4$ (Fig. 2) to $\mu_1 = 2$ (Fig. 3) and $\mu_1 = 2.5$ (Fig. 4) implies that an improvement in concept detection accuracies after $\mu_1 = 2$ is of less value in enhancing the time series based recognition due to the adaptive capabilities of recognition methods in the presence of erroneous concept detections.

According to the above experiments, the affecting factors on concept-based time series recognition can be summarized as:

- **Recognition Methods.** As we can see from Figs. 3 and 4, recognition methods play the dominant role in obtaining different performances, as depicted by two distances of d and h for low and high concept detection accuracies respectively. In both figures, $d > h$ holds which means recognition performances differ more significantly at lower concept detections. Therefore, how to better classify high-level events/activities streams based on noisy semantic attributes needs to be addressed. This is especially important for cases where we then build upon the detected concepts such as using them to infer activities, complex events or behaviours. For such applications, there may be further challenges because of the diverse range of usable concepts, or the noisy nature of the multimedia data.
- **Concept Detection.** Accuracies of all recognition methods climb while improving concept detection accuracies (increasing μ_1) in Figs. 3 and 4. Despite recent progress, automatic concept detectors are still far from perfect. The climbing rate is especially significant when original concept detections are not good enough, say, on the left side of the blue dashed lines in Figs. 3 and 4. Given the fact that current state-of-the-art concept detectors are still far from perfect despite recent progress, improving original concept detections for enhanced time series recognition represent another possible research area.
- **Concept Selection.** The benefits of appropriate concept selection can also be demonstrated in our experiments. To eliminate the effects of the above two factors of recognition methods and concept detection, we can focus on the extreme performances on two datasets, which are 0.95 and 0.69 in Figs. 3 and 4 respectively. In addition to the inherent nature of two datasets, the sophisticated selections of more appropriate concepts also lead to a difference in performances. For the lifelog dataset as shown in Fig. 3, topic-related semantic concept selections are investigated in [12] to choose concepts in terms of user experiments and semantic networks.

As shown in Table 1, our lifelog dataset uses more concepts than the ADL dataset (85 versus 45) to characterize activity time series. To further validate the effect of concept selections on the finial recognition performances, we carried out further experiments by randomly selecting n concepts ($n \leq 85$) from the lifelog dataset and performing time series recognition based on the selected concept subsets. Results for $n = \{20, 40, 60, 85\}$ are shown in Figs. 5 and 6 for two levels of concept detection accuracies ($\mu_1 = \{2.0, 5.5\}$).

Taking three recognition methods for example, Figs. 5 and 6 show the same trends in accuracies when increasing the number of selected concepts. When more appropriate concepts are utilised to characterize the dynamic evolution of time series based concepts, their recognition performances are enhanced accordingly. This is more obvious when the original concept detections are less satisfactory in Fig. 5 (@$\mu_1 = 2.0$). In this case, when concept annotations are more noisy, introducing more concepts can counteract such noise and enhance recognition performances. For example, increasing the number of selected concepts from 20

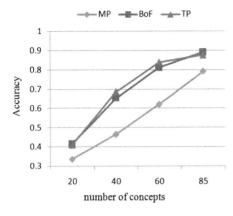 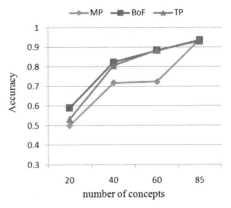

Fig. 5. Performance comparison for randomly selected concepts at lower concept detection accuracies ($\mu_1 = 2.0$).

Fig. 6. Performance comparison for randomly selected concepts at higher concept detection accuracies ($\mu_1 = 5.5$).

to 40 leads to nearly 0.3 accuracy improvement in Fig. 5, compared to 0.2 in Fig. 6. These results support our hypothesis that concept selection is another factor affecting recognition performance of time series based concepts.

From Figs. 5 and 6, we also note that the performance enhancement is less significant when the number of selected concepts are high, say, when $n \geq 60$. While increasing values of n, the slopes of performance curves of BoF and TP decline gradually in both figures. This implies that introducing more redundant concepts is of less value to characterize the time series since these concepts are not independent of each other. This can be captured through various types of correlations among concepts including co-occurrence patterns and ontological relationships. In other words, the semantic space spanned by concepts in the lexicon can be projected to a more compact space with lower ranks since concepts are not independent. This characteristic can also be utilised to enhance concept detection accuracies [22], classify time series [21], etc.

6   Conclusions

Though acceptable results can be achieved using state-of-the-art concept detection methods for narrow domains and for concepts for which there exists enough annotated training data, concept detectors are still far from perfect especially for those related to activities, events or behaviour which have a temporal aspect and how time series based recognition methods for such concepts interact with noisy semantic attributes is unclear. To validate the adaption capability of event/activity recognitions to concept detections, we carried out experiments on two datasets with various concept detection accuracies using typical time series based recognition methods. Results show that concept-based time series recognition is feasible when built on the premise of noisy underlying concept detections. In the experiment, we also explored the nature of the affecting factors on time series analytics. Besides recognition methods which are the focus of current research, concept detection and concept selection are also pointed out to have direct influence on analytics performances. This work can provide an analysis framework and guidance for time series recognition based on concept attributes.

References

1. Gupta, A., Srinivasan, P., Shi, J., Davis, L.: Understanding videos, constructing plots learning a visually grounded storyline model from annotated videos. In: CVPR 2009, pp. 2012–2019, June 2009
2. Pirsiavash, H., Ramanan, D.: Detecting activities of daily living in first-person camera views. In: CVPR 2012, pp. 2847–2854, June 2012
3. Tan, C.C., Jiang, Y.-G., Ngo, C.-W.: Towards textually describing complex video contents with audio-visual concept classifiers. In: ACM Multimedia 2011, pp. 655–658 (2011)
4. Yu, Q., Liu, J., Cheng, H., Divakaran, A., Sawhney, H.: Multimedia event recounting with concept based representation. In: ACM Multimedia 2012, pp. 1073–1076 (2012)
5. Liu, J., Yu, Q., Javed, O., Ali, S., Tamrakar, A., Divakaran, A., Cheng, H., Sawhney, H.: Video event recognition using concept attributes. In: IEEE Winter Conference on Applications of Computer Vision, pp. 339–346 (2013)
6. Merler, M., Huang, B., Xie, L., Hua, G., Natsev, A.: Semantic model vectors for complex video event recognition. IEEE Trans. Multimedia 14(1), 88–101 (2012)
7. Bhattacharya, S., Kalayeh, M., Sukthankar, R., Shah, M.: Recognition of complex events exploiting temporal dynamics between underlying concepts. In: CVPR 2014, June 2014
8. Hill, M., Hua, G., Huang, B., Merler, M., Natsev, A., Smith, J.R., Xie, L., Ouyang, H., Zhou, M.: IBM research TRECVid-2010 video copy detection and multimedia event detection system (2010)
9. Cheng, H., Liu, J., Ali, S., Javed, O., Yu, Q., Tamrakar, A., Divakaran, A., Sawhney, H.S., Manmatha, R., Allan, J., Hauptmann, A., Shah, M., Bhattacharya, S., Dehghan, A., Friedland, G., Elizalde, B.M., Darrell, T., Witbrock, M., Curtis, J.: SRI-Sarnoff AURORA system at TRECVid 2012: Multimedia event detection and recounting. In: NIST TRECVID Workshop, December 2012
10. Doherty, A.R., Caprani, N., O'Conaire, C., Kalnikaite, V., Gurrin, C., O'Connor, N.E., Smeaton, A.F.: Passively recognising human activities through lifelogging. Comput. Hum. Behav. 27(5), 1948–1958 (2011)
11. Wang, P., Smeaton, A.F.: Using visual lifelogs to automatically characterise everyday activities. Inf. Sci. 230, 147–161 (2013)
12. Wang, P., Smeaton, A.: Semantics-based selection of everyday concepts in visual lifelogging. Int. J. Multimedia Inf. Retrieval 1(2), 87–101 (2012)
13. Aly, R., Hiemstra, D., de Jong, F., Apers, P.: Simulating the future of concept-based video retrieval under improved detector performance. Multimedia Tools Appl. 60, 1–29 (2011)
14. Guo, J., Scott, D., Hopfgartner, F., Gurrin, C.: Detecting complex events in user-generated video using concept classifiers. In: CBMI 2012, pp. 1–6, June 2012

15. Laptev, I., Marszalek, M., Schmid, C., Rozenfeld, B.: Learning realistic human actions from movies. In: CVPR 2008, pp. 1–8, June 2008
16. Jaakkola, T.S., Haussler, D.: Exploiting generative models in discriminative classifiers. In: NIPS 1999, pp. 487–493 (1999)
17. Sun, C., Nevatia, R.: ACTIVE: activity concept transitions in video event classification. In: ICCV 2013, pp. 913–920, December 2013
18. Li, W., Yu, Q., Sawhney, H., Vasconcelos, N.: Recognizing activities via bag of words for attribute dynamics. In: CVPR 2013, pp. 2587–2594, June 2013
19. Liu, J., Yu, Q., Javed, O., Ali, S., Tamrakar, A., Divakaran, A., Cheng, H., Sawhney, H.: Video event recognition using concept attributes. In: WACV 2013, pp. 339–346 (2013)
20. van der Maaten, L.: Learning discriminative fisher kernels. In: ICML 2011, pp. 217–224 (2011)
21. Assari, S., Zamir, A., Shah, M.: Video classification using semantic concept co-occurrences. In: CVPR 2014, pp. 2529–2536, June 2014
22. Wang, P., Smeaton, A.F., Gurrin, C.: Factorizing time-aware multi-way tensors for enhancing semantic wearable sensing. In: He, X., Luo, S., Tao, D., Xu, C., Yang, J., Hasan, M.A. (eds.) MMM 2015, Part I. LNCS, vol. 8935, pp. 571–582. Springer, Heidelberg (2015)

Ten Research Questions for
Scalable Multimedia Analytics

Björn Þór Jónsson[1,2]([✉]), Marcel Worring[2], Jan Zahálka[2], Stevan Rudinac[2],
and Laurent Amsaleg[3]

[1] School of Computer Science, Reykjavik University, Reykjavík, Iceland
bjorn@ru.is
[2] Informatics Institute, University of Amsterdam, Amsterdam, The Netherlands
{m.worring,j.zahalka,s.rudinac}@uva.nl
[3] IRISA, Rennes, France
laurent.amsaleg@irisa.fr

Abstract. The scale and complexity of multimedia collections is ever increasing, as is the desire to harvest useful insight from the collections. To optimally support the complex quest for insight, multimedia analytics has emerged as a new research area that combines concepts and techniques from multimedia analysis and visual analytics into a single framework. State of the art multimedia analytics solutions are highly interactive and give users freedom in how they perform their analytics task, but they do not scale well. State of the art scalable database management solutions, on the other hand, are not yet designed for multimedia analytics workloads. In this position paper we therefore argue the need for research on *scalable multimedia analytics*, a new research area built on the three pillars of visual analytics, multimedia analysis and database management. We propose a specific goal for scalable multimedia analytics and present several important research questions that we believe must be addressed in order to achieve that goal.

1 Introduction

In the last decade, we have witnessed a revolution in all aspects of computing technology as the human ability to produce, store and share data has truly exploded. This data contains very valuable information for individuals, enterprises and society, and as a result we have seen a sharp rise in interest in big data analytics and related fields. Big data is typically characterised using "the three Vs"—Volume, Velocity, and Variety—which indicate respectively that the data is bountiful, is produced continuously, and contains a large variety of information.

While big data analytics has focused on relatively structured data, such as business data and transaction logs, much of the information explosion has taken the form of multimedia, in particular images and videos, along with user-generated annotations and automatically generated metadata, for example from

© Springer International Publishing Switzerland 2016
Q. Tian et al. (Eds.): MMM 2016, Part II, LNCS 9517, pp. 290–302, 2016.
DOI: 10.1007/978-3-319-27674-8_26

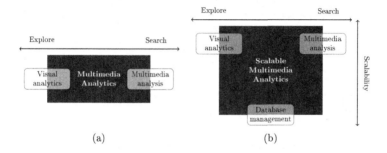

Fig. 1. Transition from (a) multimedia analytics to (b) scalable multimedia analytics

the capturing device or a social media service. Current and potential applications involving large multimedia collections are numerous, ranging from personal applications (e.g., life-logging) through societal (e.g., digital heritage) and scientific (e.g., biology, astronomy, and medicine) to business applications (e.g., marketing and profiling). What many of these media collections have in common is that they have the potential to significantly change the world in some way if we can manage to extract the knowledge and insight that they encode.

Unfortunately, existing big data analytics methods are not directly applicable to the multimedia domain. First, since the data is multimedia, automatically understanding the content and context of the data must be done at various levels of abstraction and, because it is very difficult, it is best done in collaboration between man and machine. This collaboration requires systems to learn in real-time the intention of users, the patterns they indicate, and support their interactions with the data. Developing general methods and tools for harvesting information from multimedia collections therefore represents a significant unsolved challenge.

1.1 Multimedia Analytics

Enter the new field of *multimedia analytics*. This field, which combines visual analytics with multimedia analysis as depicted in Fig. 1(a), has been developing over the last half decade [2]. The goal of multimedia analytics is to produce the processes, techniques and tools to allow users to efficiently and effectively analyze multimedia collections in order to gain insight and knowledge [2]. While multimedia analysis should be stressed to its limits to extract information automatically from the media files, visual analytics proposes an interactive process that must involve the user, through data selection and visualization techniques.

Understanding semantics of multimedia content brings many difficult challenges. Machines cannot match humans, both in the richness of semantics extracted and the speed of the extraction, while humans have great difficulty processing large multimedia collections. Combining the strengths of human and machine while alleviating the weaknesses, and providing interactive experience for a variety of analytics tasks, is the major challenge of multimedia analytics.

Addressing the analytic challenge is already daunting when collections are of moderate size. As the collections grow in size and scope, supporting the analytics process efficiently and effectively through data management becomes increasingly important. State of the art multimedia analytics solutions are highly interactive and give users freedom in how they perform their task, but they do not scale well. State of the art scalable data management solutions, on the other hand, are not designed for interactive analysis.

1.2 Scalability Challenges

In big data analytics, the requirements for data management are often described using the three Vs—volume, velocity and variety.[1] For multimedia analytics, since the nature of multimedia collections and applications leads to some specific requirements, the following major axes of scalability must be addressed:

Volume: The size of the collection is obviously the first scalability axis. Due to the large file sizes of multimedia items, some collections are enormous in their sheer size, making any manipulation of such collections very difficult. While storage capacity and cost is generally not an issue anymore, user time—and therefore processing time—is. The main interest of volume is therefore the impact on remaining axes.

Variety: Current multimedia analytics research projects generally focus on solving a particular problem. Data may have complex structure and arise from many sources, but significant effort is spent on reducing both data quantity and complexity to make the analytics process more manageable. We predict, on the other hand, that the aim will become to understand and analyze whole domains, with a variety of multimedia data coming from sources that may not yet be fully understood, such as the "Internet of Things." This requires gathering data for future use and retaining much more of the potentially useful data. Such data will be even more voluminous, but in particular more complex, requiring more effort to manage and analyze.

Velocity: As these large-scale collections grow fast and are long-lived, data is added incrementally and continuously, and data sources may come and go, so we must necessarily support incremental analysis. Furthermore, multimedia represents a world that is changing fast, both literally and in terms of its representation in the multimedia collections. Many different users will work with each collection over a long period of time and they must contribute their knowledge to the collection for the benefit of concurrent and future users, who will then add to this knowledge or even negate it.

Visual Interaction: Users are not considered a scalability axis for big data analytics, as big data tends to be used by few expert users. For multimedia analytics, however, users and user interactions play perhaps the most important role. In some domains, the number and diversity of users may

---

[1] The requirements are sometimes given as four, six or even seven Vs. The additional Vs—veracity, validity, viability and value—are not important for this discussion.

be very significant, as aspects of the collection they work on may be separated depending on their objectives, roles, location and time. Because visual interfaces are mandatory, we call this axis visual interaction.

Note that the first two axes stem from the multimedia collection itself, while the latter two stem from the user interaction with the multimedia collection.

1.3 Scalable Multimedia Analytics

Based on these scalability axes, we propose the following definition of the main goal of *Scalable Multimedia Analytics*:

> to produce the processes, techniques and tools to allow *many diverse* users to efficiently and effectively analyze *large* and *dynamic* multimedia collections over a *long period of time* to gain insight and knowledge.

We argue that scalable multimedia analytics must rest on the three pillars shown in Fig. 1(b). Visual Analytics must still contribute advanced methods for presenting information and interacting with users, while Multimedia Analysis must contribute increasingly accurate and rich methods for analysing multimedia to add semantic information about its content. However, in order to scale gracefully to very large collections, Database Management must contribute advanced methods for managing storage of and access to the large multimedia collections.

Large collections obviously require advanced storage management due to their size. Their dynamic nature—rapid growth, rapid development of file formats and capturing technology, and rapid evolution of sharing models and analysis of this information—also requires techniques for supporting multimedia analytics on dynamic collections, especially when analyzing recently added material.

Furthermore, the multimedia analytics process may span a long period of time, possibly requiring the cooperation of several analysts, which must then share the insight and knowledge extracted from the collection. For this purpose a data model is required that can seamlessly integrate the insight and knowledge into the information extracted from the existing multimedia collection. Such a data model must persistently keep track of and structure the relationships between data, knowledge and context.

Finally, query processing must be supported in the analytics process, with potentially different requirements depending on the context of the analytics application, such as whether the scope of the analysis is wide or narrow. Maintaining real-time performance in this environment will require managing resources very effectively, using a range of techniques for knowledge representation, database management, and computation management.

Database techniques to address the above challenges have been proposed to support either visual analytics or multimedia analysis, but the techniques used in each case are very different. All of the work to date has thus only addressed the combination of two of the pillars, leaving a wide gap in the middle of Fig. 1(b). It is clearly necessary to focus on all three pillars at the same time, if we are to make progress towards truly scalable multimedia analytics.

1.4   Contributions of this Paper

The contributions of this position paper are to (i) propose and elaborate on the above definition of the goal of scalable multimedia analytics, (ii) briefly review the work in the fields of multimedia analytics and database management which could help in reaching this goal, and (iii) present some important research directions that we believe must be addressed in order to achieve progress in this field.

2   Multimedia Analytics

In this section we discuss the current state of the art in multimedia analytics. We first consider the specific requirements of multimedia, before describing the multimedia analytics process. We conclude by considering the current approach in the field to achieving scalability.

2.1   From Multimedia Analysis: Multimedia Representation

For structured and numeric data the interpretation of the data items themselves is always at the same level of abstraction. Analytics comes about when aggregating the data and studying patterns through statistics. In contrast, an individual image or video can be interpreted in many different ways. Depending on the nature of analytics task, as illustrated in Fig. 2, the multimedia content may have to be interpreted at different semantic levels, associated with: (a) low-level visual features (e.g., colour histograms), (b) semantic concepts (e.g., objects, settings and events), (c) semantic theme (e.g., physics, immigrants or cultural identity) and (d) complex human interpretation, including factors such as sentiment and aesthetic appeal.

Recent advances in multimedia analysis have opened the door to enabling search and exploration at all semantic levels. However, while the features extracted from the content are getting increasingly descriptive, the size of the resulting feature collection is still prohibitively large for real-time user interactions, a key aspect of multimedia analytics. Furthermore, there is no "universal" feature representation satisfying relevance criteria in a wider range of analytics tasks. For example, a particular video search query may require a simple query-by-example matching based on low-level visual features, while utilizing concept detectors or automatic speech recognition may yield better results in other cases [10]. In the past several attempts have been made to standardising multimedia content descriptions, as well as multimedia items and user interactions with them (e.g., the MPEG standards). Some of the main reasons for their limited adoption were the insufficient effectiveness of early content analysis approaches and the inflexibility of description schemes with regard to accommodating emerging information-rich content sources (e.g., social multimedia portals) and complex user interaction modes.

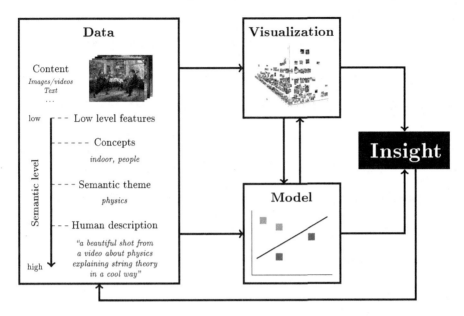

Fig. 2. Multimedia analytics process [13], adapted from the visual analytics diagram by Keim et al. [5]

2.2 From Visual Analytics: Multimedia Analytics Process

A large body of work related to the constituent parts of multimedia analytics has been surveyed by Zahálka and Worring [13]. The objective of multimedia analytics is *user insight*, a complex understanding of the analyzed data accumulated over time using all or most of the relevant data at hand [8]. The concept of insight is quite familiar in the field of visual analytics: the conceptual diagram by Keim et al., instantiated for multimedia analytics in Fig. 2, is one of the cornerstones of the field. In contrast with visual analytics, however, a media item is more complex than a data point and the analyst cannot fully understand it before seeing it; there is thus a trade-off between giving an overview of the collection and showing the individual media items in detail.

The palette of tasks leading towards insight is quite colourful. Nevertheless, all multimedia analytics tasks have a key common characteristic: they consist of a certain proportion of exploration and search. Hence, an *exploration-search axis*, shown in Fig. 3, has been proposed as the task model for multimedia analytics [13]. The analysts tilt back and forth on this axis during their quest for insight, and hence a multimedia analytics system should support the entire axis. *Analytic categorization*, i.e., maintaining a set of analyst-defined categories based on the semantic and metadata content of the multimedia collection and updating it as the analyst progresses towards insight, has been proposed as an umbrella task for the exploration-search axis [13].

Many challenges for multimedia analytics arise due to the need for overcoming two gaps. The *semantic gap* is defined as the disproportion between the

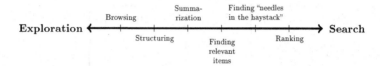

Fig. 3. The exploration-search axis with example multimedia analytics tasks [13]

semantics extractable by the humans on the one side and the machines on the other. This longstanding challenge aiming at understanding the content of a single multimedia item is increasingly being addressed using deep learning based on huge amounts of training data. Yet the semantic gap is as prevalent at the (sub-) collection level at which it has hardly been addressed. The *pragmatic gap*, defined as the gap between the highly adaptable mental categorical model of the user on the one side and the strict, bounded definition of categorization in the machine world on the other, comes into play when exploring a multimedia collection in its context [13]. In order to close the pragmatic gap, the data model of the analytics system must fully adapt to user intent and understanding, as it varies over the duration of the analytics process, so that it accurately represents the view of the user at each time.

2.3   Scalability Considerations

Multimedia analytics state of the art, however, has up to now to the best of our knowledge not explicitly considered the issue of data management, despite aiming for large-scale analytics. This applies both to the model described above, as well as to the pioneer systems conceived so far (e.g., [1,11]). As Fig. 4(a) indicates, data is only present in main memory, which limits the scale of the systems both in data volume and duration and makes the data processing ad hoc for each respective multimedia analytics system.

As illustrated in Fig. 4(b), a modern multimedia database should support search and exploration through optimal utilization of available query analysis and retrieval algorithms and ideally eliminate the need for constructing a separate framework for each analytic task from features to user interaction models. Tightly integrating a suitable data model and query processing techniques with multimedia analytics has the possibility to increase the scale of multimedia analytics and truly utilize multimedia collections as knowledge bases, rather than individual datasets. Effective data models and query languages must be able to support exploration and search based on relevance criteria defined at various semantic levels [9]. More particularly, the query model should facilitate user intent analysis and the definition of complex relevance criteria.

As the size and heterogeneity of multimedia collections increase, analysing them in their entirety becomes infeasible. The data model should thus facilitate efficient filtering of parts of the collection needed in a given analytics session. It should further enable translation of features into representations allowing for seamless interactions with the system. Finally, both data model and query language should facilitate the dynamic choice of optimal retrieval algorithms.

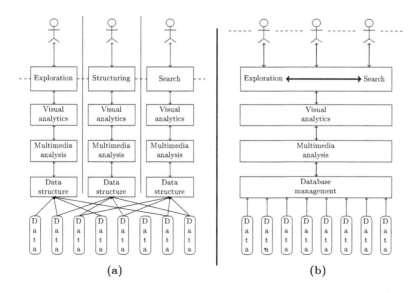

Fig. 4. Transition from (a) ad hoc to (b) scalable multimedia analytics systems

3   Database Support

In this section we review the existing support for multimedia search on one hand and analytics workloads on the other. We conclude that existing work is indeed insufficient to cope with scalable multimedia analytics, as defined above, and discuss some techniques that can pave the way forward.

3.1   Multimedia Search

High-dimensional indexing has been studied for decades, but it is only during the last decade that some breakthroughs have been made: it is now possible to run efficient similarity searches at a truly large scale. Some of those approaches are main-memory based [3], while others adapt gracefully to disk-based collections [6], but they all employ some sort of approximation—either in the descriptor generation or during query processing—to trade efficiency for accuracy.

Recently, researchers have started applying big data analytics tools, such as Hadoop and Spark with their map-reduce programming model, to high-dimensional indexing [7]. The conclusion of this work so far is that while these big data tools can support very large collections and provide excellent throughput, they come nowhere close to providing the interactive query response times that are required for multimedia analytics workloads. Furthermore, none of these tools provide a data model that can represent any of the complexities of multimedia analytics collections adequately.

3.2   Analytics Workloads

Analytics workloads have mostly been considered in two domains: business analytics and the more general big data analytics. As discussed above, current big data analytics tools are not suitable for multimedia analytics due to their response time. In the business analytics domain, however, data warehouses are used to extract data from their sources and stored in a database schema that keeps sufficient information to facilitate obtaining useful and meaningful insights, yet in a format that is sufficiently simple to do so interactively. Of course, business analytics is concerned with numerical data which is much simpler than multimedia data and, unlike multimedia data, supports easy aggregation of information.

A data model for multimedia analytics must provide sufficient semantics to facilitate long-term accumulation of data and insights, yet be simple enough to allow relatively simple query formulation and efficient query processing and optimization. The relational data model, for example excels at the latter, but handles complex relationships poorly. Ad hoc data structures can handle any relationships, but query formulation then amounts to low-level programming. The multi-dimensional model of OLAP applications seems to represent a good middle ground for those applications; similarly a good middle ground must be developed for multimedia analytics workloads. This work seems to provide a direction for going forward towards multimedia analytics and indeed recent proposals for data models for multimedia analytics have been based on this work, including Multimedia PivotTables [12] and the O^3 data model [4].

3.3   Database Management

Other aspects of database management are also relevant in the multimedia analytics domain. While space constraints prevent both complete coverage and full citations, the following list indicates the range of techniques that could be used: Transaction support ensures data integrity by enforcing the ACID properties of atomicity, consistency, isolation and durability; Query optimization dynamically chooses the best query processing algorithms and access paths, depending on query, data and hardware characteristics; Caching is used to dynamically retain as much data as possible in memory and process this data while fetching the remainder, to hide the latency of the underlying collection; Parallel and distributed processing are used to divide the workload to as many computing cores or computers as possible; Approximation and sampling is used to reduce the work needed to produce a first answer, which may then be incrementally updated if more time becomes available. A complete database management solution for multimedia analytics must undoubtedly draw on all of these aspects. In some cases, tried and tested techniques will be directly applicable (e.g., for transaction management) while in other cases entirely new methods must be developed based on the data model and associated query model.

Table 1. Ten research questions for scalable multimedia analytics

RQ1	How can database management techniques facilitate multimedia analytics for increasingly large-scale collections of multimedia items and metadata?
RQ2	Is a novel multimedia query language needed for the database to fully support multimedia analytics, or is an extension of classic query languages sufficient?
RQ3	Can the database management system support the different workloads that arise along the exploration-search axis?
RQ4	Can database management techniques be utilized to improve the quality of modality fusion?
RQ5	How can the multimedia analytics system facilitate up-to-date interactive analysis of collections that are dynamically and rapidly evolving?
RQ6	Can persistence of the machine learning model be used to improve the quality and efficiency of the analysis?
RQ7	How can insight history and the context of insight be represented as the insight develops through time?
RQ8	How could long-term learning be leveraged to improve the extraction of high-level semantics?
RQ9	How can database management techniques be leveraged to improve the user's interactive experience?
RQ10	How can the multimedia analytics system best support dynamic discovery of new analytic categories that lead to insight and knowledge?

4   Research Questions

In this section, we discuss several research directions arising from the discussion above that we believe must be addressed in order to make progress towards scalable multimedia analytics. This list, summarized in Table 1, expands on the multimedia analytics research agenda of [13], focusing on issues related to scalability. For clarity, we divide the research questions into the four axes—or Vs—of scalability described in Sect. 1.2, but of course a complete solution must address all of these issues.

4.1   Volume

With the ever increasing volume of multimedia collections in virtually all application domains, all components of multimedia analytics must handle large-scale data more efficiently. Managing increasingly large-scale data is not specific to multimedia analytics. With the advances in camera and smartphone technology, however, individual images and videos have increasingly higher resolution and detail, thus increasing not only the number of multimedia items, but also the size per item. These challenges are reflected in **RQ1**.

4.2   Variety

The variety of both data and tasks within the multimedia domain presents an interesting potential for information gain, but also brings many processing challenges. As mentioned in Sect. 2.1, multimedia data makes common database aggregations and data operations difficult or impossible. The limitations of current query languages with respect to semantics inspire **RQ2**.

Analysts conduct a variety of tasks in the multimedia domain. These tasks, modelled by the exploration-search axis, require different data to be presented to the user, which could be possibly handled by database management techniques, leading to **RQ3**.

A second aspect of variety is the variation in the multimedia data itself. Fusion of individual modalities is required to truly utilize the heterogeneous nature of multimedia. Efficient fusion at the database level could positively impact semantic quality, inspiring **RQ4**.

4.3   Velocity

The main challenge of velocity is that of handling collections that are growing rapidly, allowing the analysis of up-to-date information and merging these with existing information; this is represented by **RQ5**.

Furthermore, database persistence has a tremendous potential to improve incremental analysis not only by persisting data, but also by persisting elements of the analysis itself: the machine learning model used by the systems and the history of insight as the user develops it over time. Persisting and reusing the elements of the analysis reduces the start-up time of the analysis. Moreover, this enables the use of multimedia data as a true knowledge base: instead of starting the analysis every time anew, analysts are able to continue where they or their colleagues left off in previous sessions. Longitudinal analysis of the stored data might thus very well improve the accuracy of high-level semantic concepts and their boundaries, improving semantic quality in general. These considerations lead to research questions **RQ6** through **RQ8**.

4.4   Visual Interaction

The interactivity requirement of multimedia analytics places a rather stringent requirement on performance throughout the entire pipeline. Maintaining and improving interactivity with database management is a research challenge of its own, as witnessed by **RQ9**.

As mentioned in Sect. 2.2, analytic categorization was introduced as an umbrella model for user tasks [13]. Whether it is a sufficient model, however, remains an open question. The degree of categorization support from the database management component of the system is an open question as well. Moreover, analytic categorization involves enabling the user to discover new categories as she progresses towards the insight. These categorization-related concerns inspire the final research question **RQ10**.

5   Conclusions

In this paper we have argued that research is needed at the boundary of multimedia analytics and database management, and in fact that database management should be integrated as the third pillar of *scalable multimedia analytics*. We have presented a list of important research challenges that relate to the scalability of the multimedia analytics process. This list is no doubt incomplete and as these issues are addressed new will arise. What is important, however, is that the research community immediately starts tackling these research questions so that we can start harvesting the information encoded in today's and tomorrow's multimedia collections.

Acknowledgments. This work was in part supported by the CNRS PICS grant "MMAnalytics" and by sabbatical support from Reykjavik University. Thanks go to the anonymous reviewers, for comments that helped improve the paper.

References

1. Burtner, R., Bohn, S., Payne, D.: Interactive visual comparison of multimedia data through type-specific views. In: Wong, P.C., Kao, D.L., Hao, M.C., Chen, C., Healey, C.G. (eds.) SPIE Conference on Visualization and Data Analysis (VDA), Burlingame, CA, USA, pp. 86540M–86540M-15 (2013)
2. Chinchor, N.A., Thomas, J.J., Wong, P.C., Christel, M.G., Ribarsky, W.: Multimedia analysis + visual analytics = multimedia analytics. IEEE Comput. Graph. Appl. Mag. **30**(5), 52–60 (2010)
3. Jégou, H., Douze, M., Schmid, C.: Product quantization for nearest neighbor search. IEEE Trans. Pattern Anal. Mach. Intell. **33**(1), 117–128 (2011)
4. Jónsson, B.Þ., Tómasson, G., Sigurþórsson, H., Eiríksdóttir, Á., Amsaleg, L., Lárusdóttir, M.K.: A multi-dimensional data model for personal photo browsing. In: He, X., Luo, S., Tao, D., Xu, C., Yang, J., Hasan, M.A. (eds.) MMM 2015, Part II. LNCS, vol. 8936, pp. 345–356. Springer, Heidelberg (2015)
5. Keim, D.A., Kohlhammer, J., Ellis, G., Mansmann, F. (eds.): Mastering The Information Age-Solving Problems with Visual Analytics. Eurographics (2010)
6. Lejsek, H., Jónsson, B.Þ., Amsaleg, L.: NV-Tree: nearest neighbours at the billion scale. In: De Natale, F.G.B., Del Bimbo, A., Hanjalic, A., Manjunath, B.S., Satoh, S. (eds.) ACM International Conference on Multimedia Retrieval (ICMR), Trento, Italy, pp. 54:1–54:8 (2011)
7. Moise, D., Shestakov, D., Guðmundsson, G.Þ., Amsaleg, L.: Indexing and searching 100M images with Map-Reduce. In: Jain, R., Prabhakaran, B., Worring, M., Smith, J.R., Chua, T.S. (eds.) ACM International Conference on Multimedia Retrieval (ICMR), Dallas, TX, USA, pp. 17–24 (2013)
8. North, C.: Towards measuring visualization insight. IEEE Comput. Graph. Appl. Mag. **26**(3), 6–9 (2006)
9. Rudinac, S., Larson, M., Hanjalic, A.: Learning crowdsourced user preferences for visual summarization of image collections. IEEE Trans. Multimedia **15**(6), 1231–1243 (2013)

10. Snoek, C.G.M., van de Sande, K.E.A., de Rooij, O., Huurnink, B., van Gemert, J.C., Uijlings, J.R.R., He, J., Li, X., Everts, I., Nedovic, V., van Liempt, M., van Balen, R., Yan, F., Tahir, M.A., Mikolajczyk, K., Kittler, J., de Rijke, M., Geusebroek, J.M., Gevers, T., Worring, M., Smeulders, A.W.M., Koelma, D.C.: The MediaMill TRECVID 2008 semantic video search engine. In: Over, P., Awad, G., Rose, R.T., Fiscus, J.G., Kraaij, W., Smeaton, A.F. (eds.) TRECVID Workshop, Gaithersburg, MD, USA (2008)
11. Viaud, M.L., Thièvre, J., Goëau, H., Saulnier, A., Buisson, O.: Interactive components for visual exploration of multimedia archives. In: Luo, J., Guan, L., Hanjalic, A., Kankanhalli, M.S., Lee, I. (eds.) ACM International Conference on Image and Video Retrieval (CIVR), Niagara Falls, Canada, pp. 609–616 (2008)
12. Worring, M., Koelma, D.C.: Insight in image collections by multimedia pivot tables. In: Hauptmann, A.G., Ngo, C.W., Xue, X., Jiang, Y.G., Snoek, C., Vasconcelos, N. (eds.) ACM International Conference on Multimedia Retrieval (ICMR), Shanghai, China, pp. 291–298 (2015)
13. Zahálka, J., Worring, M.: Towards interactive, intelligent, and integrated multimedia analytics. In: Chen, M., Ebert, D.S., North, C. (eds.) IEEE Conference on Visual Analytics Science and Technology (VAST), France, Paris, pp. 3–12 (2014)

Shaping-Up Multimedia Analytics: Needs and Expectations of Media Professionals

Guillaume Gravier[1], Martin Ragot[3], Laurent Amsaleg[1(✉)], Rémi Bois[1], Grégoire Jadi[4], Éric Jamet[3], Laura Monceaux[4], and Pascale Sébillot[2]

[1] CNRS, IRISA and Inria Rennes, Rennes, France
`laurent.amsaleg@irisa.fr`
[2] INSA Rennes, IRISA and Inria Rennes, Rennes, France
[3] Univ. Rennes 2, CRPCC, Rennes, France
[4] Univ. Nantes, LINA, Nantes, France

Abstract. This paper is intended to help clarifying what multimedia analytics encompasses by studying users expectations. As a showcase, we focus on the very specific family of applications doing search and navigation of broadcast and social news content. This paper is first describing what professional practitioners working with news currently do. Thanks to extensive conversations with media professionals, mockup interfaces and a human-centered design methodology, we analyze the perceived usefulness of a number of functionalities leveraging existing or upcoming technologies. This analysis helps (i) determining research directions for the technology underpinning the very recent field of (multi)media analytics and (ii) understanding how multimedia analytics should be defined. In particular, dependency to the domain is discussed: are multimedia analytics tasks domain-specific or can we find general definitions?

1 Introduction

From the early days of DARPA and NIST evaluations, news search and navigation have been widely studied. The fundamental technologies needed to search and navigate news have been designed and evaluated in a number of different, though closely related, contexts such as broadcast news transcription, topic detection and tracking, automatic content extraction, etc. Several systems and prototypes originated from there. Early technology-driven systems such as Broadcast News Navigator [13], Informedia [8] and Físchclár news [12] primarily targeted indexing for search based on transcription, topic segmentation and entity extraction. Departing from the index and search philosophy, pioneering work on topic threading aimed at organizing news collections with explicit links to offer event-oriented navigation capabilities, e.g., [9,21]. Exploratory news visualization and exploration interfaces were also designed in lab settings [7].

However, no commercial news navigation product is widely operational today for media professionals. Practitioners mainly rely on general public search systems, synthesizing and organizing search results by themselves. For example, press agents in state offices are often asked to make a brief on a given topic,

© Springer International Publishing Switzerland 2016
Q. Tian et al. (Eds.): MMM 2016, Part II, LNCS 9517, pp. 303–314, 2016.
DOI: 10.1007/978-3-319-27674-8_27

e.g., Charlie's killings in Paris, mentioning the main dates and the chronology of the event, the causes and consequences as well as the main characters. To do so, they search media databases to complement their personal knowledge. Obviously, tools to organize, navigate, synthesize and extract information from news collections would be beneficial. This is even more true now that information sources have been significantly multiplied, in particular with social networks and user-generated news and comments.

Multimedia analytics exactly match this description: organize data collections and provide tools to extract knowledge by interacting with the data. Practitioners have at hand a large amount of multimedia material that they need to understand and explore in order to gain insight. Taking advantage of this insight, they will, e.g., create some new multimedia material for the general public to have a better understanding of the society they live in, and/or provide other practitioners with another source of knowledge useful to get insight, etc. While multimedia analytics is a recent field, there are already several technological solutions available to facilitate working and understanding multimedia collections. In particular, the sheer number of content description and search tools published in the literature clearly shows that multimedia analysis has reached maturity. So why haven't we real-world, commercial systems, facilitating multimedia analytics over news data?

To better understand why practitioners have not adopted the latest technology and to foster new research directions in media analytics targeting users' needs, this paper studies current practices and expectations in the news media business. In contrast to many other papers, the study here is not technologically oriented but rather user-oriented. It does not expose users to the vast bestiary of existing tools, asking them to imagine what they could be doing with these tools. The study in this paper goes the other way around and adopts a general human-centered design approach for interactive systems. We use here an ergonomic analysis based on interviews of practitioners, pursuing multiple goals via media analytics. First, we want to gain a better understanding of potential users and of their professional activity to identify relevant functionalities to develop. Second, we want to measure the perceived usefulness of a number of functionalities to prioritize research and development. The idea is to measure how users perceive the utility of functionalities that can be developed based on existing and upcoming technology, before actually developing a prototype system. Finally, we also want to gather suggestions and new ideas from discussions with future users, so as to shape the future of multimedia analytics.

2 Methodology

2.1 The UCD Approach

The approach taken to better understand how media professionals work and how media analytics can assist them borrows from human-centered design (UCD). The ISO 9241-210 standard identifies six basic principles governing a human-centered design approach, leading to four main activities in application design:

(*i*) understanding and characterizing the context of use: skills or habits of potential users, tasks to perform with the system, analysis of the environment in which the tasks will be performed; (*ii*) identifying users' demands; (*iii*) producing design solutions (scenarios, mock-ups or prototypes) based on technical knowledge; (*iv*) assessing these solutions in relation to the demands. With the goal of defining major trends for (multi)media analytics and shaping the related research directions, we consider here the first three activities applied to the design of an interface to search and explore a collection of documents related to the news domain so as to gain insight and extract information.

The scenario considers large-scale collections of newswires and online newspapers, radio podcasts and videos along with comments and reactions from the web, e.g., via blogs, or on social networks. We explicitly envision two distinct uses of the application. The first case is a standard *information extraction* usage where the interface is used to search for a precise piece of information, be it a simple fact such as a date, a location, or the reaction of public persons to an event. The second case considers *information exploration* (or analytics) where the interface is typically provided to apprehend a subject or an event, gain insight and get a global understanding, e.g., to make a synthesis of an event. This is typically what press agents do when they're asked a press kit or a brief on a subject. Contrary to the search scenario, we believe that advancing the state of the art in Multimedia Analytics is needed to address this task and develop better analytics tools.

The target population for such a navigation interface is typically that of media professionals—journalists, press agents, news-related community managers and website designers/editors, politicians and their press assistant, etc.— rather than the general public whose needs are typically very heterogeneous.

2.2   Interview Stage #1: Knowing the Crowd

Over a period of one month, we interviewed 13 media professionals representing three different professions: journalists, press agents, and community managers. All analyze the press on a daily basis though with different purposes. The professions were chosen to provide a variety of practices for which we believe technology can help. Each interview was divided in four main parts. A questionnaire was first used to gain a better understanding of the interviewee's profile, in particular regarding his relation to the press and his degree of acquaintance with Internet and multimedia technology. The interview itself, led by an ergonomist, started with an analysis of the current activity and practices. Each interview lasted from 1 h to 1:30. Thirteen professionals were selected for these user tests. This sample seems adequate compared to studies on this subject. Indeed, a user test conducted using a sample between 5 and 10 persons is enough to reveal the majority of usability problems [15].

The 13 interviewed persons were mostly male (9/13): 4 press agents, 7 journalists and 2 community managers. Average age is 32 with an average professional experience of 5 years. We measured several aspects of their technology profile. Personal innovativeness, i.e., willingness to test innovative technology [1], gets a score around 5 on a Likert scale from 0 (innovation reluctant) to 10 (innovation victim), except for community managers having an average score of 8.5. Skills in Internet browsing and information retrieval range from average for press agents (6 on a scale from 0 to 10) to high for community managers (10) with journalists in between. All practitioners make heavy use of social networks, mostly Facebook (100 %) and Twitter (91.7 %), with 58 % of the persons spending more than 5 h daily on social networks. Search engines on the Internet are frequently used for all groups.

We also measured how traditional media sources are used. Surprisingly, TV news are not frequently used by any of the professional categories. Newspapers are regular sources for press agents and journalists who however mostly rely on the radio news, with many of them listening to radio news several times a day. News aggregation sites are not usual sources of information except maybe for some journalists where 3 out of 7 visit such sites at least once a day.

One interesting outcome of the study of practices is that media aggregation interfaces are not used (and not trusted) by media professionals. We see that as a sign that search engines and content aggregation hardly make analytics! With this in mind, we explore functionalities that could be useful for media analytics.

2.3 Interview Stage #2: Assessing Design Decisions

The second stage of the interview addressed the assessment of the design solution, focusing on the acceptability of various functionalities. Acceptability refers to an individual's perception of the value of a system or a technology. According to the technology acceptance model [3], the two most influential theoretical constructs in terms of acceptability are the perceived usefulness of a technology (i.e., the degree to which a person believes that using a given system will improve performance) and its perceived ease of use (i.e., the degree to which a person believes that using a given system will require little or no effort). In this study, we focus on perceived usefulness to evaluate more specifically each of the proposed functionalities. Finally, ranking the functionalities was proposed to conclude the interview and global suggestions were collected.

From a practical point of view, evaluation of the functionalities was done using a psychometric Likert scale from 0 (strongly disagree with the proposition) to 10 (strongly agree with the proposition). In addition, we analyzed verbatim transcripts of the interviews for a better understanding of the judgments. Note that ranking the functionalities was done interactively with the interviewees who were allowed to ask for details. We observed that this interaction was needed for persons to get a good understanding of non-conventional functionalities, thus fully justifying the use of interviews as opposed to a online questionnaire. In preliminary experiments, the latter was deemed not adequate for complex functionalities and usage significantly departing from standard practices. We shall note here that this situation might arise on a regular basis in the assessment of what multimedia analytics can do and how it can help, with foreseen practices departing from the standard search philosophy.

3   Perceived Usefulness of Analytics Functionalities

During the interviews, the navigation interface was globally described to a person as a web-based system to explore the various types of media data used in his/her professional activity, offering a number of functionalities to search information, have a synthetic view of documents, see links between documents in the collection, explain those links, etc. It is important to stress again that no actual system was used at this stage of the design process. The understanding of the functionalities is thus based on explanations where the interviewee imagines how he would use the functionality rather than based on experience. Following the UCD approach, we built mockups screens such as the one illustrated in Fig. 2 to help describing a wide range of functionalities. Some of the functionalities obviously build on existing technology while others are clearly innovative if not futuristic, e.g., generating a short text that explains the links between two documents. Functionalities were grouped in four broad categories, respectively related to text and keywords, social networks and opinions, links and recommendation, and content abstraction and fast access (details are given below).

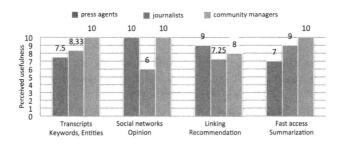

Fig. 1. Perceived usefulness of each group of functionalities.

A synthetic view of the trends is given in Fig. 1, where perceived usefulness is reported for each professional profile and for each group of functionalities. Globally, for press agents, social network analysis and linking/recommendation are the two most useful functionalities. For journalists, fast access to information ranks first along with transcripts, keywords and entities. Community managers find everything useful, with a weaker interest for links and recommendation.

3.1   Transcripts, Keywords and Entities

The first group gathers fairly classical tools highlighting salient information in a document or in the collection based on language data, mostly from automatic speech transcription for radio and video content. Showing the transcript appears as the most straightforward possibility. Keywords as well as entity extraction and characterization can further be used to present key information to users.

Six persons judged displaying transcripts useful and 3 mentioned this functionality would help them save time. Displaying word clouds instead of the whole transcript was globally perceived negatively, 3 persons mentioning that there is no interest in doing so, 2 mentioning that word clouds are not practical. Highlighting proper names was judged useful or interesting by 5 persons, but only one found interest in having direct access to the corresponding biography (if not from Wikipedia). Highlighting locations was surprisingly judged of no use by 3 while 3 mentions that using a map would be more appropriate. In the free suggestion phase, 5 persons for whom classifying keywords according to relevance and frequency would be nice, better than displaying word clouds.

From the technology point of view, this set of functionalities builds on existing tools (automatic speech recognition, named entity detection, term and keyword extraction, etc.) and common practices in the field of interfaces (word cloud, Google map localization, etc.). Comments during the interviews however highlighted the importance of the accuracy of those techniques: accuracy and robustness of the underlying content extraction tools involved should still be improved to make them fully acceptable in media analytics. Automatic processing quality assessment, i.e., saying whether or not the outcome of content analysis can be trusted, is needed. Comments on Wikipedia also highlight that external knowledge sources must be trustworthy to be of interest for media analytics.

3.2 Social Networks and Opinions

A number of functionalities related to opinion mining in social networks were also considered, ranging from standard and existing opinion mining techniques targeting valence (positive, negative, neutral) to upcoming fine-grain characterization techniques. In particular, recent work in opinion mining addresses the identification of the emotions (anger, surprise, fear, etc.) expressed in addition to valence [6] and it is expected that these techniques will be mature in the near future. The identification of the aspects on which people are reacting also appears as a promising functionality. Typically, on a news item such as the Strauss-Kahn Sofitel scandal, people reacted on several aspects: the main offender, the director of the International Monetary Fund, the candidate to the French presidential election, etc. After identifying the different aspects targeted in social networks, opinions and feelings can be analyzed, displayed and synthesized for each aspect individually to have a better synthetic view of social reactions.

The group of functionalities was globally judged interesting and useful, with 5 persons mentioning usefulness and 3 time gain. Perceived usefulness is particularly high for press agents, while journalists are more doubtful. The ability to analyze the evolution of opinions over a period of time was well perceived in terms of usefulness, with 5 persons mentioning this functionality. Extracting opinions and feelings, whether globally on a subject or more precisely for an event, was also found interesting. Looking at feelings appears more interesting at a global level while opinions seem more interesting for a precise event. Unsurprisingly, opinion analysis in traditional media (as opposed to social media) is judged of limited interest. The lack of confidence in the outcome of automated opinion analysis was again mentioned on several occasions. Finally 5 persons suggested a functionality implementing filters to sort comments (by social network, by keywords, by number of retweets, likes, followers, etc.).

There clearly is a high demand for social network analysis to extend the classical sources of information of media analytics. There is apparently a clear interest to go further into the characterization of opinions and feelings, beyond valence. This is an ongoing research topic and confidence in the result indicates that progress are still required for this technology to be considered as mature for social media analytics: better ressources (lexicons, tree-banking of social texts, etc.), fine grain lexical and syntactic analysis on degraded language are among the elements that will shape the future. As for the first group of functionalities, being able to measure the confidence in the outcome of processing modules is vital. Explaining the decision, which is something out of reach of the current technology, appears like a good option to help with this issue.

3.3 Links and Recommendation

Creating explicit links between documents in a collection provides a set of functionalities to group similar items, recommend content related in some way to a document or provide a chronological thread of an event. We also consider explicit linking as a potential prerequisite for analytics, exploiting generic graph analytics techniques and knowledge propagation.

Links can be established on a number of grounds—e.g., same content, same event, same topic, same person involved—and organized in various ways: chronological order to show the evolution of an event, reactions to an event, causes and consequences. To help better apprehend why links are established, link characterization can further be used to inform users of the interface of the meaning of the link. The mockup screen in Fig. 2 illustrates these functionalities.

Providing links and recommendations with documents do not appear as a priority. We asked people to rank the types of links according to perceived usefulness and obtained the following ranking: by topic, keywords, date, location, etc. Linking by dates and/or locations was however mentioned several times as useful during the interviews. Regarding link explanation, journalists found fine-grain explanation of why the link is provided more useful than a simple coarse grain explanation such as a type (average rating of 8.5 vs. 6.3). Limiting redundancy by grouping highly similar documents wasn't judged as necessary. However, press agents rank the functionalities 'grouping similar documents' and 'highlight key/central documents' with a perceived utility of 10. Interestingly, these two features can easily be implemented using graph analytics should the data be organized as a graph. A suggestion, made by two persons, is to emphasize links to raw information sources, i.e., not processed by (other) journalists.

While the technology to compare content exists for all modalities, little has been done to create links on a large scale in collections. Evaluations on video hyperlinking over the past few years, e.g., [5], are typically first step in this direction. Clearly, analytics cannot satisfy itself of documents taken regardless of the collection. The question of knowing what usage for what types of links and structure (k-nn graphs, threads, etc.) thus appears as a crucial one for analytics.

Fig. 2. Mockup user interface illustrating the functionalities related to links and recommendation. The current item appears in the top-left corner. Below are links to items on the same subject; The right panel gives a timeline related to the current item. Explanation of a link is illustrated with the text 'Consequence of the Greek elections'.

Fine grain semantic links are still missing, along with the ability for the machine to explain why a link was established, to facilitate navigation.

3.4   Fast Access to Information

The last group of functionalities gathers a number of features whose goal is to provide a more efficient access to the information and to ease a global view of the collection (or of part of the collection). Apart from the classical table of content and search engine, several rapid access features were considered: showing a timeline of the collection or of a selected event, providing a clickable word cloud depicting the whole collection, grouping similar documents along with a summary, or highlighting key documents.

While fast access to information was globally judged as a useful set of functionalities, we were able to gain limited insight on the detailed functionalities. Topic segmentation was globally perceived as useful and time saving. Links from topic segments to the corresponding (transcription of the) content were suggested when topic segmentation was mentioned. Several persons also found interesting displaying a chronological record of an event. In contrast, summarizing documents or groups of similar documents was judged of no interest: one reason given is that machine interpretation of content cannot be trusted by professionals.

Most of these functionalities rely on existing technology and interface design. Summarization, in particular in the context of multiple documents from multiple modalities, still needs improvement in order to be accepted as part of an interface for media analytics. Yet, the perceived usefulness of these functionalities remain limited though we believe they are all relevant for access and fast selection of relevant information, an important feature of multimedia analytics. Progress in interface design to accommodate these functionalities might be considered for users to find utility in them.

4   Lessons for Multimedia Analytics

Multimedia analytics could be defined as the *process of organizing multimedia data collections and providing tools to extract knowledge, gain insight and help make decisions by interacting with the data.* This definition clearly goes beyond traditional multimedia analysis whose goal is to describe content, usually for indexing purposes, and requires organizing collections of data. Studying users practices and expectations in media analytics sheds light on the design of such applications, on technology requirements and on research directions to pursue.

Social Networks. A first important result that was highlighted by the interviews is the importance of social networks, at least in the news domain. While this does not come as a surprise, for news as for any other domain [22], this state of fact calls for tools to characterize opinions and feelings in social networks at a fine grain: better resources are required for fine grain semantic characterization; improved word representations (e.g., derived from word embeddings [14,20]) along with the corresponding learning machinery; etc. Apart from opinons and sentiments, on which we focused in this paper, a huge amount of information can be extracted from social signals, e.g., for event detection (see, e.g., [19], to cite a famous example), fact checking, etc.

Heterogeneity of Information Sources. Regardless of the scientific challenges for NLP and machine learning, the importance of social media in the news domain (and in many other domains) points out to the variety of information sources that need to be ingested and connected within multimedia analytics applications. In turn, this raises a number of scientific challenges regarding heterogeneous and distributed information integration. There are no clear model to do so as of today and mostly ad-hoc solutions are adopted, often with probabilistic modeling. Mathematical tools for the joint analysis of modalities and/or sources are still poor as of today. We will come back to this point later in the discussion with the notion of *knowledge.*

Reliablity of Sources. With the variety of sources of information comes the notion of reliability of the information, which is clearly a crucial point. This is obviously true for media professionals who prefer unedited sources, who need to know who published and who read, etc. We believe that this is also true in most application domains and that analytics interfaces must implement a trust mechanism of some sort. From a multimedia modeling point of view, we see a number of implications. First, information automatically extracted need be reliable. At the very least, an automatic decision should come with a confidence. Confidence measures have been used for a long time in some domains, e.g., speech recognition [10], but need be generalized to all content-based analysis tools. Another possible option is the ability for a classification system to explain on what ground the decision was made. Early attempts in this direction, e.g., recounting [4] or evidence [16], are encouraging and, again, should be generalized. But the current trends in big data machine learning goes in a different direction with neural models acting as black boxes. Second, content-based analysis can be used to validate information. This is clear from current trends in data journalism and fact checking [17]. These are recent research areas, mostly unexplored, implementing analytics on a limited scale. Here again, trust and confidence are crucial.

Contextualization. Another point that appears meaningful is the need for contextualization in general. Showing and analyzing a document must take into account the metadata context (authors, number of views, etc.) and the context of the collection: tweets often make sense in the context of more global threads, video fragments can hardly be understood outside the chronology of events they illustrate. Links between parts of documents form a nice basis to keep track of the context. From the technology point of view, organizing collections according to links materializing the relations between items within the collection as discussed in Sect. 3.3 is convenient for context-aware analysis, navigation and visualization. While the links themselves are not judged relevant by professionals for recommendation, they could easily be used for contextual display, a feature that was judged as highly relevant.

Dynamic Management of Knowledge. With the goal of analytics being to gain insight and knowledge, there is a need for research on knowledge representation in tight cooperation with content analysis and interpretation. We believe that multimedia modeling and collection organization should evolve as more knowledge is gained. To do so, we need knowledge representation mechanisms that can serve content-based analysis and, conversely, content-based analysis able to handle evolving knowledge. While knowledge representation and content-based modeling have mostly been two separate fields in the past, we can anticipate that they will converge, to some extent. Recent work on image processing [2] and natural language processing [18] hint in this direction. Multimedia analytics certainly goes with a trend towards using knowledge available as linked open data. Note that information trust remains an issue. Finally, as knowledge evolves with the process of multimedia analytics, a constant back and forth between data description and organization, on the one hand, and knowledge extraction via analytics on the other hand, is required. This in turn requires multimedia models and content-based description algorithms to adapt and evolve, e.g., relying on active learning techniques.

Conclusion. To conclude this discussion, let us highlight a few points that were left out of the study we performed (and thus out of the discussion) but that we judge very relevant. First of all, the dynamicity (aka velocity) of the data has been totally ignored. Many collections evolve at a very fast pace, in particular on social media. How do applications adapt to the constant stream of data and, more generally, to increasing knowledge? This question remains largely open and few multimedia modeling tools can today handle dynamic collections and knowledge evolutions. Second, user interfaces have also been disregarded to focus on functionalities regardless of their inclusion in an actual interface. All interviewees in our study insisted on the ease of use of media analytics functionalities, calling for research on interfaces and navigation in multimedia databases, e.g., to handle knowledge-aware multi-faceted view of heterogeneous and multimodal data. Recent work such as [11] on photo browsing goes in this direction and require generalization to multimodal and heterogeneous data, a challenge for the database world.

As we can see from this discussion, the future of media analytics goes beyond the sole multimedia community, standing at the intersection of multiple domains including databases, knowledge discovery, or human-computer interaction. Targeted users should also be actively involved in the process of shaping-up multimedia analytics. In the near future, efforts should be made to put all these actors together.

Acknowledgments. This work was funded via the CominLabs excellence laboratory financed by the National Research Agency under reference ANR-10-LABX-07-01.

References

1. Agarwal, R., Prasad, A.: A conceptual and operational definition of personal innovativeness in the domain of information technology. Inf. Syst. Res. **9**(2), 204–215 (1998)
2. Atif, J., Hudelot, C., Bloch, I.: Explanatory reasoning for image understanding using formal concept analysis and description logics. IEEE Trans. Syst. Man Cybern. Syst. **44**(5), 552–570 (2014)
3. Davis, F.D., Bagozzi, R.P., Warshaw, P.R.: User acceptance of computer technology: a comparison of two theoretical models. Manage. Sci. **35**(8), 982–1003 (1989)
4. Ding, D., Metze, F., Rawat, S., Schulam, P.F., Burger, S., Younessian, E., Bao, L., Christel, M.G., Hauptmann, A.: Beyond audio and video retrieval: towards multimedia summarization. In: Proceedings of the ACM International Conference on Multimedia Retrieval (2012)
5. Eskevich, M., Jones, G.J.F., Aly, R., et al.: Multimedia information seeking through search and hyperlinking. In: ACM International Conference on Multimedia Retrieval (2013)
6. Fraisse, A., Paroubek, P.: Toward a unifying model for opinion, sentiment and emotion information extraction. In: International Conference on Language Resources and Evaluation (2014)
7. Ghoniem, M., Luo, D., Yang, J., Ribarsky, W.: NewsLab: exploratory broadcast news video analysis. In: IEEE Symposium on Visual Analytics Science and Technology, pp. 123–130 (2007)
8. Hauptmann, A.G., Witbrock, M.J.: Informedia: news-on-demand multimedia information acquisition and retrieval. In: Intelligent Multimedia Information Retrieval, pp. 215–239 (1997)
9. Ide, I., Mo, H., Katayama, N., Satoh, S.: Topic threading for structuring a large-scale news video archive. In: Enser, P.G.B., Kompatsiaris, Y., O'Connor, N.E., Smeaton, A.F., Smeulders, A.W.M. (eds.) CIVR 2004. LNCS, vol. 3115, pp. 123–131. Springer, Heidelberg (2004)

10. Jiang, H.: Confidence measures for speech recognition: a survey. Speech Commun. **45**(4), 455–470 (2005)
11. Jónsson, B.T., Eiríksdóttir, Á., Waage, Ó., Tómasson, G.,Sigurthórsson, H., Amsaleg, L.: M3+ p3+ o3= multi-d photo browsing. In: MultiMedia Modeling, pp. 378–381 (2014)
12. Lee, H., Smeaton, A.F., O'Connor, N.E., Smyth, B.: User evaluation of Físchlárnews: an automatic broadcast news delivery system. ACM Trans. Inf. Syst. **24**(2), 145–189 (2006)
13. Merlino, A., Morey, D., Maybury, M.: Broadcast news navigation using story segmentation. In: ACM International Conference on Multimedia, pp. 381–391 (1997)
14. Mikolov, T., Yih, W.-T., Zweig, G.: Linguistic regularities in continuous space word representations. In: HLT-NAACL, pp. 746–751 (2013)
15. Nielsen, J., Landauer, T.K.: A mathematical model of the finding of usability problems. In: Proceedings of the INTERACT 1993 and CHI 1993 Conference on Human Factors in Computing Systems, CHI 1993, pp. 206–213. ACM, New York (1993)
16. Poignant, J., Bredin, H., Barras, C.: Multimodal person discovery in broadcast tv at mediaeval 2015. In: Working Notes Proceedings of MediaEval 2015 Workshop (2015)
17. Ratkiewicz, J., Conover, M., Meiss, M., Gonçalves, B., Flammini, A., Menczer, F.: Detecting and tracking political abuse in social media. In: ICWSM (2011)
18. Rizzo, G., Troncy, R.: NERD: a framework for unifying named entity recognition and disambiguation web extraction tools. In: Conference of the European Chapter of the Association for Computational Linguistics (2012)
19. Sakaki, T., Okazaki, M., Matsuo, Y.: Earthquake shakes twitter users: real-time event detection by social sensors. In: Proceedings of the 19th International Conference on World Wide Web, pp. 851–860. ACM (2010)
20. Turian, J., Ratinov, L., Bengio, Y.: Word representations: a simple and general method for semi-supervised learning. In: Proceedings of the 48th Annual Meeting of the Association for Computational Linguistics, pp. 384–394 (2010)
21. Wu, X., Ngo, C.-W., Li, Q.: Threading and autodocumenting news videos: a promising solution to rapidly browse news topics. IEEE Signal Process. Mag. **23**(2), 59–68 (2006)
22. Zeng, D., Chen, H., Lusch, R., Li, S.-H.: Social media analytics and intelligence. IEEE J. Intell. Syst. **25**(6), 13–16 (2010)

Informed Perspectives on Human Annotation Using Neural Signals

Graham F. Healy[✉], Cathal Gurrin, and Alan F. Smeaton

Insight Centre for Data Analytics, Dublin City University, Glasnevin, Ireland
me@grahamhealy.com

Abstract. In this work we explore how neurophysiological correlates related to attention and perception can be used to better understand the image-annotation task. We explore the nature of the highly variable labelling data often seen across annotators. Our results indicate potential issues with regard to 'how well' a person manually annotates images and variability across annotators. We propose such issues arise in part as a result of subjectively interpretable instructions that may fail to elicit similar labelling behaviours and decision thresholds across participants. We find instances where an individual's annotations differ from a group consensus, even though their EEG signals indicate in fact they were likely in consensus with the group. We offer a new perspective on how EEG can be incorporated in an annotation task to reveal information not readily captured using manual annotations alone. As crowd-sourcing resources become more readily available for annotation tasks one can reconsider the quality of such annotations. Furthermore, with the availability of consumer EEG hardware, we speculate that we are approaching a point where it may be feasible to better harness an annotators time and decisions by examining neural responses as part of the process. In this regard, we examine strategies to deal with inter-annotator sources of noise and correlation that can be used to understand the relationship between annotators at a neural level.

Keywords: Brain-computer interface · EEG · Hci · Information retrieval · Semantic

1 Introduction

In recent years there has been a focus in the multimedia analytics community towards extracting semantically meaningful value from multimedia content. Mining the semantic content of multimedia data to extract visual features is an application of content-based image retrieval (CBIR) [7]. In a naive implementation, low-level image features such as colour, texture, shape, local features or their combination could represent images [2]. However, it was noticed that the problem of the Semantic Gap arises where an individual's interpretation of multimedia content can be different from that of a machine. As a consequence in recent years higher-level semantic extraction (typically based on deep

© Springer International Publishing Switzerland 2016
Q. Tian et al. (Eds.): MMM 2016, Part II, LNCS 9517, pp. 315–327, 2016.
DOI: 10.1007/978-3-319-27674-8_28

learning) has become popular [6]. In order to be effective, such deep learning approaches require a significant amount of training data, which naturally poses a large human-factor overhead in terms of the selection of appropriate training data.

One convenient solution that has gained favour is the integration of non-expert annotations via a process of crowd-sourcing, in which contributions are solicited from a large group of people, usually from an online community. Services such as Amazons Mechanical Turk [12] provide the facility to support this. Crowd-sourcing has been applied to generate a variety of multimedia analytics datasets (and subsequently tools) such as food labelling [9], machine translation [1] and concept detection in digital images [6]. The widespread adoption of crowd-sourcing has substantially benefited the multimedia analytics community, yet it relies on a human component that is not well controlled or even well understood. Jia et al. [6] ask how can one trust the labels obtained from such services? They propose an algorithm that reduces the number of labels required, and thus the total cost of labelling, while keeping error rates low on a variety of datasets.

What is the clear from such trends is the important need for human annotators although they are potentially unreliable and in many instances agreement between non-expert annotators can be low. Since it can be difficult to solely interpretthe factors that might be affecting the way humans annotate from the annotation data and/or annotator post-task questionnaires, we propose the use of a neurally combined perspective.

Given a specified data annotation task, we explored relationships between behavioural (manual annotations) and neurally derived predictions from an EEG device, for eight annotators annotating a image collection for the presence of semantic concepts. Our findings suggest that many annotators who partake in the activity may do so in a careless manner, neglecting to annotate images of interest. We can make this observation based on the analysis of the EEG signals understood to correlate with attentional-orientation processes (i.e. *interest*). In the subsequent sections we describe our experimental procedure, discuss the results of our experiment, and frame our findings in terms of future perspectives on the application of EEG in understanding annotation tasks as they can be performed on a crowd-sourced level. We end this paper by making a list of suggestions that could be employed in future annotation activities to reduce the risk of incomplete or incorrect annotations being generated.

2   Background to EEG in Annotation Tasks

The P300 ERP (Event-Related Potential) is a neural signal present in response to stimuli (such as images) that significantly capture or engage a participant's attention [10]. While there are a large number of measures that can be extracted from EEG, the 'P300' is typically understood to provide a measure of attentional allocation/orientation to a stimulus and subsequently has been the focus of a wide variety of BCI (Brain-computer Interface) application research including

EEG-image labelling tasks [3–5,8]. Although similar approaches of using EEG for image/stimulus annotations tasks have been explored by others before, we extend upon this prior work by examining how these techniques can allow us to better understand annotators and potential underlying factors relating to quality of the annotations.

In this work we explore an alternative and complementary approach by integrating EEG (Electroencephalography) to derive multimedia dataset annotations in a controlled laboratory environment. We do this by examining neural signals present at the time a stimulus was first seen (and subsequently encoded) with respect to later annotations provided by the user/annotator as to which particular visual stimuli significantly captured their attention. We propose we should find 'P300' activity that should ultimately later correspond with the ground-truth labelled data from the annotator later in the experiment.

We use Lifelog images as our dataset because our research interest is directed at understanding and decoding neural responses to semantically rich imagesets i.e. in this task, places and artefacts that the participant would be familiar with. Although in this work we shape our focus and approach around attention orientation related processes, other authors have used implicit responses and information contained in other ERP time windows that are specific to other neural processes such as face perception/processing for instance [11].

3   User Annotation Experiment

In order to assess the level to which ERP responses could be used in a image annotation task (and understand the information they provide about how a person completes the annotation task) eight participants were recruited from the research staff/postgraduate body. Each subject took part in three phases of experimentation, during a single session. In the first phase (the pay-attention phase) participants viewed previously unseen lifelog images without a specific task to accomplish; this is to ensure that particular images that capture their attention are not as a result (e.g. semantic/visual similarity) of target image categories they will be seeking during the model-training phase (phase II). In the second phase (model-training phase) the annotator is shown images and they must identify specific objects of interest; in this phase we already have the ground truth from the attention phase. We used this second phase to build prediction models that use EEG signals and then applied these on phase I data. Finally in phase III (annotation phase), the annotators annotate the images they seen in phase I so that we can better understand the relationship between their explicit (manual) and their more implicit (EEG-based) annotations.

The eight participants recruited were academic researchers working in a computer science department; hence they would understand the importance of the annotation process and the need for accurate annotations to be provided.

3.1   EEG Setup and Configuration

The EEG data was recorded using an ActiCHamp 32-channel EEG system at 10-20 electrode locations Fp1, Fz, F3, F7, FT9, FC5, FC1, C3, T7, TP9, CP5, CP1, Pz, P3, P7, O1, Oz, O2, P4, P8, TP10, CP6, CP2, Cz, C4, T8, FT10, FC6, FC2, F4, F8 and Fp2. Signals were bandpassed between .2 Hz to 20 Hz and epochs of 50 ms to 750 ms relative to stimulus onset were extracted. Epochs were baselined to the 200 ms directly prior to stimulus presentation. Impedance was kept below 5 kOhm across channels. Noisy channels were removed and interpolated. No trial rejection strategy was used. EEG signals were rereferenced to common average reference. Independent Component Analysis was used to removed activity related to eye movements such as blinks.

Features were extracted from epoch windows generated on averaged clusters of localised channels corresponding to typical topographic mappings of related ERP phenomena: (O1 + O2 + Oz), (P3 + P7 + O1), (P4 + P8 + O2), (Pz + CP1 + CP2 + Cz), (Cz + C3 + C4 + CP1 + CP2 + FC1 + FC2), (Cz + Fz + FC1 + FC2), (Fz + Fp1 + Fp2), (F3 + F4 + Fz + FC1 + FC2), (F3 + F7 + Fp1), (F4 + F8 + Fp2), (C3 + T7 + FC5 + CP5) and (C4 + T8 + FC6 + CP6). This yielded 12 (pseudo-) channels of 50 features per channel (600 features per trial). All of these considerations were based on our experience of using EEG devices over a number of years for BCI-related applications.

3.2   Experiment Outline

Participants were seated approximately 60 cm from the computer screen and given an overview of the nature of the task (i.e. high speed image search). Following this introduction, they begin the experimental session. Participants were asked to refrain from physical movement to avoid any chance of missing an image and also to reduce movement-related noise in the EEG signal. The total experiment took about approximately 50 min per participant (including setup time). The purpose of the experiment was not explained to participants until they had completed all 3 phases of experiments.

Phase I - Pay-Attention Phase. In the first phase of the experiment participants viewed a stream of lifelog images captured from the perspective of a person walking through Dublin city centre; the participants were seeking to identify images that captured their attention. There were 470 images chosen from an archive of images approximately 1800 images after filtering. These images were manually filtered for those containing occluded images by clothing and such. Examples of the lifelog images employed are shown in Fig. 1.

These 470 images (consistent across participants) were presented in a randomised order each time to participants via a RSVP (Rapid Serial Visual Presentation) protocol at 5 Hz (high speed). Participants were instructed to look-out-for images that captured their interest/attention. This phase was completed first so as to minimize any carry-over effects from image content to be used in

(A) Top-ranked images across participants using manual annotations

(B) Top-ranked images across participants using EEG-based model

Fig. 1. Top-ranked images across participants using manual annotations (A) and EEG-based annotations (B)

the training of phase II, such as attention being orientated towards a particular image not because it is of interest itself but that it might appear to be related to an image (visually/semantically) used in the training block. In effect as this block is completed first across all participants, the application of our later trained classification models (from phase II) to this block are not compromised by non-related attentional artefacts as a result of recognising images potentially related to the training blocks.

Phase II - Model Training Phase. In phase II we collect data to be able to train EEG-prediction models that can be used to annotate the images from phase I. In each of these blocks a participant was required to count the collective number of occurrences of 4 predefined target concepts/categories. Prior to the start of phase II each participant was shown target images for these concepts and allowed time to become familiar with them before they felt they could accurately recognise them in a 5 Hz presentation.

Each target category was of a recognisable building/object which could be easily recognised from different camera angles/views. Each target category had 2 related images (e.g. 2 different photos containing the same building/artefact). Target categories were selected to be different across participants but balanced across 10 possible categories across participants. That is each participant had a unique combination of targets to search for compared to other participants with some overlap.

During this phase, 4 training blocks (80 sec each) were completed containing 80/300 target/non-targets (totalling 320/1200 target/non-targets), providing a means of capturing ground truth labelled data that in turn could be used to train EEG-prediction models. Each block contained an equal number of each of the 4 target concepts to be searched for. Participants were required to count occurrences of target concepts so as to tune their attention to the task and validate the participant was capable of performing the task. Target categories used were images of: 1. Starbucks shop, 2. a pub (The Earl), 3. Temple Bar, 4. a restaurant (Bull & Castle), 5. a government building, 6. a game shop, 7. a pub (Grogan's), 8. the outside of a public shopping mall, 9. a cafe (Bewley's) and 10. a statue in a public park.

Following completion of each training block, total counts were reported by the participants. The requirement to count targets in this way during a RSVP visual search task is known to stop issues of wandering/lapsing attention during a task. For our classification model training we used data available over all 4 training blocks.

A Bayesian Ridge classifier was used to train a model for each participant using the training data available in blocks from phase II (4 blocks of 80/280 target/non-targets). A randomised cross-validation grid search was used for parameter tuning of the model with the final accuracy of the chosen parameters validated on a subset of the available training data. Each model's 4 hyper-parameters (lambda 1, lambda 2, alpha 1, alpha 2) were sampled from a log space distribution of -20,20 (base 10) with 10 values per hyper-parameter. 20 iterations (random parameter selection) were used in estimating optimal parameters for each participant's model. A withheld test set from the training process (stratified) of 20 % (64/224 targets/non-targets) was used to ascertain the EEG classifier model's accuracy on training data as shown in Table 1 (Training Accuracy).

The model trained on blocks available from phase II was then applied to the EEG data captured in phase I to derive 'interest' predictions for each image. ROC (Receiver operating characteristic)-AUC(Area Under Curve) metric was used to measure accuracy alongside a Mann-Whitney U statistical test.

Phase III - Annotation Phase. Following completion of the training/calibration blocks participants were asked to label (via keystrokes) the 470 images shown in phase 1 into 3 categories: (A) I do not remember this image from the first block, (B) I remember this image but there was nothing particularly interesting about it that captured my attention and (C) I remember this image capturing my attention in a significant way. This was done so that we could evaluate the agreement between predictions from the phase II EEG-based model when applied to phase I, to the participant's own manual annotations of the phase 1 image content. These labels were later remapped to new labels hereafter referred to as targets (C) and non-targets (A&B) for the rest of text.

In Table 1 we show the counts for targets across annotators and the (model) accuracy of the phase 2 learned EEG model applied to phase 1 data compared to participant's annotations that were acquired in phase III. Accuracy is measured using ROC-AUC so as to allow comparison between varying target/non-target counts across participants. Mann-Whitney U p-values are also presented to assist in assessing accuracy.

We refer to EEG-prediction scores for the remainder of this text as the respective annotator's phase II learned model applied to their phase I EEG data.

Table 1. * indicates significant results @ $\alpha < .05$ (via bootstrap resampling randomization test). # Targets: Number of targets (i.e. 'interesting' images) labelled per annotator during phase III. Model Accuracy: ROC-AUC accuracy of phase II model applied to phase I EEG-data with respect to labels obtained in phase III. Training Accuracy: ROC-AUC of phase II models applied to an independent test set of data kept in phase II i.e. not part of parameter selection or training. MW-U: P-Value (Mann-Whitney U) comparing EEG-prediction scores (obtained from phase II prediction models applied to phase I data) for targets v non-targets for each annotator (as obtained in phase III).

ID	# Targets ('Interesting')	Model Accuracy	Training Accuracy	Bootstrap AUC	MW-U
1	71	.64 *	.92	.56	< .00
2	19	.41	.92	.61	.042
3	12	.64 *	.85	.61	.043
4	73	.53	.90	.57	.178
5	27	.62 *	.91	.59	.009
6	20	.54	.93	.62	.035
7	122	.45	.85	.55	.267
8	34	.49	.89	.57	.157

4   Results of the User Annotation Experiment

In Table 1 it can be seen that, for each participant, it was possible to discriminate target from non-target concepts with high accuracy for phase II data. Surprisingly, however, the accuracy of the phase II model when applied to phase I data using participant's manual annotations as ground truth labels (captured in phase III) dramatically decreases (measured via ROC-AUC). A bootstrapped ROC-AUC statistic reveals that for 3 of these participants (annotators 1,3 and 5) the model does perform above a chance level ($\alpha < .05$). As ROC-AUC might not be a sensitive/suitable enough as a test we use a Mann-Whitney U test that also reveals a similar pattern of relationships (also shown in Table 1). There is indication here too that annotators 2 & 6 might also display an albeit weak relationship between their EEG-predictions and behavioural responses.

In order to help understand the underlying relationships present between annotations and EEG predictions across annotators, in Fig. 3 we show univariate Mann-Whitney U tests (a measure we use for cross prediction accuracy) for each annotator's labels predicting each other person's EEG-based scores. Here, we can see for instance EEG-predictions for annotators 4, 7 and 8 are not well predicted by other annotator's labels (and nor the annotator's own labels). Conversely, we can see for instance labelling data from annotators 7 & 8 seemingly predict other annotator's EEG-predictions scores better than their own EEG-prediction scores. Such evidence indicates the presence of shared target image relationships across annotators. Comparatively, in Fig. 4 we can see underlying patterns of correlated (via Spearman's rho) EEG-prediction scores across annotators for phase I images. The presence of such relationships is further confirming shared responses across annotators irrespective of how they later label. A PCA

(Principal Component Analysis) of annotator EEG-prediction scores across images in phase I indicate via bootstrapping statistic (testing % variance accounted for in first component against distribution generated using randomised image - prediction score mappings within annotators) reveals PCA is finding significant patterns of co-varying activity between annotator's EEG-predictions (p < .05).

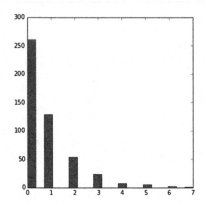

Fig. 2. Histogram showing distribution of summed manual annotation scores across all annotators for all images (N = 460). For example a score of 6 indicates 6 annotators marked the image as a target.

Examining Spearman's rho correlation on the consensus annotation counts (from phase III) and average EEG-prediction scores (from phase II trained model applied to phase I data) across all annotators we find a near significant correlation (Spearman's rho = 0.087, α = 0.06, N = 460). When only data from annotators 1, 3 and 5 are examined in this way we find greater correlation (spearman's rho = 0.204 α < .000001 N = 460). Comparatively, when only considering the remaining annotators (2, 4, 6, 7 and 8) we find no significant pattern of correlation in this way. This reaffirms our earlier observation of significant ROC-AUC scores for annotators 1, 3 and 5, and indicates such patterns between annotators might be a strong source of potential correlated activity that needs to be considered when examining relationships between group-level (combined across annotators) manual annotations and EEG-based predictions.[1] In effect some of these could be considered 'good' annotators.

Importantly, such results indicate (insofar as we can measure) that similar images are arousing similar responses across annotators, it's just that annotators 4, 7 and 8 (and less so 2 and 6) may not be accurately reporting these in a sense we can effectively measure. We find further evidence of this examining Spearman's rho correlation of the averaged manual annotation scores (for annotators

---

[1] As we are using non-parametric (rank based) statistics in our analysis, there is no difference between averaging a subset of EEG-prediction scores or taking their sum i.e. the respective underlying ranking remains the same.

1, 3 and 5) with the averaged EEG-prediction scores for annotators 2, 4, 6, 7 and 8 (Spearman's rho = .095, α = .039, N = 470). Such relationships can be further teased out examining Fig. 4 and Fig. 3. Taken in tandem these results (from PCA and examining correlative based measures) support the conclusion similar images - at least in part - captured annotator's attention in a similar way although more complex underlying relations do seem to be present.

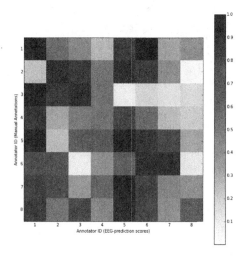

Fig. 3. Mann-Whitney U (p-values) evaluating the relationship between manual and EEG-prediction scores across participants, that is we are comparing distributions of EEG-prediction scores (for annotator on x axis) using annotator labels (y axis) for target v non-targets. P-values are inverted (i.e. 1 - p-value). That is higher values (more red) indicate more significant relationships. (Color figure online)

Although we could theoretically use the annotations provided in phase III as potential inputs in a machine learning scheme to learn models directly from - and apply to - phase I data, this presents a problem as there is high variability across user's annotation counts for the different possible response types (e.g. interesting (target) vs non-interesting (non-target)). This can be seen in Table 1 ('# Targets'). An implication of this is that there can be too few training examples in many instances to adequately train and evaluate a machine learning model. Moreover, the fact a subset of annotators annotate a larger number of images of 'interest' is likely indicating too that annotators are applying different thresholds in how they decide which images to report as having caught their attention (i.e. 'of interest'). Analysis in the presence of such an unequal distribution of label counts across participants could likely be improved by using a likert scale and normalising scores as one method to improve comparability i.e. the annotator rates from 1 to 10 how interesting they found the image. Inherent too in this approach, however, is the fact that annotators may equally fail in providing consistent labelling judgements over time. We find evidence of this in our labelling (phase III) task, that is the distribution of target labels made

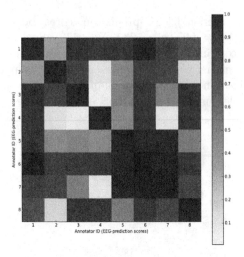

Fig. 4. Spearman-r correlation p-values between participant's for EEG-based annotations. P-values are inverted (i.e. 1 - p-value)

by annotators tend to be made in a way during the annotation task that does not fit with an expected uniformally-tending distribution across time. As the presentation order of images to be annotated in phase III is randomised we should expect judgements to be made in such a way that labels of a similar type should not be clustered together in time any more than by chance. It would seem annotators are not annotating consistently over time, however, indicating a fundamental problem in how we might naively interpret these labels as being a reliable ground truth. Using a bootstrapping process with a chi-square statistic examining the expected distribution of target labels during the labelling task we find for annotators 4, 5, 6, 8 this effect of non-uniform distribution is present ($\alpha < .05$). We involve a bootstrapping process as some assumptions for the chi square statistic are violated due to low target counts for some annotators. Comparing how much individual annotators display this effect with respect to each is compromised by the fact of these low target counts meaning in instances with lower target counts and our bootstrapping process, we are increasingly likely to make a type II error.

In Fig. 2 we can see there becomes relatively fewer images on which there is shared consensus as the consensus count increases across annotators. We define consensus here as the number of annotators that indicate the image was a target image. Here we sought to investigate whether manual annotations for images with high consensus as being targets across annotators might be indicating those who label such images as non-interesting might be doing so from a neurally driven perspective 'wrongly'. In order to investigate this we use a similar strategy as described above except we examine Mann-Whitney U p-values after applying masks (that is including or excluding certain annotators) during the averaging process for EEG-based annotations scores to be examined on a group level. In Table 2 we show consensus counts and the number of annotators in agreement

Table 2. * indicates significant results @ $\alpha < .05$ (via bootstrap resampling randomization test). Consensus N Threshold: The number of annotators that must at minimum agree on an image as a target for it to be included by the threshold as part of the target distribution. Non-Masked MW-U: P-value (Mann-Whitney U test) testing for differences between average EEG-prediction scores (across annotators) for each image i.e. target distribution contains average EEG-prediction scores for all annotators on images where consensus $> N$. Similarly, non-target distribution comprises average EEG-prediction scores over all images where consensus $N = 0$. Masked MW-U: P-value (Mann-Whitney U test) as before except only annotators who select the image as a target are included in generating average EEG-prediction scores. Inverted MW-U: P-value (Mann-Whitney U test) except only using averages of EEG-prediction scores for annotators who do not select image as a target.

Consensus N Threshold	Non-Masked MW-U	Masked MW-U	Inverted Mask MW-U	# Pool target counts	# Incremental target counts
1	0.063	*0.004	0.194	212	122
2	0.04	*0.007	0.168	90	53
3	0.024	*0.002	0.16	37	22
4	*0.0003	*0.0002	*0.008	15	7
5	0.037	0.059	*0.009	8	5
6	0.242	0.162	0.225	3	2
7	0.056	0.064	*0.043	1	1
8	n/a	n/a	n/a	n/a	n/a

for increasing consensus counts using all annotators. Important to remember in interpreting results here is that there are uneven contributions from annotators thus averages derived from lower consensus counts tend to be increasingly dominated by annotators with higher target counts.

As consensus increases we find fewer sample points available and for this reason in our analysis here (Table 2) we pool inputs to our test (Mann-Whitney U) across incremental thresholds of increasing N (e.g. all examples above or equal to a consensus of 3). For each target image ($<$ consensus N) we calculate an average of EEG-prediction scores of only those annotators who labelled the image as a target (masked), the average prediction scores of those annotators who did not select the image as a target (inverted mask), and an average of all annotator prediction scores. Our comparison distribution (same process) in all cases were images where no annotator selected a target ($N = 249$).

Here a clear relationship emerges in examining pooled targets, masked pooled targets and inverted pooled targets where we can see as consensus increases greater significantly detectable effects that shared annotator consensus on an image indicates for those who labelled it oppositely (non-target), their EEG-prediction scores indicate otherwise. Where consensus $N = 4$ and 5 we can see from the inverted masked analysis high consensus on an image results in greater prediction scores for annotators who did not actually annotate the image as a target. As discussed, as the consensus N increases we have fewer (target) examples which in turn increases our likelihood of a type II error explaining weaker/non-existent significance at higher N. Similarly for low consensus N we do not see this relationship present.

5   Perspectives on the Experiment and Suggestions

In our experiment with 8 researchers, it seems only 3 of them were careful enough to translate recognition of an object of interest into an actual significantly detectable annotations (as per ROC-AUC). It is our conjecture that in a crowd-sourced environment where annotators are being paid to annotate as many images or items as possible, then the annotator effort and accuracy is likely to be low. This suggests the need for significant annotator performance assessment before relying on the coverage of the annotations and in this respect EEG could be a useful tool in understanding the user's annotation process and how it relates to neurophysiological correlates of perception and attention.

Another suggestion is the potential for the application of EEG-based annotation that does not require the physical motion of an annotator to press a keyboard or mouse. The observation of the EEG signals is that it is feasible to annotate images at a rate of 5 per second using an EEG-based approach. In experimentation, we have found that such an annotation task can last for approximately 30 min (with rest periods) before the average annotator gets too fatigued. However, the application of EEG sensing can easily be done incorrectly such as failing to acquire clean signals or failing to remove confounding/detrimental artefacts from the signal in preprocessing stages such as applying ICA (Independent Component Analysis) or band-pass filtering. These confounding sources of noise due to eye blinks for instance can provide discriminative information in a prediction model but are not of neural origin so are not considered in our work.

These suggestions have the potential to de-risk the annotation process and potentially to significantly improve the speed of annotation and the coverage of the annotations. We suggest that this has implications for the use of crowd-sourced annotations and we have made a number of suggestions about the potential for enhanced annotators and EEG-based annotation. Although our preliminary results here are specifically tied to a single dataset and task type, we use these results as a basis to further advocate and discuss the use of neural measures in multimedia research and particularly in tasks involving annotation.

References

1. Ambati, V.: Active learning and crowdsourcing for machine translation in low resource scenarios (2012). aAI3528171
2. Deselaers, T., Keysers, D., Ney, H.: Features for image retrieval: an experimental comparison. Inf. Retrieval **11**(2), 77–107 (2008)
3. Gerson, A.D., Parra, L.C., Sajda, P.: Cortically coupled computer vision for rapid image search. IEEE Trans. Neural Syst. Rehabil. Eng. **14**(2), 174–179 (2006)
4. Healy, G., Smeaton, A.: Eye fixation related potentials in a target search task. In: Engineering in Medicine and Biology Society, EMBC, 2011 Annual International Conference of the IEEE, pp. 4203–4206, August 2011
5. Healy, G., Gurrin, C., Smeaton, A.F.: Lifelogging and EEG: utilising neural signals for sorting lifelog image data. Quantified Self Europe Conference, 10–11 May 2014, Amsterdam, Netherlands (2014)

6. Jia, Y., Shelhamer, E., Donahue, J., Karayev, S., Long, J., Girshick, R., Guadarrama, S., Darrell, T.: Caffe: Convolutional architecture for fast feature embedding, pp. 675–678 (2014). http://doi.acm.org/10.1145/2647868.2654889

7. Lew, M.S., Sebe, N., Djeraba, C., Jain, R.: Content-based multimedia information retrieval: state of the art and challenges. ACM Trans. Multimedia Comput. Commun. Appl. (TOMM) **2**(1), 1–19 (2006)

8. Mohedano, E., Healy, G., McGuinness, K., Giró-i Nieto, X., O'Connor, N., Smeaton, A.: Improving object segmentation by using EEG signals and rapid serial visual presentation. Multimedia Tools and Applications, pp. 1–23 (2015). http://dx.doi.org/10.1007/s11042-015-2805-0

9. Noronha, J., Hysen, E., Zhang, H., Gajos, K.Z.: Platemate: Crowdsourcing nutritional analysis from food photographs, pp. 1–12 (2011). http://doi.acm.org/10.1145/2047196.2047198

10. Polich, J.: Updating P300: an integrative theory of P3a and P3b. Clin. Neurophysiol. **118**(10), 2128–2148 (2007)

11. Shenoy, P., Tan, D.: Human-aided computing: utilizing implicit human processing to classify images. In: CHI 2008 Conference on Human Factors in Computing Systems (2008)

12. Welinder, P., Perona, P.: Online crowdsourcing: rating annotators and obtaining cost-effective labels, pp. 25–32, June 2010

Demo Session Papers

GrillCam: A Real-Time Eating Action Recognition System

Koichi Okamoto and Keiji Yanai$^{(\boxtimes)}$

The University of Electro-Communications - Tokyo, 1-5-1 Chofugaoka, Chofu-shi,
Tokyo 182-8585, Japan
okamoto-k@mm.inf.uec.ac.jp, yanai@cs.uec.ac.jp

Abstract. In this demo, we demonstrate a mobile real-time eating action recognition system, GrillCam. It continuously recognizes user's eating action and estimates categories of eaten food items during mealtime. With this system, we can get to know total amount of eaten food items, and can calculate total calorie intake of eaten foods even for the meals where the amount of foods to be eaten is not decided before starting eating. The system implemented on a smartphone continuously monitors eating actions during mealtime. It detects the moment when a user eats foods, extract food regions near the user's mouth and classify them. As a prototype system, we implemented a mobile system the target of which are Japanese-style meals, "Yakiniku" and "Oden". It can recognize five different kinds of ingredients for each of "Yakiniku" and "Oden" in the real-time way with classification rates, 87.7 % and 80.8 %, respectively. It was evaluated as being superior to the baseline system which employed no eating action recognition by user study.

Keywords: Mobile food recognition · Eating action recognition · Food recording system

1 Introduction

In recent years, due to a rise in healthy thinking on eating, many people take care of eating and foods, and some people record daily diet regularly. To assist them, many mobile applications for recording everyday meals have been released so far. Some of them employ food image recognition, which enable users to record daily foods only by taking photos.

We has proposed a mobile food recording system which has a 100-kind food recognition engine so far [1]. Since the recognition engine employed the state-of-the-art image recognition method, Fisher Vector and liner SVM classifiers, the classification rate was relatively high. The top-5 classification rate for 100 classes was 79.2 %. However, it required taking a meal photo before eating, and all the foods taken as a photo had to be served before eating.

Then, we proposed a mobile food recognition application, GrillCam, which was applicable for the case that the amount of food eaten by one person was

© Springer International Publishing Switzerland 2016
Q. Tian et al. (Eds.): MMM 2016, Part II, LNCS 9517, pp. 331–335, 2016.
DOI: 10.1007/978-3-319-27674-8_29

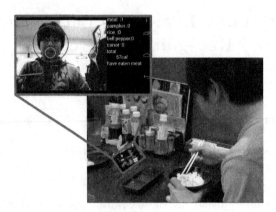

Fig. 1. A typical usage of the proposed system. A smartphone on which the proposed system is running is put toward a user's face. It continuously detects eating action and estimates food categories of eaten foods.

not decided before eating such as sharing large dishes or barbecue-style meal (grilling meats and vegetables while eating) before [2]. After that, we improved the proposed system in terms of the recognition accuracy and target meals.

2   System Overview

Figure 1 shows a typical scene when we use the proposed system implemented on a Android smartphone. In this scene, the user is having a "Yakiniku" meal which is a Japanese-style barbecue of grilling thin-sliced beef and vegetables on the hot plate. When we use it, we stand it with slight tilt in front of a user who is eating so that the built-it inner camera of the smartphone faces to the user's face. Although it is common for image recognition application for a smartphone to use a backside camera, we use an inner camera to record user's action by the camera and show information of eaten food items to the user at the same time.

The screen-shot image of the UI of the proposed application is shown in Fig. 2. A user can always check what the system is recognizing at the time regarding face, mouth and chopsticks. A blue circle, a yellow circle and a green line shown in the detected result area represent detected face region, mouth region and detected chopsticks region, respectively. On the right side of the screen, total calorie intake and the number of eaten items are displayed. A user can check how many calories he/she has taken while eating.

The proposed system performs eating action recognition according the following processing flow:

1. Detect a user's face and mouth.
2. Detect and track chopsticks.
3. Segment out a region candidate region (bounding box) around the tip of the chopsticks at the moment when the tip of the chopsticks is approaching the mouth.

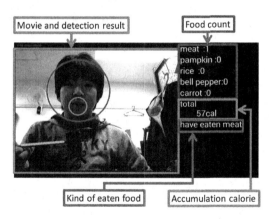

Fig. 2. The screen of the proposed mobile application, GrillCam.

4. Recognize a food item category for the segmented region.
5. Calculate and accumulate food calorie intake, and display it on the screen.
6. Repeat this procedure until the meal is finished.

In the current implementation, to detect face and mouth, we used a standard face and mouth detector in the OpenCV Library. To detect chopsticks, we used background subtraction and the Hough transform.

To segment out a food candidate region, in the previous implementation, we assumed a food item is located on the tip of the detected chopsticks, which was relatively a straight way. If lines of chopsticks were detected as being longer than their actual length, a food region was also estimated incorrectly.

Therefore, we improved a step to estimate a food candidate bounding box by taking into account the center of the mass of binarized images as follows (Fig. 3):

1. Segment out a large rectangular region around the tip of the chopsticks.
2. Binarize the region.
3. Calculate the center of mass in the region.
4. Extract a bounding box around the estimated center.

To recognize a food category of the food item being picked by chopsticks. In the previous implementation [2], we used a Bag-of-Features representation of the ORB [3] features and HSV color histogram. However, the accuracy was not enough.

In the current implementation, we adopt ORB and HSV color histogram, and each of the local descriptors are represented by Fisher Vector, and one-vs-rest linear SVM as a classifier for rapid recognition.

To calculate the total calorie intake, we use the pre-defined standard calorie values on each food item. To estimate food calorie intake precisely, we will estimate the volume of the detected food items in the future work.

Table 1. The classification rate at Yakiniku and Oden.

Yakiniku [2]	Yakiniku	Oden
74.8 %	87.7 %	80.8 %

Table 2. The mean and standard deviation of the five-step evaluation score.

Baseline	GrillCam
2.36 ± 1.12	4.36 ± 1.41

3 Experiments

We have implemented the proposed system as an Android application. In the experiments, we used Google Nexus 5 (2.3 GHz QuadCore, Android 4.4) as a target smartphone. We chose "Yakiniku" and "Oden" meal as target domains for the demo system implementation. "Yakiniku" is a Japanese-style barbecue meal to eat baked sliced meats and vegetables, while "Oden" is a Japanese winter dish consisting of several ingredients in a light, soy-flavored dashi broth. In the experiments, we evaluate classification accuracy and the system usability.

For the implementation, we selected the following five typical food items in Yakiniku: "meat", "rice", "pumpkin", "bell pepper" and "carrot". Further, we selected the following five typical food items in Oden: "radish", "egg", "Hanpen (Boiled Fish Cake)", "Konjac" and "chiikuwa (Grilled Fish cake)". Figure 4 shows kind of Yakiniku, and Fig. 5 shows Oden. We stored typical calorie values on the above food items on the system to calculate total calories of eaten food items.

For training, we prepared more than one hundred images for each of the food items. As shown in Table 1, the result of the classification rate at Yakiniku and Oden. In the previous implementation, the system was able to recognize only Yakiniku. The classification rate by the previous system with a conventional bag-of-features was 74.8 %, while the rate by the proposed system employing Fisher Vector is 87.7 %. The rate was improved by 12.9 points.

We made a simple user study as well. For comparison, we prepared a baseline system which had no image recognition function and instead required users to

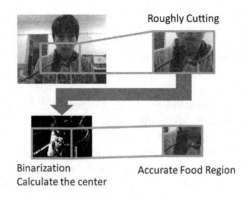

Fig. 3. Food region extraction

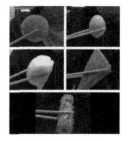

Fig. 4. The target food items for "Yakiniku".

Fig. 5. The target food items for "Oden".

touch food item buttons on the screen every time they eat a food item. We asked eleven users to eat five kinds of food items in front of both the proposed system and the baseline system, and to evaluate both systems in five-step score regarding their usability. As shown in Table 2, the score of the proposed system outperformed the score of the baseline greatly. This indicates the effectiveness of the proposed system.

4   Conclusions

In this paper, we proposed a new-style food recording system, GrillCam, which employs real-time eating action recognition and food categorization for meal scene. The system continuously monitors user's eating action and estimates the calorie intake in a real-time way.

This enables us to estimate total calories of the food the amount of which is undecided before eating. This special feature is totally different from existing image-recognition-based food calorie estimation systems which requires taking meal photos before starting to eat.

Although in the current implementation the kinds of mean scenes and the number of food items are still limited, we believe the proposed system, GrillCam, will be more practical by extending it so as to recognize various meal scenes such as sharing large platters, conveyor-belt-style sushi and hot-pot-style meals like sukiyaki. To do that, we will extend it so as to detect fork, knife and hands as well as chopsticks.

References

1. Kawano, Y., Yanai, K.: Real-time mobile food recognition system. In: Proceedings of CVPR International Workshop on Mobile Vision (IWMV) (2013)
2. Okamoto, K., Yanai, K.: Real-time eating action recognition system on a smartphone. In: Proceedings of ICME Workshop on Mobile Multimedia Computing (2014)
3. Rublee, E., Rabaud, V., Konölige, K., Bradski, G.: ORB: an efficient alternative to SIFT or SURF. In: Proceedings of IEEE International Conference on Computer Vision (2011)

Searching in Video Collections Using Sketches and Sample Images – The Cineast System

Luca Rossetto[✉], Ivan Giangreco, Silvan Heller, Claudiu Tănase,
and Heiko Schuldt

Databases and Information Systems Research Group, Department of Mathematics
and Computer Science, University of Basel, Basel, Switzerland
{luca.rossetto,ivan.giangreco,silvan.heller,
c.tanase,heiko.schuldt}@unibas.ch

Abstract. With the increasing omnipresence of video recording devices and the resulting abundance of digital video, finding a particular video sequence in ever-growing collections is more and more becoming a major challenge. Existing approaches to retrieve videos based on their content usually require prior knowledge about the origin and context of a particular video to work properly. Therefore, most state of the art video platforms still rely on text-based retrieval techniques to find desired sequences. In this paper, we present Cineast, a content-based video retrieval engine which retrieves video sequences based on their visual content. It supports Query-by-Example as well as Query-by-Sketch by using a multitude of low-level visual features in parallel. Cineast uses a collection of 200 videos from various genres with a combined length of nearly 20 h.

1 Introduction

From all commonly used types of media, video offers the largest variety to express information. In the previous years, the amount of videos and thus the sheer size of video data accessible online has seen a significant growth. The same is true for the number of both video related usages and users. This development is expected to continue with even increased speed in the coming years. A result of this development is that finding a specific video (or a scene within a video) in this data deluge usually requires the collection to be properly annotated. However, it is not realistic to assume that every video creator is willing to provide or even capable of adding the required annotations during the upload process. Automating this process would require a highly sophisticated software which is able to understand the semantics and the context of any arbitrary video sequence, and it would have to anticipate any subsequent use of this video. While this is –especially constrained to certain situations and depending on the context– a non-trivial task for a human, it is highly difficult or even impossible for a machine. One possibility to overcome this problem is to employ content-based retrieval techniques. In what follows, the term "content" is used to refer to the raw information such as pixels within frames, samples within audio tracks, and letters within subtitles

© Springer International Publishing Switzerland 2016
Q. Tian et al. (Eds.): MMM 2016, Part II, LNCS 9517, pp. 336–341, 2016.
DOI: 10.1007/978-3-319-27674-8_30

as they are encoded, and not to the semantic meaning instilled within a video as interpreted by a human being.

Content-based search in video collections requires sophisticated retrieval support. This includes both the back-end of a video retrieval system and the user interaction at the front-end. In terms of the retrieval back-end, the content of the videos has to be effectively reflected by means of characteristic features and efficient search mechanisms have to be provided that are able to identify the most similar videos or video shots for a given query. In terms of the user interface, different query paradigms need to be supported that help users expressing their information need, i.e., to specify (parts) of the contents of the videos or shots a user is looking for.

In this paper, we present Cineast [2], a content-based video retrieval engine which retrieves video sequences based on their visual content. Cineast is built as a general-purpose video retrieval engine, with main usage in known-item search. It supports multiple query paradigms such as Query-by-Example (*QbE*) and Query-by-Sketch (*QbS*), as well as relevance feedback.

This paper briefly introduces the architecture of the Cineast system (Sect. 2) and discusses in Sect. 3 how the system can be used in a Query-by-Sketch and/or in a Query-by-example mode.

2   Cineast

2.1   Architecture

From a conceptual point of view, the Cineast system can be divided into an on-line and an off-line part [2]. The on-line part deals with retrieval while the off-line part handles the feature extraction. Both parts are architecturally very similar. The primary component for both parts is the *feature module runtime* which manages the individual feature modules. These modules, in turn, are responsible for extraction and retrieval based on their individual feature vectors.

During feature extraction which is performed off-line, the input video is decoded and segmented into shots[1]. For every shot, a set of shot descriptors is generated. These shot descriptors only contain a part of the information present within the shot and serve as input for the feature modules. Shot descriptors include, among others, pixel-wise average or median over all frames, a most representative keyframe (which is the one frame with the smallest distance to the average image), the set of all frames of the shot, the set of all subtitle elements present during the shot, etc. The feature modules then use one or multiple of these descriptors to produce one or multiple feature vectors which are then stored in ADAM, a database system capable of managing feature data [1].

During retrieval, a set of shot descriptors is produced from the query provided by the user. The individual modules use these descriptors in a similar way as in the off-line extraction case to produce query vectors which are used to generate

---

[1] A shot is a continuously recorded sequence of frames. All frames within one shot are visually similar to each other.

a list of results for every module. These result lists are combined into a final ordered list of results by the feature runtime using a weighted average over the similarity scores.

2.2   Features

Currently, Cineast implements over three dozen features. These include global features, such as the average/median color, dominant shot colors, chroma and saturation, and a color histogram. Furthermore, it provides regional features, e.g., for color (color moments, color layout descriptor, color element grids), and for edges (partitioned edge image, edge histogram descriptor, dominant edge grid). In addition to the visual low-level features, Cineast also comes with motion features, i.e., directional motion histograms and regional motions sums, and text features applied to the subtitles. The features are grouped into categories that subsume multiple features.

2.3   User Interaction

Figure 1 shows the browser-based user interface of Cineast with the retrieved results (right hand side) for the given sketch. The left-hand side of the UI is used for query specification.

Query Specification: The left hand side of the UI contains two input canvases which can either be used for sketching or for providing query images acquired,

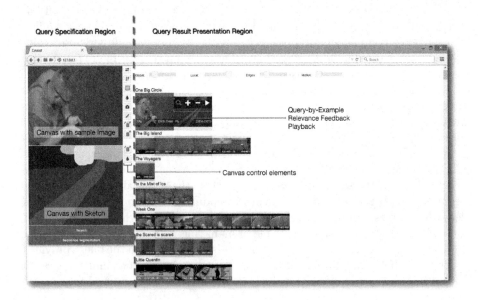

Fig. 1. Screenshot of the Cineast user interface with sequence segmentation

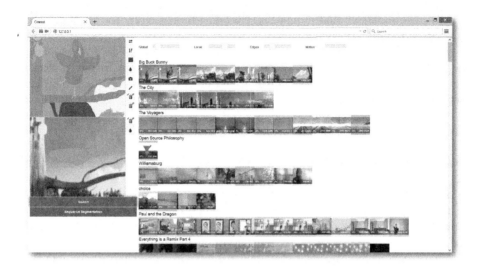

Fig. 2. Screenshot of the Cineast user interface with results grouped by video (Color figure online)

for instance, from a webcam or the file system. This gives the user a wide range of input possibilities. Multiple canvases enable the user to specify multiple subsequent shots of a particular video sequence which opens up the possibility to search for longer, more visually diverse sequences while simultaneously increasing the chances of the correct video being high up in the result list. Both canvases additionally offer the possibility for a flow-field input to specify motion within a shot. Cineast also supports hybrid queries that consist of a sketch in one canvass and a sample image in the other. Moreover, sketches and sample images can even be combined in a single canvass by superimposing a sketch and a sample image (e.g., for adding further objects by means of a sketch on top of a query image).

Query Result Presentation: The right-hand side of the UI is used to display the retrieved video sequences. The smallest unit of information which is retrieved by Cineast is a shot, but since a query from multiple input canvases may return multiple shots, the results are grouped by video of origin as shown in Fig. 2. The shots are ordered chronologically within each video while the displayed videos are ordered by the average score of their shots. The similarity score for each shot is displayed in the bottom left-hand corner as well as visually indicated by the color of its border.

Because certain features used for retrieval take more time for evaluation than others, partial results become available quite some time before the complete retrieval process is complete. Partial results are streamed to the front-end as soon as they become available. The front-end will then start displaying results as soon as they become available and re-order them as necessary which highly increases the usability of the interface. A loading icon in the lower left corner of the screen indicates that a query is still being processed by the back-end and more results might still be coming.

A click on the Sequence Segmentation button re-arranges the results. Shots are no longer grouped simply by video but a video is broken apart into multiple sequences as depicted in Fig. 1. This is useful in cases where the query matches a certain video multiple times.

For each shot in the results, the user has the possibility to play the video at the corresponding position.

Query Refinement: In addition to displaying the retrieved video shots, a user can also take these shots as input for a next, refined query. This can be done in one of three ways. A shot can be used as a direct query input by clicking on the looking-glass symbol which appears when the mouse cursor enters the thumbnail. This will cause Cineast to search for shots which are similar to the one selected. It is also possible to drag a thumbnail from the results into one of the canvases and modify it by sketching to make it more closely resemble the shot one is looking for. Lastly, multiple shots can be selected from the result set to serve as positive and negative examples for relevance feedback, which is then triggered using a dedicated button.

System Configuration/Parameter Setting: The four sliders at the top of the result area control how the shot similarity is computed. Cineast uses many features with as many different measures of similarity. For reasons of usability, these features are grouped into four categories; global and local color similarity, edge similarity, and motion similarity. Depending on the properties of the query, a user might want to adjust these sliders to better reflect the intent of the search. Manipulating the sliders will re-rank the results in real time without involvement of the retrieval back-end.

2.4 Implementation

Cineast is written in Java and provides a network-accessible JSON-API. The user interface is browser-based. The server-side which is implemented in PHP serves as a proxy between the browser and Cineast as well as a server for static content such preview images. The client interface uses a JSON-streaming library[2] for the asynchronous processing of retrieved results.

3   Cineast in Action

In order to show the Cineast system in action, two independent and unconnected machines should be used. The first one hosts the OSVC1 dataset [3], a collection of 200 creative commons videos. A user is able to explore the roughly 20 h of video by browsing to select a video sequence to be searched for.

---

[2] http://oboejs.com/.

The second machine, which has to offer a touch- and pen-based input, runs Cineast. With this machine, users are able to search for the video sequence they have previously selected as their target.

Users of Cineast may sketch one or multiple scenes using color- and/or line-sketches and draw motion to describe their selected scene. Each canvas thereby describes a different shot. In case the scene in question was not among the results, a user has the possibility to refine a query using parts of the results. This can either be done using a retrieved shot directly as an input, thereby switching from a *Query-by-Sketch* to a *Query-by-example* modality, or by using it as a base for a new query image by dragging the preview into one of the canvases and modifying it using the sketching tools. Additionally, multiple results can be tagged as positive or negative examples and used for relevance feedback.

This set-up allows to demonstrate the highly efficient and effective query support for large video collections provided by the Cineast system. A video of the system in action can be found at: http://www.youtube/NjDqToONeZc.

4   Conclusion

In this paper, we have presented the Cineast video retrieval engine and we have shown how it can be interactively used to retrieve sequences of video based on sketches, sample images, and motion information.

Acknowlegements. This work was partly supported by the Swiss National Science Foundation in the context of the CHIST-ERA project IMOTION (20CH21_151571).

References

1. Giangreco, I., Al Kabary, I., Schuldt, H.: ADAM – a database and information retrieval system for big multimedia collections. In: Proceedings of the IEEE International Congress on Big Data (BigData 2014). IEEE, Anchorage (2014)
2. Rossetto, L., Giangreco, I., Schuldt, H.: Cineast: a multi-feature sketch-based video retrieval engine. In: 2014 IEEE International Symposium on Multimedia (ISM), pp. 18–23. IEEE (2014)
3. Rossetto, L., Giangreco, I., Schuldt, H.: OSVC - Open Short Video Collection 1.0. Technical report CS-2015-002, University of Basel (2015)

LoggerMan, a Comprehensive Logging and Visualization Tool to Capture Computer Usage

Zaher Hinbarji[(✉)], Rami Albatal, Noel O'Connor, and Cathal Gurrin

Insight Centre for Data Analytics, Dublin City University, Dublin, Ireland
zaher.hinbarji@insight-centre.org
https://www.insight-centre.org

Abstract. As we become increasingly dependent on our computers and spending a major part of our day interacting with these machines, it is becoming important for lifeloggers and Human-Computer Interaction (HCI) researchers to capture this aspect of our life. In this paper, we present LoggerMan, a comprehensive logging tool to capture many aspects of our computer usage. It also comes with reporting capabilities to give insights to the data owner about his/her computer usage. In this work, we aim to address the current lack of logging software in this domain, which would help us, and other researchers, to build datasets for HCI experiments and also to better understand computer usage patterns. Our tool is published online (http://loggerman.org/) to be used freely by the community.

Keywords: Logging tool · Computer usage · Lifelogging · Human-computer interaction · User modeling

1 Introduction

Lifelogging is the process of gathering personal data generated by the activities of an individual. Most of the related work on lifelogging is focused on gathering and analysing visual and physical activity data [1]. In this work we take this further by including HCI tracking and analysis. In our modern society computers have become a core part of our professional and personal environments, which makes capturing HCI activities a rich source of information to mine for generating better insights into our life. It is our conjecture that capturing HCI data is supporting technology for several applications, ranging from simple automatic memory/diary generation, to the discovery of user's interests, skills, preferences and mood. Personalisation, profiling and security applications are common examples of how such lifelog data can be utilised. We believe that the passive continuous tracking of HCI can reveal important information about the user, which can be used to perform advanced analytics and to build various personalised services. In this work, we present a comprehensive logging tool to capture a wide range of our human-computer interactions.

© Springer International Publishing Switzerland 2016
Q. Tian et al. (Eds.): MMM 2016, Part II, LNCS 9517, pp. 342–347, 2016.
DOI: 10.1007/978-3-319-27674-8_31

2   Background

Although one can easily find a commercial spyware/surveillance software providing the ability to monitor and record the computer activities, it is not easy to find dedicated and comprehensive tools for logging computer usage for analysis purposes. In [2], authors discussed using surveillance software for conducting HCI experiments. They described several issues related to time stamps, mouse clicks, web usage and some other missing functionality in such software. Other currently available software is designed to track interaction with one application, for example, *WOSIT* [3] can be used to observe user actions on one application's UI in UNIX systems. *OWL* [4] is another application-specific logging tool that tracks Microsoft Word usage.

Several works have utilised HCI logging for various applications. For example, the use of keystroke logging tools in the domain of cognitive writing research has created new possibilities by providing detailed information about the writing process that was not accessible previously. Particularly, key-loggers are utilised in studies on writing processes, description of writing strategies, children writing development, writing and spelling difficulties analysis [5–7]. Inputlog is an example of such logging tool [8]. Keyboard logging also can be integrated in educational fields for programming and typing skills [9]. In addition, detailed timing of keystrokes can be employed for security purposes [10,11] and for stress [12] and emotion detection [13]. Screenshots are another example of possible logging technique that can be used to capture and analyse the content consumed or produced by the user. Screenshots were applied for search and task automation [14], and for building user friendly help manuals and demonstrations [15].

Furthermore, intrusion detection and authentication are common applications based on HCI logging. In [16], mouse curves are logged and used as a signature to authenticate users. Several mouse usage features are used to build user profiles [17,18]. Log files, resource and command usage are used to profile users behaviour [19,20]. In [21], user identification was explored via logging various GUI interactions such as windows switching frequency, time between new windows and number of opened windows.

Motivated by the beneficial potential and applications of HCI logging, and by the lack of generic and comprehensive HCI tracking and visualisation tools, we present *LoggerMan* a generic, passive and application-independent logging solution for researchers and lifeloggers. LoggerMan provides a robust tool to assist data gathering, visualisation and analytics for research community.

3   LoggerMan Overview

LoggerMan helps researchers and lifeloggers to collect interaction data produced during normal computer usage. The main goal of LoggerMan is to work passively in the background, intercept usage events and store them for later analysis. It gathers wide range of keyboard, mouse and UI actions. LoggerMan is available online (LoggerMan.org) for interested researchers/lifeloggers. In the next sections, we will describe the system main components and functionality.

3.1   Privacy and Technical Specification

LoggerMan works under Mac OSX 10.7 or later. To avoid any privacy concerns among users, all captured data is stored locally on the computer. This provides the user with a full control of the data. Buffering techniques are used to store the stream of events efficiently to the hard disk. All modules are designed to be Unicode compatible. Thus, the tool can log texts of different languages properly. Loggerman can be configured to run automatically after a system restart. To maintain user privacy, LoggerMan does not log data typed in secured fields (password fields).

3.2   Modules

LoggerMan has been designed to be flexible and modular. Our tool consists of multiple logging modules which can be switched on/off independently (Fig. 1). Each logging module captures a different type of data and every log-entry is timestamped to the current time.

Fig. 1. LoggerMan menu

Keyboard-Related Modules. These modules are responsible for capturing keystrokes events and storing them in local files. They provide two levels of detail: word based and key based. Word-based mode (visible as "Keyboard" menu item on Fig. 1) segments and concatenates the events stream into words. Whereas, key-based mode (visible as "Keystrokes" menu item on Fig. 1) tracks each key separately and stores it associated with its time stamp. This particularly important for the domain of keystrokes dynamics [10].

Mouse Module. All mouse actions are traced by this module: move, right/left up/down mouse buttons events, scroll and drag-and-drop actions. We previously utilised the data stored by this module to build an authentication system based on mouse curves [16].

Screenshots Module. This will capture a screenshot of the current active window. The user can select one of the shooting intervals (every 5, 10, 30 s) or a smart-shooting option, which takes a screenshot: after every window transition (from app to app or even different windows in the same app) and at fixed 1 min intervals also. This helps to ensure a reasonable trade-off between capture frequency and storage usage. The screenshot can then provide valuable information about the content that the user is consuming or non-texual content that the user is creating.

Apps Module. This module is designed to track apps transitions. It only logs the currently used app regardless of how it is being used, and as such it offers a simple overview of app usage, without any potential privacy issues of actually capturing the screens.

Clipboard Module. Clipboard module is responsible for tracking copy-paste operations. Any text the user copy to the clipboard is captured and logged.

3.3 Reporting and Insight

The current version of LoggerMan comes with a local reporting tool to visualise and give insights to the data owner about his/her computer usage. Specifically, the reporting tool presents three tag clouds: used apps, typed words and windows titles. These clouds are constructed by computing the normalised frequencies of items occurrence during the selected date range by the user (Figs. 2 and 4). Information about mouse clicks and typed keystrokes during a time frame is also presented as a graph (Fig. 3). In addition, LoggerMan organises the screenshots and the app usage data in a zoom-able timeline to provide a dyanmic and up-to-date view of computer usage (Fig. 5).

Fig. 2. Apps cloud

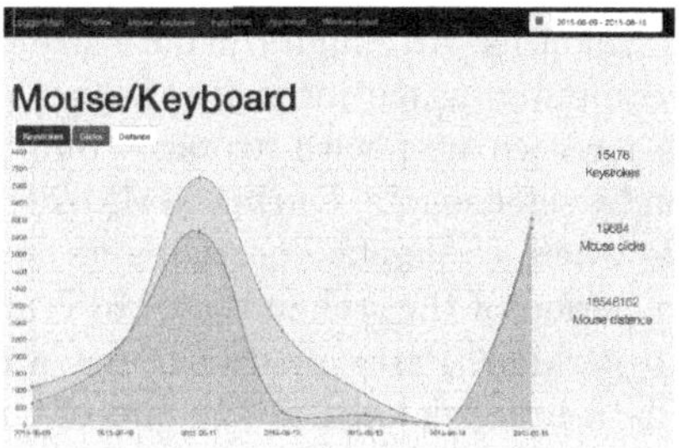

Fig. 3. Mouse/keyboard usage

4 Demonstration and Evaluation

LoggerMan is designed to be easy to install and run by any user. The user is all set by just double clicking the app icon for the first time. After that, the app will keep working automatically and passively in the background. Three subjects have been using LoggerMan to capture their computer usage for more

Fig. 4. Windows titles cloud

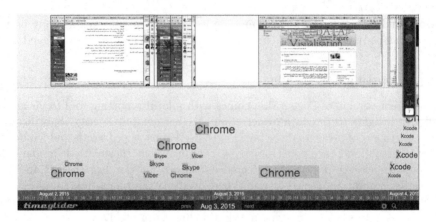

Fig. 5. Apps & screenshots timeline

than 2 months. Although the data size is dependent of the running modules and the computer usage intensity, LoggerMan produces around 1 GB of data per week for each user with average computer usage. Most of this space (98 %) is taken by screenshots. During MMM 2016, our demo will be conducted in a demo mode where conference participants will be able to log their interactions for a short period of time, then they can visualise their data by the reporting module of LoggerMan. Participants can get a copy of their interaction log files before the data being deleted at the end of the session. A personalized report will be generated for each user to show their uniqueness compared to other users.

5   Conclusions

There is a lack of a useful logging tools in the domain of human-computer interaction. In this paper, we presented LoggerMan, a logging and visualisation tool to capture computer activities and to give some insight based on this data. LoggerMan can easily support a wide range HCI related research such as user identification, personalisation, recommendation and cognition.

Acknowledgments. This publication has emanated from research conducted with the financial support of Science Foundation Ireland (SFI) under grant number SFI/12/RC/2289.

References

1. Gurrin, C., Smeaton, A.F., Doherty, A.R.: LifeLogging: personal big data. Found. Trends Inf. Retr. **8**(1), 1–125 (2014)
2. Liffick, B.W., Yohe, L.K.: Using surveillance software as an HCI tool. In: Information Systems Education Conference (2001)

3. The MITRE corporation. Widget observation simulation inspection tool. http://www.openchannelfoundation.org/projects/WOSIT. Accessed 07 Aug 2015

4. Frank, L.: OWL: A recommender system for it skills

5. Sullivan, K.P.H., Lindgren, E.: Computer Key-Stroke Logging and Writing: Methods and Applications. Studies in Writing, vol. 18. Elsevier, Oxford (2006)

6. Leijten, M., Van Waes, L.: Keystroke logging in writing research: using inputlog to analyze and visualize writing processes. Writ. Commun. **30**(3), 358–392 (2013)

7. Stromqvist, S., Holmqvist, K., Johansson, V., Karlsson, H., Wengelin, A.: What keystroke-logging can reveal about writing. computer key-stroke logging and writing: methods and applications. Stud. Writ. **18**, 45–72 (2006)

8. Leijten, M., Van Waes, L.: Inputlog: a logging tool for the research of writing processes. Technical report

9. Rodrigues, M., Gonçalves, S., Carneiro, D., Novais, P., Fdez-Riverola, F.: Keystrokes and clicks: measuring stress on e-learning students. In: Casillas, J., Martínez-López, F.J., Vicari, R., De la Prieta, F. (eds.) Management Intelligent Systems. AISC, vol. 220, pp. 119–126. Springer, Heidelberg (2013)

10. Gunetti, D., Picardi, C.: Keystroke analysis of free text. ACM Trans. Inf. Syst. Secur. **8**(3), 312–347 (2005)

11. Shimshon, T., Moskovitch, R., Rokach, L., Elovici, Y.: Clustering di-graphs for continuously verifying users according to their typing patterns. In: 2010 IEEE 26th Convention of Electrical and Electronics Engineers in Israel (IEEEI), pp. 000445–000449 (2010)

12. Vizer, L.M.: Detecting cognitive and physical stress through typing behavior. In: CHI 2009 Extended Abstracts on Human Factors in Computing Systems, pp. 3113–3116. ACM (2009)

13. Kolakowska, A.: A review of emotion recognition methods based on keystroke dynamics and mouse movements. In: 2013 the 6th International Conference on Human System Interaction (HSI), pp. 548–555 (2013)

14. Yeh, T., Chang, T.-H., Miller, R.C.: Sikuli: Using GUI screenshots for search and automation. In: Proceedings of the 22nd Annual ACM Symposium on User Interface Software and Technology, UIST 2009, pp. 183–192. ACM (2009)

15. Yeh, T., Chang, T.-H., Xie, B., Walsh, G., Watkins, I., Wongsuphasawat, K., Huang, M., Davis, L.S., Bederson, B.B.: Creating contextual help for GUIs using screenshots. In: Proceedings of the 24th Annual ACM Symposium on User Interface Software and Technology, UIST 2011, pp. 145–154 (2011)

16. Hinbarji, Z., Albatal, R., Gurrin, C.: Dynamic user authentication based on mouse movements curves. In: He, X., Luo, S., Tao, D., Xu, C., Yang, J., Hasan, M.A. (eds.) MMM 2015, Part II. LNCS, vol. 8936, pp. 111–122. Springer, Heidelberg (2015)

17. Ahmed, A.A.E., Traore, I.: A new biometric technology based on mouse dynamics. IEEE Trans. Dependable Sec. Comput. **4**(3), 165–179 (2007)

18. Pusara, M., Brodley, C.E.: User re-authentication via mouse movements. In: Proceedings of the 2004 ACM Workshop on Visualization and Data Mining for Computer Security, pp. 1–8. ACM (2004)

19. Anderson, J.P.: Computer Security Threat Monitoring and Surveillance. James P. Anderson Co. (2002)

20. Yeung, D.Y., Ding, Y.: Host-based intrusion detection using dynamic and static behavioral models. Pattern Recognit. **36**, 229–243 (2003)

21. Goldring, T.: User profiling for intrusion detection in windows nt. Comput. Sci. Stat. **35** (2003)

E^2SGM: Event Enrichment and Summarization by Graph Model

Xueliang Liu[1], Feifei Wang[1], Benoit Huet[2], and Feng Wang[3]([✉])

[1] Hefei University of Technology, Hefei, China
`liuxueliang@hfut.edu.cn`, `ahfywff@gmail.com`
[2] EURECOM, Sophia Antipolis, France
`huet@eurecom.fr`
[3] East China Normal University, Shanghai, China
`fwang@cs.ecnu.edu.cn`

Abstract. In recent years, organizing social media by social event has drawn increasing attentions with the increasing amounts of rich-media content taken during an event. In this paper, we address the social event enrichment and summarization problem and propose a demonstration system E^2SGM to summarize the event with relevant media selected from a large-scale user contributed media dataset. In the proposed method, the relevant candidate medias are first retrieved by coarse search method. Then, a graph ranking algorithm is proposed to rank media items according to their relevance to the given event. Finally, the media items with high ranking scores are structured following a chronologically ordered layout and the textual metadata are extracted to generate the tag cloud. The work is concluded in an intuitive event summarization interface to help users grasp the essence of the event.

1 Introduction

With thousands of photos and videos being uploaded every minute to social media sharing websites, there is an increasing need to automatically uncover the structure of online media, so as to help users grasp the gist at a glance. It is often the case that among user contributed media, much content is captured during real-life events and conveys various interesting information about them [3]. However, in most cases users fail to provide detailed annotations about the media before sharing them online. Efficiently and effectively understanding the heterogeneous event semantic behind these multimedia content and summarizing them succinctly for consumer consumption, is a significant challenge in the research community [1, 4].

In this work, we address the event enrichment and summarization problem and propose a framework to summarize events by photos/videos from a large media collection. The problem is solved in three steps. At first, considering the huge amount of media in the dataset, we employ a coarse retrieval method to obtain a summarization candidate set, so as to reduce computation cost. Then, in order to find the most relevant medias for each event, a graph ranking method

© Springer International Publishing Switzerland 2016
Q. Tian et al. (Eds.): MMM 2016, Part II, LNCS 9517, pp. 348–353, 2016.
DOI: 10.1007/978-3-319-27674-8_32

is proposed to rank them by the relevance to the event, and the ones with high ranking score are kept for summarization purpose. Finally, we provide a vivid interface to users. A graph layout, where all of the media items are ordered in chronological order, is used to visualize and structure the summarized event. The framework of the proposed approach is illustrated in Fig. 1.

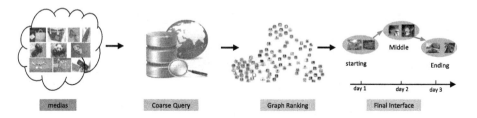

Fig. 1. The framework of the proposed method.

2 Our Solution

The main goal of the proposed event-based media summarization system is to find the relevant media from a big multimedia dataset for a given event and represent it with an informative and engaging layout. In this section, we reveal the details of the proposed approach.

2.1 Dataset

The demonstration is built on the YF100MCC dataset which contains 100M photos and videos compiled from Yahoo! Flickr [5]. For each photo or video, many metadata, such as the owner, taken/uploading time, title, description, location, and the url are provided.

The efficiency of the textual, temporal and spatial feature in event based research has been proven in previous work [2]. Unfortunately the spatial feature is optional in this dataset. Therefore, only the textual and temporal features are selected to model the media data. Specifically, the tf-idf method is used to extract the textural feature and 1000D textual feature is obtained for each media item and the media taken time, which is represented as the number of pas seconds from Unix epoch, is used as the temporal feature.

2.2 Coarse Query

To reduce the computation cost when summarizing an event, we build a text indexing system on the dataset and use the first sentence in an event description as the textual query. The results are further filtered by temporal span to remove the irrelevant ones according to the event taken time.

Considering there are different types of events, we design different time spans to filter the results. For the global multi-day event, we keep the media items that are taken in some days before an event. For the local single-day event, we only keep the media items that are taken on the same day of the event timestamp. In the dataset, there are some media without taken time attribute, and we use the upload time instead.

2.3   Graph-Based Ranking

In the coarse query result, the amount of media is still too large, and their importance on representing an event is not evaluated. The users still can not get the event gist directly and quickly by looking at them. In our framework, we propose a graph ranking method to rank the media data and obtain a small subset which are highly relevant to an event.

Graph ranking method [6] is popularly used in media analysis, where the vertices and edges represent medias and their relations respectively. The graph ranking method is derived from the local similarity assumption. Obviously, the images with the similar content should be assigned the similar relevant score on a given event. Mathematically, given a media collection $D = \{x_1, x_2, \cdots, x_n\}$, we denote r an ordering of theses media data, and $r(i)$ be the score of media x_i under the ordering. The similarity between two media can be computed based on Gaussian function,

$$W_{ij} = exp(-\frac{\|x_i - x_j\|^2}{\sigma^2}) \tag{1}$$

where σ is the radius parameter of the Gaussian function.

In addition, the relevance of an image should depend on its "closeness" to the event. In this problem, the event is described using both textual and temporal modality. Hence the images which are closer in the two modalities to an event should be assigned a higher relevant score.

There have been several works addressing the similarity problem. For simplicity, we use Cosine similarity and Gaussian similarity method to measure the textual and temporal similarities respectively.

The similarity of an event e and an image x_i with textual feature can be formulated as follows.

$$\mathbf{Sim}(e, x_i) = \frac{\|Text(e)Text(x_i)\|}{\|Text(e)\| \, \|Text(x_i)\|} \tag{2}$$

where $Text(o)$ is the textual feature extracted from o.

The time similarity can be formulated according to the time span between the image taken time and event held time.

$$\mathbf{Span}(e, x_i) = exp^{-\frac{\|Time(e)-Time(x_i)\|}{\sigma_t}} \tag{3}$$

Based on these assumptions, we formulate the event based media ranking problem in a graph learning framework as follows.

$$\Omega(\mathbf{r}) = W_{ij} \left(\frac{\mathbf{r}(i)}{\sqrt{D(i)}} - \frac{\mathbf{r}(j)}{\sqrt{D(j)}} \right)^2 + \lambda_1 \sum_{i=1}^{n} (\mathbf{r}(i) - \mathbf{Sim}(e, x_i))^2 + \lambda_2 \sum_{i=1}^{n} (\mathbf{r}(i) - \mathbf{Span}(e, x_i))^2. \tag{4}$$

where $\mathbf{r}(i)$ is the relevance score of x_i, and $D(i) = \sum_{i=1}^{n} W_{ij}$. There are three parts in the above objective function. The first part is the graph smoothness term which means that the similar medias could achieve similar score, the second and third parts mean that the relevance score should be larger if a media is closer to the given event measured by textual or temporal similarity.

Equation 4 can be optimized by derivative method, and the closed form solution could be derived as follows.

$$\mathbf{r}^* = \frac{\lambda_1 \mathbf{Si} + \lambda_2 \mathbf{Sp}}{(\lambda_1 + \lambda_2)I + L} \tag{5}$$

The media are sorted by r^* and the top items which are assigned with high values are used in the following illustration steps.

2.4   Illustration

A collection of high relevant media is obtained by the graph based ranking method as proposed in Sect. 2.3. To help the users understand the evolution of an event, we represent the data by a graph method. In other words, we link the image pairs according to their similarity and show them according to the chronological order. In detail, we measure the similarity of a image pair by Eq. 1. For each image, only two nearest neighbors are connected by an edge to encourage the sparsity of the graph and to provide a neatly-arranged visual to the users. When the graph is built, all the image vertices are presented along the timeline according to their chronological order. Figure 2 illustrates an example to demonstrate the rationale of this illustrating approach.

Besides visual illustration, we use a tag cloud to organize the textual data to provide a nice and meaningful visualization. Tag cloud is a form of histogram which can represent the amplitude of over a hundred items. In tag cloud, the importance of each tag is shown with different format. This format is useful for quickly perceiving the most prominent terms. For each media illustrating an event, we segment the text description into tags, count the frequency of each tag and generate the tag cloud with tags in different font sizes and colors. By tag cloud, the key concept of an event could be perceived effectively.

We design a pilot system for event summarization to help users visualize and understand the event. For more details, please visit the project website[1].

3   Implement and Result

3.1   Implement Configuration

In the experiment, the YF100MCC dataset is managed by MongoDB[2]. To assist the coarse search, we create the index on taken time and text attributes. The

---

[1] http://lmc.hfut.edu.cn/eventgraph.
[2] http://www.mongodb.org.

first sentence and the timestamp in event description is used as the query to retrieve the relevant media roughly. Specifically, we use the first event description sentence to do the full-text search which is supported by MongoDB. As to the temporal attribute, we design different rule to different events. For the global and multi-day events, we query the media which is taken 5 days before the timestamp. For the one-day events, it is reasonable consider the relevant media from the ones that were taken on the same as the event.

The coarse query candidates are ranked by the proposed graph ranking algorithm. The more relevant a media is to an event, the higher score it is assigned by the model. In our experiments, the parameters λ_1 and λ_2 are both set to 0.1 experimentally.

To provide a readable result to users, we keep the top 50 media items to illustrate an event. The 50 media items of each event are connected by a graph, where each media is linked by two successive items in chronological order. The textual metadata from each media item is extracted and used to generate the tag cloud according to their frequencies.

3.2 Results

In this section, we evaluate the quality of the two types of summarization result. Here, we take an query as example: the event "Occupy Movement" at 1 Nov 2011 12:00 am UTC. The event summarization is visualized in Fig. 2. As shown in this figure, there are two parts in this summarization, while the left part is the visual graph and the right part is the tag cloud.

In the visual graph, there are 50 medias to illustrate this event. These medias are linked to another item according to their similarity by textual metadata, which provides a clue on the event evolution in the past days. These medias are presented along the timeline according to their chronological order. Thanks to

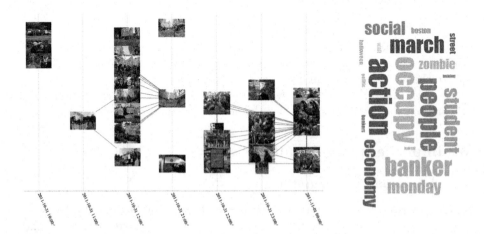

Fig. 2. The Visual Example of the Event "Occupy Movement" at timestamp 1 Nov 2011 12:00 am UTC

such a graph, the user is able to find out how the event updates at each short period, so that they can comprehend the event gist at a glance.

Besides the visual graph, we also use the tag cloud to present the gist of the event. In the tag cloud, we can find out many terms such as "occupy", "banker", "economy", "action". With these words, it could be perceived that this event relates to bankers and economy.

4 Conclusions

In this paper, we address the event enrichment and summarization challenge and propose a system that can help users obtain the gist of an event at different timestamps. In the proposed demonstration, the event-relevant candidates are first queried by coarse search approach and are ranked by the proposed graph ranking method to find the most relevant samples for summarizing an event. Finally the event summary is presented by a media graph in chronological order to show the event trend, and by tag cloud to present the gist of the event.

Acknowledgments. This research is supported in part by the National High Technology Research and Development Program of China(Grant No. 2014AA015104) and in part by the National Nature Science Foundation of China under Grant 61502139, Grant 61125106, Grant 61272393, Grant 61322201, and Grant 61432019.

References

1. Hong, R., Tang, J., Tan, H.-K., Ngo, C.-W., Yan, S., Chua, T.-S.: Beyond search: event-driven summarization for web videos. ACM Trans. Multimedia Comput. Commun. Appl. **7**(4), 35:1–35:18 (2011)
2. Liu, X., Huet, B.: Heterogeneous features and model selection for event-based media classification. In: Proceedings of the 2nd ACM International Conference on Multimedia Retrieval, April 2013
3. Mezaris, V., Scherp, A., Jain, R., Kankanhalli, M.S.: Real-life events in multimedia: detection, representation, retrieval, and applications. Multimedia Tools Appl. **70**(1), 1–6 (2014)
4. Petkos, G., Papadopoulos, S., Mezaris, V., Troncy, R., Cimiano, P., Reuter, T., Kompatsiaris, Y.: Social event detection at mediaeval: a three-year retrospect of tasks and results. In: Proceedings of ACM ICMR 2014 Workshop on Social Events in Web Multimedia (2014)
5. Thomee, B., Shamma, D.A., Friedland, G., Elizalde, B., Ni, K., Poland, D., Borth, D., Li, L.-J.: The new data and new challenges in multimedia research. arXiv preprint arXiv:1503.01817 (2015)
6. Wang, M., Yang, K., Hua, X.-S., Zhang, H.-J.: Towards a relevant and diverse search of social images. IEEE Trans. Multimedia **12**(8), 829–842 (2010)

METU-MMDS: An Intelligent Multimedia Database System for Multimodal Content Extraction and Querying

Adnan Yazici[1]($\boxtimes$), Saeid Sattari[1], Turgay Yilmaz[2], Mustafa Sert[3],
Murat Koyuncu[4], and Elvan Gulen[5]

[1] Multimedia Database Laboratory, Department of Computer Engineering,
METU, Ankara, Turkey
yazici@ceng.metu.edu.tr
[2] Command Control and Combat Systems, HAVELSAN Inc., Ankara, Turkey
[3] Department of Computer Engineering, Baskent University, Ankara, Turkey
[4] Department of Information System Engineering, Atilim University, Ankara, Turkey
[5] C + E Management, Microsoft Corporation, Redmond, WA, USA

Abstract. Managing a large volume of multimedia data, which contain various modalities (visual, audio, and text), reveals the need for a specialized multimedia database system (MMDS) to efficiently model, process, store and retrieve video shots based on their semantic content. This demo introduces METU-MMDS, an intelligent MMDS which employs both machine learning and database techniques. The system extracts semantic content automatically by using visual, audio and textual data, stores the extracted content in an appropriate format and uses this content to efficiently retrieve video shots. The system architecture supports various multimedia query types including unimodal querying, multimodal querying, query-by-concept, query-by-example, and utilizes a multimedia index structure for efficiently querying multi-dimensional multimedia data. We demonstrate METU-MMDS for semantic data extraction from videos and complex multimedia querying by considering content and concept-based queries containing all modalities.

1 Introduction

With the increasing amount of multimedia data production favored by technological advances and decreasing prices of digital devices, people are exposed to a very large volume of multimedia data in daily life. However, searching such a large volume of data efficiently is a real challenge. Although managing and retrieving textual content (metadata, tag) are relatively straightforward, users are usually interested in the semantic content (i.e., concept) such as objects and events, which are obtained from the various modalities of multimedia data [1]. In addition, complex properties of video data (i.e., multi-dimensional nature, uncertainties and temporal/spatial aspects) present the need for specialized MMDSs to efficiently access and retrieve videos based on their semantic content.

© Springer International Publishing Switzerland 2016
Q. Tian et al. (Eds.): MMM 2016, Part II, LNCS 9517, pp. 354–360, 2016.
DOI: 10.1007/978-3-319-27674-8_33

In this study, we develop a competent MMDS to address the following issues:

– Semantic content extraction: The semantic content in videos is very valuable for querying purposes. However, it is not feasible to extract the semantic content of large amounts of digital videos manually. Thus, the automatic extraction of semantic content is a vital issue.
– Multimodal processing: Multimedia data usually has a complex structure containing multimodal information (i.e., visual, audio and textual). During the extraction process, combining the information from different modalities is crucial to improve the extraction performance.
– Modeling & storing data: The data model for MMDS handles low-level and high-level features, information obtained from different modalities (multi-modality), and relations (e.g. inheritance) between extracted objects. It is also important to handle the uncertainty existing in multimedia data.
– Powerful querying: In order to satisfy user requirements, the MMDS provides appropriate querying mechanisms like unimodal/multimodal querying, query-by-content (based on low-level features), query-by-concept (based on high-level features) and fuzzy querying (based on fuzzy measures).
– Efficient access structures: Since multimedia data contains a huge amount of information and exists in large sequential file formats, relatively slow responses to user queries are inevitable. Therefore, an efficient multi-dimensional access structure is used in MMDS. For this purpose, both low and high level contents are indexed for efficiently retrieving various types of queries.

Despite several studies exist [1–3], the need for a complete MMDS that combines all required capabilities in a single multimedia database system has not yet been fulfilled [4,5]. Hence, a well-defined system architecture for an integrated multimedia database system covering these requirements is vital.

In this demo paper, we present METU-MMDS to address all of the issues mentioned above. The rest of this paper gives an overview of METU-MMDS system architecture (Sect. 2) and a brief description of our demonstration (Sect. 3) and finally the conclusion (Sect. 4).

2   System Architecture

METU-MMDS is composed of two major components: (i) The semantic content extraction subsystem, and (ii) the storage and retrieval subsystem (Fig. 1).

2.1   Semantic Content Extraction

The semantic content extraction subsystem enables extraction of the semantic contents of raw videos and passes them to the storage and retrieval subsystem. The extraction process starts with the shot-boundary detection of the given videos. The shot-boundary detection is performed by using a Canny Edge Detection algorithm based solution, which calculates Edge Change Ratio between

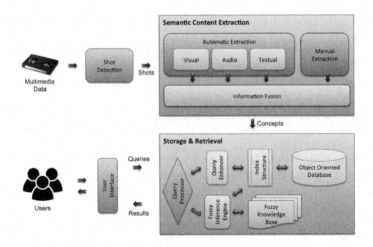

Fig. 1. The architecture of METU-MMDS

video frames [6]. After shot-boundary detection, available modalities (visual, audio and textual) are processed for extracting the semantic content. Once the processing of available modalities are completed, a multimodal fusion process is applied on the extracted content to improve the extraction accuracy. After performing semantic content extraction process, the extracted contents are stored in the storage and retrieval subsystem. For this purpose, the outputs of all modalities (visual, audio and text) and the fusion results are stored in the database separately to provide users with more flexibility in querying the system.

Visual Concept Extraction. The visual semantic content extraction is performed by employing a Class Specific Feature (CSF) [7] and a Support Vector Machine (SVM) based classification approach for object extraction. The object categories are defined with the CSF model and 8 different MPEG-7 descriptors (color, shape and texture based) are used to classify objects using image segments. Prior to the feature extraction and object classification steps, the key-frames of video are partitioned spatially into segments, by employing JSEG segmentation algorithm [8]. After extracting low-level features of these segments, the SVM classifier is utilized to classify these segments into potential objects. The classified objects are then passed to the storage & retrieval subsystem as the visual concepts, as well as the multimodal fusion module.

Audio Concept Extraction. The main task of this module is to extract meaningful audio content from the video data. The module first divides a given audio clip into 1 s consecutive segments and extract low-level audio features from each of these segments. Then, an acoustic classification procedure is carried out to label each of the segments into predefined classes. Then we apply two smoothing rules to reduce possible prediction errors. We use an SVM classifier and a joint audio feature set due to their success as described in [9]. Finally, the classified segments are aligned with the corresponding video shots.

Textual Concept Extraction. With an intention to combine the advantages of the text modality, i.e., video subtitles, we execute a rule-based named entity recognizer presented in [10] on the video texts. The recognizer extracts named entities of the three basic types, namely, person, location, organization names. Assuming that the keywords meaningful for querying are the named entities occurring the video text, we accept the named entities as textual concepts.

Multimodal Information Fusion. After obtaining the visual, audio and textual concepts, SVM-based multimodal fusion process [11] is applied. The purpose of the fusion module is to integrate the observations belonging to the same class and to utilize the interactions and relations between the observations of different classes captured from independent modalities. This process results in increasing the detection accuracy of the concepts by obtaining new information from exploiting the relations of the concepts.

2.2   Storage and Retrieval

The storage and retrieval subsystem stores the extracted data and answers various types of queries. It consists of an object-oriented database, an indexing structure, a query enhancer, a fuzzy inference engine and a fuzzy knowledge-base. When a query is received by the system, it is firstly processed by the query processor. Based on the type of the query, it is forwarded to the query enhancer or to the fuzzy inference engine. If the query is a unimodal or multimodal query, based on the retrieval of particular concepts, the retrieval is performed by using the index structure and database.

The Database and Data Model. The low-level (feature) and the semantic (concepts) contents, which are extracted by the semantic content extraction subsystem, are stored in an object-oriented database. The object-oriented database design is based on a conceptual video data model developed by considering the special characteristics of multimedia data such as the relations between videos, shots, frames, concepts, low-level features [12]. The database is also capable of storing the content with uncertainty measures for handling fuzzy querying.

Fuzzy Inference and Knowledge Base. The fuzzy inference engine manages the interoperability between the object-oriented database and the fuzzy knowledge-base. The fuzzy knowledge-base is used to define and store knowledge about the video domains. Instead of storing some implicit data in the database, the fuzzy knowledge-base includes some (fuzzy) logical rules that infer new information from the existing data.

Multimedia Index Structure. We make use of a multi-dimensional index structure for efficiently retrieving multimedia data from the database. Our index structure indexes both content and concept descriptors of the multimedia data. This index structure is an adaptation of FOOD-Index [13] in which a spatial indexing technique is incorporated in order to handle both concept-based and content-based queries. The index structure uses a multi-dimensional scaling approach [14] that reduces the retrieval problem to a spatial-indexing task to lower the search space, and thus, querying the multimedia databases is much more efficient than using distance based indexing methods.

Multimodal Query Enhancer. The query enhancer improves the querying and retrieval capability of the system and helps increase the retrieval performance of the system by performing a query-level fusion process. The query-level fusion is based on the idea of exploiting the correlation between different concepts, and provides retrieval of the videos that have correlated concepts in addition to the queried ones [15]. The developed system enables partial match capability for retrieval by utilizing Vector Space Model (VSM). In this context, the query-level fusion exploits the co-occurrence of the VSM terms within a modality and across different modalities [16]. Co-occurrence metrics (calculated by using the Pearson correlation and Canonical Correlation Analysis methods) are used to update the weights of terms in the query.

3   Demonstration Details

Our demonstration allows conference attendees to use METU-MMDS for (i) extracting semantic content from videos, and (ii) querying multimedia data using various types of queries.

For semantic content extraction, the scenario includes uploading a video to the video server, performing shot-boundary detection on the video, key-frame extraction of all shots, visual segmentation of key-frames, visual concept recognition, audio segmentation, audio concept classification, textual concept extraction (named entity recognition) and multimodal information fusion. The defined process is usually performed as a batch job, yet a manual intervention on the visual/audio/textual concept extraction tasks, which enables the system to include human analysis when required and provides a hybrid (manual and automatic) extraction process. The hybrid annotation functionality enables interactive segmentation by defining or selecting automatically calculated minimum bounding rectangles for visual objects, automatic prediction of semantic concepts and reviewing the prediction results. After semantic content extraction, both the low-level features and the high-level semantic concepts are stored in the multimedia database, and indexed by the multi-dimensional index structure.

The demonstration of querying capabilities of the system basically includes queries based on (i) the semantic concepts (query-by-concept) and (ii) the low-level content (query-by-content or query-by-example). The system has an integrated architecture including components for different modalities. Thus, for the query-by-concept, both unimodal querying (i.e. retrieving videos based on a concept related to a single modality like images, audio clips, or textual documents), and multimodal querying (i.e. retrieving videos based on multiple concepts, each of which belongs to a different modality, like a query of retrieve videos including cars and car horn) are enabled. In addition, for the video documents, queries can be made on fused annotations by using a single query concept. For the query-by-example query type, the query can be performed by giving an example multimedia document (image, audio clip, etc.). The system also supports querying by both specifying a concept and an example document.

Another important capability of the developed system is the improved querying capability provided by the multimodal query enhancer and the query-level fusion mechanism. Such a capability enables comparing the retrieval performance of the improved querying capability with the base multimodal querying functionality. This capability is also included in the demonstration.

4   Conclusion

In this paper, we present METU-MMDS, an intelligent MMDS, for automatic semantic content extraction and querying multimedia data. We overview the architecture and the modules of the system, then briefly explain how the system extracts semantic content from different modalities. We also describe the types of queries supported by the system. We demonstrate the user interfaces to extract semantic content online and its ability to query multimedia data stored in the multimedia database, which is indexed using a multi-dimensional index structure for efficiently retrieving multimedia data from the database.

Acknowledgments. This work is supported by the research grant from TUBITAK with the grant number 114R0182. We also thank to all of the previous researchers of Multimedia Db. Lab. at METU who have contributed to this study.

References

1. Rashid, U., Bhatti, M.A.: Exploration and management of web based multimedia information resources. In: Elleithy, K. (ed.) Innovations and Advanced Techniques in Systems, Computing Sciences and Software Engineering, pp. 500–506. Springer, The Netherlands (2008)
2. Brendan, J., Hongzhi, L., et al.: Structured exploration of who, what, when, and where in heterogeneous multimedia news sources. In: ACM MM, pp. 357–360 (2013)
3. Stefanidis, K., Koutrika, G., Pitoura, E.: A survey on representation, composition and application of preferences in database systems. J. TODS **36**, 19–45 (2011). ACM
4. Meng, T., Shyu, M.L.: Leveraging concept association network for multimedia rare concept mining and retrieval. In: ICME, pp. 860–865. IEEE, Melbourne (2012)
5. Smith, J.R.: Riding the multimedia big data wave. In: SIGIR, pp. 1–2. ACM (2013)

6. Aydinlilar, M., Yazici, A.: Semi-automatic semantic video annotation tool. In: Gelenbe, E., Lent, R. (eds.) International Symposium on Computer and Information Sciences, pp. 303–310. Springer, Paris (2012)

7. Yilmaz, T., Yazici, A., Yildirim, Y.: Exploiting class-specific features in multi-feature dissimilarity space for efficient querying of images. In: Christiansen, H., De Tré, G., Yazici, A., Zadrozny, S., Andreasen, T., Larsen, H.L. (eds.) FQAS 2011. LNCS, vol. 7022, pp. 149–161. Springer, Heidelberg (2011)

8. Deng, Y., Manjunath, B.S.: Unsupervised segmentation of color-texture regions in images and video. IEEE J. TPAMI **23**(8), 800–810 (2001)

9. Okuyucu, C., Sert, M., Yazici, A.: Audio feature and classifier analysis for efficient recognition of environmental sounds. In: ISM, pp. 125–132. IEEE, USA (2013)

10. Kucuk, D., Yazici, A.: Exploiting information extraction techniques for automatic semantic video indexing with an application to Turkish news videos. J. Knowl.-Based Sys. **25**(6), 844–857 (2011)

11. Gulen, E., Yilmaz, T., Yazici, A.: Multimodal information fusion for semantic video analysis. J. IJMDEM **3**(4), 52–74 (2012)

12. Kucuk, D., Ozgur, N.B., Yazici, A., Koyuncu, M.: A fuzzy conceptual model for multimedia data with a text-based automatic annotation scheme. J. IJUFKS **17**(1), 135–152 (2009)

13. Yazici, A., Ince, C., Koyuncu, M.: Food index: a multidimensional index structure for similarity-based fuzzy object-oriented database models. J. IEEE Trans. Fuzzy Sys. **16**(4), 942–957 (2008). IEEE

14. Arslan, S., Yazici, A., Sacan, A., Toroslu, I.H., Acar, E.: Comparison of feature-based and image registration-based retrieval of image data using multidimensional data access methods. J. TKDE **86**, 124–145 (2013). Elsevier

15. Safadi, B., Sahuguet, M., Huet, B.: When textual and visual information join forces for multimedia retrieval. In: ICMR, pp. 265–272. ACM (2014)

16. Yu, J., Cong, Y., Qin, Z., Wan, T.: Cross-modal topic correlations for multimedia retrieval. In: International Conference on Pattern Recognition, pp. 246–249. IEEE, Japan (2012)

Applying Visual User Interest Profiles for Recommendation and Personalisation

Jiang Zhou, Rami Albatal, and Cathal Gurrin[✉]

Insight Centre for Data Analytics, Dublin City University, Dublin, Ireland
jiang.zhou@dcu.ie, cgurrin@computing.dcu.ie
https://www.insight-centre.org

Abstract. We propose that a visual user interest profile can be generated from images associated with an individual. By employing deep learning, we extract a prototype visual user interest profile and use this as a source for subsequent recommendation and personalisation. We demonstrate this technique via a hotel booking system demonstrator, though we note that there are numerous potential applications.

1 Introduction

In this work, we conjecture that images associated with an individual can provide insights into the interests or preferences of that individual, by means of a visual user interest profile (hereafter referred to as the visual profile), which can be utilised to personalise or recommend content to that individual. Given a set of images from an individual, applying deep-learning for semantic content extraction from images generates a set of concepts that form the visual profile. Using this visual profile, it is possible to personalise various forms of information access, such as highlighting multimedia content that a user would potentially like or personalising online services to the particular interests of the user. We believe that it is even possible to develop new (or complimentary) recommendation engines using the visual profile. We demonstrate the visual profile in a real-world information access challenge, that of hotel booking systems. An individual using our prototype hotel booking engine will be presented with visual ranked lists and personalised hotel landing pages. The contribution of this work is the prototype visual profile which imposes no overhead on the user to gather, but can be employed to both recommend content (e.g. hotels to book) and to optimise content (e.g. the hotel landing page).

2 Visual User Interest Modeling

User interest modelling is a topic that has been subject to extensive research in the domain of personalisation and recommendation. While personalisation and recommendation techniques can take many forms, content-based filtering is the approach that best suits our requirements. In content-based filtering, items are recommended to a user based upon a description of the item and a profile

© Springer International Publishing Switzerland 2016
Q. Tian et al. (Eds.): MMM 2016, Part II, LNCS 9517, pp. 361–366, 2016.
DOI: 10.1007/978-3-319-27674-8_34

of the user's interests [1]. Content-based recommendation systems are used in many domains, for example, recommending web documents, hotels, restaurants, television programs, and items for sale. Content-based recommendation systems typically support a method for describing the items that may be recommended, a profile of the user that describes the types of items the user likes, and a means of comparing items to the user profile to determine what to recommend.

However, while such recommender systems operate effectively for item-item recommendation, it is our conjecture that a user profile operating at a deeper, more semantic, level than simple item-based user interest profile will capture user interest in more detail and extend recommender and personalisation functionality beyond item-item or faceted recommenders. Hence we introduce the concept of visual user interest modelling which examines media content that are known to be of interest to the user and generates a visual profile of visual concept labels that can be used to subsequently recommend and personalise content to that user. To take a naive example, an individual who regularly views (or captures) images of aircraft, food, architecture, would maintain a visual profile [aircraft, food, architecture] that can be used to highlight/personalise related content when interacting with retrieval systems.

2.1 Visual Feature Extraction

Given a set of images that are associated with an individual, it is necessary to extract the semantic content of the images for the visual profile. Mining the semantic content to extract visual features is an application of content-based image retrieval (CBIR), which has been an active research field for decades [2]. In a naive implementation, low-level image features such as colour, texture, shape, local features or their combination could represent images [3]. Yu et al. [4] investigated the weak attributes, a collection of mid-level representations, for large scale image retrieval. Weak attributes are expressive, scalable and suitable for image similarity computation, however we do not consider such approaches to be suitable for generating the visual profile. Firstly, the problem of the semantic gap arises where an individual's interpretation of an image can be different from an individual interpretation. Secondly, the performance of such conventional handcrafted features has plateaued in recent years while higher-level semantic extraction (typically based on deep learning) has gained favour [5].

Wang et al. [6] proposed a ranking model trained with deep learning methods, which is claimed to be able to distinguish the differences between images within the same category. Given that efficiency is a concern, Krizhevsky and Hinton [7] applied deep autoencoders and transformed images to 28-bit codes, such that images can be fast and accurately retrieved with semantic hashing. Babenko et al. [8] and Wan et al. [9] both proved that pre-trained deep CNNs (Convolutional Neural Networks) for image classification can be re-purposed to image retrieval problem. Babenko used the last three layers before the output layer from CNNs as the image descriptors while Wan chose the last three layers including the output layer.

In our work, we wish to model user visual interest, so our approach is similar to Babenko's and Wan's methods, which also reused a pretrained deep learning network to retrieve and rank hotels images. Our image features are extracted with a CNN [10] where the distribution over classes from the output layer is used as the descriptor for each image. This CNN produces a distribution over 1,000 visual object classes for the visual profile (see Fig. 1 for an example of the top 10 object classes for sample images). Because each dimension of the feature vector is actually a class in ImageNet, the descriptor helps to bridge the semantic gap between low-level visual features and high-level human perception.

Instead of training a CNN by ourselves, we employ a pre-trained model on the ImageNet "ILSVRC-2012" dataset from the Caffe framework [5]. Every image is forward passed through the pre-trained network and a distribution over 1,000 object classes from ImageNet is produced. This 1,000 dimension vector is regarded as the descriptor of the image and saved in the visual profile. The visual profile can contain the descriptors of many images.

3   Utilising the Visual Profile

In this work, we represent the prototype visual profile as a collection of feature vectors extracted from a set of images from the individual. Given this set of images, example-based matching between the user profile feature vectors and images in the dataset can be performed. In our case, cosine distance is applied as the similarity metric between pairs of images. The distance is computed with two vectors u and v in an inner product space as Eq. 1. The outcome is bounded between $[0, 1]$. When the images are very similar, the distance approaches to 0, otherwise the value is close to 1.

$$distance = 1 - \frac{u \cdot v}{\|u\|_2 \|v\|_2} \tag{1}$$

Figure 1 shows a simple example of the image matching from our demonstrator system, which assumes a visual profile of one image and a dataset of two images. The first row shows an image (from the user profile) and its top 10 dimensions with highest class probability values in the descriptor. As there is no swimming pool class in the pre-trained model, the model considers the content of the image to be container ship, sea-shore, lake-shore, dam, etc., with meaningful likelihoods. The second row is a similar image from the dataset which also has a swimming pool. As we can observe, classes such as the container ship, dock, boathouse, lakeside are detected in its top 10 classes as well. Not surprisingly, the Cosine distance between these two images is calculated at 0.28 which suggests visual similarity. The third row is a less-related image, for which the top 10 classes has no overlapped with the top 10 of the user profile image. Moreover, the 10 classes are even not semantically close to those from the query image. The distance of 0.98 suggests a high-degree of dissimilarity. By applying this approach on a larger scale, it is possible to select images that are more similar to the visual profile and this knowledge can then be applied to build novel personalised systems and recommendation engines.

Fig. 1. An image matching example using CNNs

4   Applications of the Prototype Visual Profile

It is our conjecture that the use of a visual profile can have many applications when personalising content to an individual. In personal photograph retrieval, the summary given to events or clusters of images can be tailored to the visual profile. In lifelogging, the key images representing an event can be personalised in a retrieval application, or images can be selected from the lifelog that support positive reminiscence. Since image content is modeled as a set of concepts, then the application of this profile can be extended to recommend non-visual content, such as in online stores or social media recommendation. Finally, the application that we find most compelling is to utilise the visual profile to provide a personalised view over, or summary of, visual data. The prototype we chose to develop was in hotel booking, where both the hotel ranked list and the hotel landing pages can be customised to suit the interests of the individual.

4.1   Hotel Booking System: An Example Application

We chose to implement a hotel image recommender system to demonstrate the visual user interest profile in operation. The idea was that a user profile would be augmented by the visual profile, which would be generated from images the user viewed when booking previous hotels, or even from social network postings or online browsing of the user. The visual profile will consist of a weighted list of concepts that occur in images that the user has shown interest in.

The demonstrator both ranked hotels based on similarity to the visual profile, but also personalised every hotel landing page based on the visual profile. In this demonstration, the visual profile is generated by either selecting several hotel images from the interface or uploading images which the user likes.

The recommendation engine will the utilise the visual profile as the searching criteria and return an ranked order hotel list; we compute the minimum distance of N images from the same hotel as Eq. 2 to be the representative distance of that hotel, such that all hotels can be reordered according to the hotel representative distances from 0 to 1 as well.

$$hotel\_distance = \min_{j \in N} \{distance_j\} \tag{2}$$

For each hotel, the images that represent that hotel are also ranked based on the similarity to the visual profile, in effect producing a personalised view of every hotel landing page for each user. Figure 2 shows a screenshot of the demonstrator system. The interface consists of three sections. Across the top of the web page is a set of example image queries to construct the visual profile. These examples cover a wide range of image types (e.g. bedroom images,entertainment systems, pools, dining, etc.). For example, adding swimming pool and pool table in the queries would rank hotels and tailor each hotel page based on these two facilities. The user can also choose to upload any images they wish, which demonstrates the flexibility of the visual profile. The middle section of the interface (Fig. 2) is the result frame, which shows the ranked list of hotels or the personalised hotel landing page. Images that are used to construct the visual profile are displayed at the bottom of the window, in which a user can remove or add images to the visual profile before executing a query by selecting 'reorder'.

Fig. 2. Applying the visual profile for a prototype hotel booking application. The personalised landing page is shown in this figure.

5   Conclusion and Future Work

In this work we presented a prototype visual user interest profile which attempts to capture the interests of a user by analysing the images that they are known to like. By employing deep learning, we can extract a visual user interest profile and use this as a source for subsequent recommendation and personalisation. This is novel in that we attempt to capture semantic user interest from visual content, and use it as a means to personalise additional content to the user. We demonstrate this technique via a hotel booking system demonstrator. The next steps in this work are to perform an evaluation of the accuracy of the proposed visual profile in various use cases and applications. We also intend to consider the temporal dimension and explore how time can impact on the weighting in the visual profile and how this impacts on real-world implementations.

Acknowledgments. This publication has emanated from research conducted with the financial support of Science Foundation Ireland (SFI) under grant number SFI/12/RC/2289.

References

1. Pazzani, M.J., Billsus, D.: Content-based recommendation systems. In: Brusilovsky, P., Kobsa, A., Nejdl, W. (eds.) Adaptive Web 2007. LNCS, vol. 4321, pp. 325–341. Springer, Heidelberg (2007)
2. Lew, M.S., Sebe, N., Djeraba, C., Jain, R.: Content-based multimedia information retrieval: state of the art and challenges. ACM Trans. Multimed. Comput. Commun. Appl. (TOMM) **2**(1), 1–19 (2006)
3. Deselaers, T., Keysers, D., Ney, H.: Features for image retrieval: an experimental comparison. Inf. Retr. **11**(2), 77–107 (2008)
4. Yu, F.X., Ji, R., Tsai, M.-H., Ye, G., Chang, S.-F.: Weak attributes for large-scale image retrieval. In: 2012 IEEE Conference on Computer Vision and Pattern Recognition (CVPR), pp. 2949–2956. IEEE (2012)
5. Jia, Y., Shelhamer, E., Donahue, J., Karayev, S., Long, J., Girshick, R., Guadarrama, S., Darrell, T.: Caffe: convolutional architecture for fast feature embedding. In: Proceedings of the ACM International Conference on Multimedia, pp. 675–678. ACM (2014)
6. Wang, J., Song, Y., Leung, T., Rosenberg, C., Wang, J., Philbin, J., Chen, B., Wu,. Y.: Learning fine-grained image similarity with deep ranking. In: 2014 IEEE Conference on Computer Vision and Pattern Recognition (CVPR), pp. 1386–1393. IEEE (2014)
7. Krizhevsky, A., Hinton, G.E.: Using very deep autoencoders for content-based image retrieval. In: ESANN. Citeseer (2011)
8. Babenko, A., Slesarev, A., Chigorin, A., Lempitsky, V.: Neural codes for image retrieval. In: Fleet, D., Pajdla, T., Schiele, B., Tuytelaars, T. (eds.) ECCV 2014, Part I. LNCS, vol. 8689, pp. 584–599. Springer, Heidelberg (2014)
9. Wan, J., Wang, D., Hoi, S.C.H., Wu, P., Zhu, J., Zhang, Y., Li, J.: Deep learning for content-based image pp. 157–166. ACM (2014)
10. Krizhevsky, A., Sutskever, I., Hinton, G.E.: Imagenet classification with deep convolutional neural networks. In: Advances in Neural Information Processing Systems, pp. 1097–1105 (2012)

Cross-Modal Fashion Search

Susana Zoghbi[✉], Geert Heyman,
Juan Carlos Gomez, and Marie-Francine Moens

KU Leuven, Leuven, Belgium
{susana.zoghbi,geert.heyman,
juancarlos.gomezcarranza,sien.moens}@cs.kuleuven.be
http://roshi.cs.kuleuven.be/multimodal_search/

Abstract. In this demo we focus on cross-modal (visual and textual) e-commerce search within the fashion domain. Particularly, we demonstrate two tasks: (1) given a query image (without any accompanying text), we retrieve textual descriptions that correspond to the visual attributes in the visual query; and (2) given a textual query that may express an interest in specific visual characteristics, we retrieve relevant images (without leveraging textual meta-data) that exhibit the required visual attributes. The first task is especially useful to manage image collections by online stores who might want to automatically organize and mine predominantly visual items according to their attributes without human input. The second task renders useful for users to find items with specific visual characteristics, in the case where there is no text available describing the target image. We use a state-of-the-art visual and textual features, as well as a state-of-the-art latent variable model to bridge between textual and visual data: bilingual latent Dirichlet allocation. Unlike traditional search engines, we demonstrate a truly *cross-modal* system, where we can directly bridge between visual and textual content without relying on pre-annotated meta-data.

1 Introduction

The Web is a multi-modal space. It is flooded with visual and textual information. The ability to natively organize and mine its heterogeneous content is crucial for a seamless user experience. This is especially true in e-commerce search, where product images and textual descriptions play key roles, because it is not feasible to physically inspect the goods. In particular this is true for items which are predominantly visual, such as fashion products.

Automatically mining fashion products, while considering their multi-modal nature, has a large potential impact on Web technologies. Globally, consumers spend billions of dollars on clothing every year. Therefore, applications that organize and retrieve fashion items would have great societal value. More specifically, in fashion e-commerce search, allowing users to query in one modality and obtain results in another is greatly beneficial for providing relevant content. For example users may write textual queries indicating the visual attributes they wish to find, and our system retrieves product images that display such

© Springer International Publishing Switzerland 2016
Q. Tian et al. (Eds.): MMM 2016, Part II, LNCS 9517, pp. 367–373, 2016.
DOI: 10.1007/978-3-319-27674-8_35

attributes, without having to rely on textual meta-data on the image to match against the textual query. Additionally, an e-commerce site may wish to automatically organize an image collection according to its visual attributes. In this case, automatically annotating the images with visual properties would facilitate this time-consuming task.

2   Functionality Overview

In this work, we demonstrate a complete system that performs two truly cross-modal search tasks in the fashion domain: **Task 1 (Img2Txt):** Given a query image without any surrounding text, our system generates text that describes the visual properties in the image; and **Task 2 (Txt2Img):** Given a textual query without any visual information, our system finds images that display the visual properties in the query. Unlike traditional search engines, these tasks are realized in a purely cross-modal way, where the query modality is completely distinct from the target. In other words, textual queries retrieve target images that have not been previously annotated, therefore, keyword matching is not possible. Likewise, image queries without any textual annotations, retrieve words that describe the image. These are challenging tasks for both computer vision and natural language processing. Figures 1 and 2 show examples of these tasks.

To build this demo, we collected a dataset consisting of 53,689 fashion products, specifically dress-like garments. Each product contains one image and surrounding natural language text. We used Amazon.com's API to query products

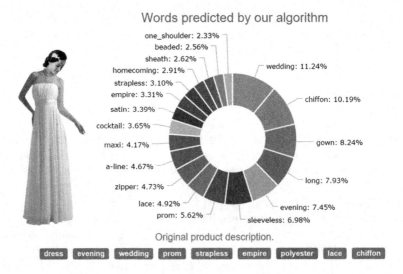

Fig. 1. Img2Txt: Given a query image (left), our system predicts words that describe the attributes of the image (right), ordered by the probability of the word occurrence. On the bottom, we show the original words from the product description and highlight in green those predicted by our algorithm(Color figure online).

Fig. 2. Txt2Img: The user may enter a textual query (e.g., 'floral print' on the top half or 'red gown' on the bottom), and our system finds images that display the attribute in the query(Color figure online).

within the Apparel category. We focused on dresses under all available categories, such as Casual, Bridesmaids, Night Out & Cocktail, Wear to Work, Wedding, Mother of the Bride, etc. From the API's output, we concatenate all the natural text that surrounds the image, such as title, features and editorial content.

While all the images contain a dress-like garment, there exist very large visual variations in terms of visual attributes, such as shapes, textures and colors[1].

3 Methodology

Text Representation. The textual descriptions are represented as a bag-of-words. We used the online glossary of www.zappos.com, an online clothing shop[2] to index terms. After lower-casing the terms, only the glossary terms are retained. The Zappos glossary contains multi-word terms (for instance 'little black dress' or 'one shoulder'), these were treated as if they were a single word.

Image Representation. We use Convolutional Neural Networks (CNNs) [5] to represent images. A CNN is a type of feed-forward artificial neural network where the individual neurons are connected to respond to overlapping regions in the image [6]. They have been extremely successful in image recognition tasks in computer vision. The training process may be interpreted as learning a template given by a set of weights, which aim to maximize the probability of the correct class for each of the training instances. These networks contain many layers. The activation weights corresponding to the last fully connected layer of CNNs may be used as image features. The CNN was pre-trained on over one million images from the famous ImageNet computer vision classification task [3]. For a full description of this representation, we refer the reader to the study of Krizhevsky et al. [5,8]. We used the Caffe implementation of CNNs [4]. Under this framework, images may be represented as 4096-dimensional real-valued vectors. Specifically, we use a 16-layer convolutional neural network from [8]. The vectors correspond to the weights of the last fully connected layer of the CNN, that is, the layer right before the softmax classifier. We may interpret each component of the CNN vectors as a visual concept or visual word. The actual value corresponding to a particular component then represents the degree to which the visual concept is present. It turns out that these visual concepts are extremely powerful to correctly classify images. Here we will show that they also outperform SIFT features for our cross-modal retrieval tasks.

Inducing a Multi-Modal Space. An attribute may be seen as a latent concept that generates different representations depending on the modality, (visual or textual). For example the word *a-line* is an attribute of the object, in particular it may describe the A-shape of a skirt. This attribute is instantiated in the image by a set of pixel values; and it is instantiated in the text by the actual textual words. In our training set, we have a collection of image and textual description pairs. From these, we wish to learn associations between the visual and textual information. We may project onto a multimodal space where we enforce that

---

[1] Examples of our data are available in http://glenda.cs.kuleuven.be/multimodal_ search under the 'Training Data' tab.

[2] http://www.zappos.com/glossary.

image and text elements that often occur together during training, are close together in this multimodal space.

There exist several approaches that we may use to find associations between visual and textual words. In particular, we focus on a state-of-the-art multimodal retrieval model: Bilingual Latent Dirichlet Allocation (BiLDA). It has been used in [7] to annotate image data from shoes and bags and it constructs a multimodal space by softly clustering textual and visual features into topics.

The BiLDA produces an elegant probabilistic representation to model our intuition that image-description pairs instantiate the same concepts (or topics) using different word modalities. To learn, the main assumptions are that each product d, consisting of aligned visual and textual representations $d = \{d^{img}, d^{text}\}$, may be modelled by a multinomial distribution of topics, $\theta = P(z|d)$; and each topic may be represented by two distinct word distributions ϕ^{img} and ϕ^{txt}, which correspond to a visual-word distribution, and a textual-word distribution (both multinomial), respectively as

$$\phi^{img} = P(w^{img}|z), \qquad \phi^{txt} = P(w^{txt}|z). \tag{1}$$

These may be interpreted as two views of the same entity: the topic z. Consequently topics become multi-modal structures, and documents may effectively be projected into a shared multi-modal space via their topical distributions. We estimate the posterior probability distributions of θ, ϕ^{img} and ϕ^{txt}, using Gibbs sampling as an inference technique, which is a simple and easy to implement method. We can infer the topic distribution of an unseen image document $\theta^{img} = P(z|d^{img})$, and the topic distribution of a previously unseen textual document $\theta^{txt} = P(z|d^{text})$, by using again Gibbs sampling, but this time we keep the word-topic distributions fixed ϕ^{img} and ϕ^{txt}, as we have already learned them, and we need only to infer the topic assignments within each unseen document. We have in-house software that implements this model. We gloss over much of the details, but more information may be found in [1,2].

Cross-Modal E-Commerce Search. Given that we can induce a multi-modal space, we may now formulate the mechanism that allows us to bridge between visual and textual content for applications in multi-modal Web search. Using Bilingual Latent Dirichlet Allocation (BiLDA), we can bridge between image and textual representations in a probabilistic manner by marginalizing over all possible topics. To generate textual words w^{text}, given a query image d^{img},

$$P(w^{text}|d^{img}) \propto \sum_z P(w^{text}|z)P(z|d^{img}) \tag{2}$$

$$\propto \sum_z \phi_z^{txt}\theta_z^{img} \tag{3}$$

The words with the highest probability are chosen as the top candidates to annotate the query image. To retrieve relevant images d^{img} given a textual query d^{txt}, we rank the images based on the similarity of their topic distributions,

$$sim(d^{img}, d^{txt}) = \sum_z P(z|d^{text})P(z|d^{img}) \tag{4}$$

$$= \sum_z \theta_z^{txt}\theta_z^{img} \tag{5}$$

the images with the highest similarity become the top candidates to satisfy the textual query. Full technical details and formal evaluation can be found in [9].

4   Conclusions

In this work we have demonstrated cross-modal search of fashion items. Given a textual query, our system retrieves relevant images of dresses, and given a picture of a dress as query, the system describes the attributes of the dress in natural language terms. Our system was trained on real Web data found at Amazon.com composed of fashion products and their textual descriptions and was evaluated on an additional set of Amazon data. We used CNN-based visual features and a controlled, commonly used fashion vocabulary. By visually inspecting the annotations our system generates, we find reasonable descriptions that capture different garment lengths, colors and textures. For example an image displaying a yellow long gown, receives these actual words as descriptions. Furthermore, we were able to find relevant images given textual queries. For example, a query indicating 'casual sleeveless floral print', retrieves garments with these characteristics. These results are extremely useful for navigating the multi-modal data on fashion search.

Acknowledgments. We greatly thank Anirudh Tomer for building the Web interface of our demonstrator. This project is part of the SBO Program of the IWT (IWT-SBO-Nr. 110067).

References

1. De Smet, W., Moens, M.-F.: Cross-language linking of news stories on the Web using interlingual topic modeling. In: Proceedings of the CIKM 2009 Workshop on Social Web Search and Mining (SWSM), pp. 57–64 (2009)
2. De Smet, W., Tang, J., Moens, M.-F.: Knowledge transfer across multilingual corpora via latent topics. In: Huang, J.Z., Cao, L., Srivastava, J. (eds.) PAKDD 2011, Part I. LNCS, vol. 6634, pp. 549–560. Springer, Heidelberg (2011)
3. Deng, J., Dong, W., Socher, R., Li, L.-J., Li, K., Fei-Fei, L.: ImageNet: a large-scale hierarchical image database. In: CVPR 2009 (2009)
4. Jia, Y., Shelhamer, E., Donahue, J., Karayev, S., Long, J., Girshick, R., Guadarrama, S., Darrell, T.: Caffe: convolutional architecture for fast feature embedding. arXiv preprint arXiv:1408.5093 (2014)
5. Krizhevsky, A., Sutskever, I., Hinton, G.E.: Imagenet classification with deep convolutional neural networks. In: Advances in Neural Information Processing Systems, vol. 25, pp. 1097–1105. Curran Associates Inc., (2012)

6. Lecun, Y., Bottou, L., Bengio, Y., Haffner, P.: Gradient-based learning applied to document recognition. In: Proceedings of the IEEE, pp. 2278–2324 (1998)

7. Mason, R., Charniak, E.: Annotation of online shopping images without labeled training examples. In: North American Chapter of the ACL Human Language Technologies, vol. 2013, p. 1 (2013)

8. Simonyan, K., Zisserman, A.: Very deep convolutional networks for large-scale image recognition. CoRR, abs/1409.1556 (2014)

9. Zoghbi, S., Heyman, G., Gomez, J.C., Moens, M.-F.: Fashion meets computer vision and natural language processing. In: Submitted to ACM International Conference on Web Search and Data Mining WSDM 2015 (2015)

Video Browser Showdown

IMOTION – Searching for Video Sequences Using Multi-Shot Sketch Queries

Luca Rossetto[1]([✉]), Ivan Giangreco[1], Silvan Heller[1], Claudiu Tănase[1],
Heiko Schuldt[1], Stéphane Dupont[2], Omar Seddati[2],
Metin Sezgin[3], Ozan Can Altıok[3], and Yusuf Sahillioğlu[3]

[1] Databases and Information Systems Research Group, Department of Mathematics
and Computer Science, University of Basel, Basel, Switzerland
{luca.rossetto,ivan.giangreco,silvan.heller,
c.tanase,heiko.schuldt}@unibas.ch
[2] Research Center in Information Technologies, Université de Mons, Mons, Belgium
{stephane.dupont,omar.seddati}@umons.ac.be
[3] Intelligent User Interfaces Lab, Koç University, Istanbul, Turkey
{mtsezgin,oaltiok15,ysahillioglu}@ku.edu.tr

Abstract. This paper presents the second version of the IMOTION system, a sketch-based video retrieval engine supporting multiple query paradigms. Ever since, IMOTION has supported the search for video sequences on the basis of still images, user-provided sketches, or the specification of motion via flow fields. For the second version, the functionality and the usability of the system have been improved. It now supports multiple input images (such as sketches or still frames) per query, as well as the specification of objects to be present within the target sequence. The results are either grouped by video or by sequence and the support for selective and collaborative retrieval has been improved. Special features have been added to encapsulate semantic similarity.

1 Introduction

In this paper we introduce the improvements made to the IMOTION system to adapt to the changed rules of the 2016 Video Browser Showdown [6] and to improve the system performance. With this version, we address the shortcomings of the 2015 edition of our system [4], especially in the textual challenges. We briefly discuss the architecture and implementation of the IMOTION system in Sect. 2 and elaborate on the changes made in this version in Sect. 3.

2 The IMOTION System

2.1 Architecture

The IMOTION system can be divided into a front-end and a back-end part. The back-end is based on the Cineast content-based video retrieval engine [3] which uses a multitude of different features in parallel to perform retrieval.

© Springer International Publishing Switzerland 2016
Q. Tian et al. (Eds.): MMM 2016, Part II, LNCS 9517, pp. 377–382, 2016.
DOI: 10.1007/978-3-319-27674-8_36

The front-end is browser-based. It communicates with the back-end through a web server which serves as a proxy for the retrieval engine while also serving static content such as preview images and videos.

In [4], we provide a more in-depth discussion of the architecture.

2.2 Implementation

The retrieval engine is written in Java and uses a customized version of PostgreSQL for storing all the feature data and meta-data. The adapted database provides various indexing techniques to index the feature data and by that decreases the retrieval time.

For object, scene, and action recognition, we train Convolutional Neural Networks (ConvNets) using the publicly available Torch toolbox [1].

3   New Functionality and User Interaction

This section outlines the various improvements we made to the system in comparison to the version which has participated to the 2015 VBS.

3.1   Multi-Shot Queries

An important new feature is the possibility to search for multiple shots in a single query. While the 2015 edition of the IMOTION system allowed only for one shot per query, the current version enables users to search for an arbitrary amount of (succeeding) shots. This greatly increases the overall expressiveness of a single query, especially when searching for heterogeneous video sequences, i.e., sequences which span several subsequent shots. In this case, separate query sketches can be provided for the different shots. Figure 1 shows a screen-shot illustrating a multi-shot query and Fig. 2 shows the corresponding results.

3.2   Object Recognition and Retrieval

To augment the visual queries with semantic information, we use an object recognition system which is trained to recognize several hundred commonly seen objects in the video. For each shot of a video, all recognized objects and their positions within the frame are stored.

Even though the focus of the visual query specification lies on sketching, we decided against using sketch recognition [7] for query specification because of time constraints during the competition. Instead, the users will be presented with a list of clip-arts representing the recognizable objects which they can add to the query image via drag and drop.

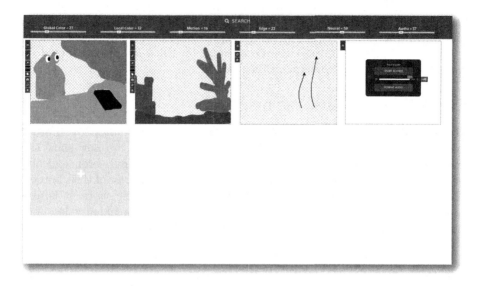

Fig. 1. Screen-shot of the IMOTION 2016 prototype UI

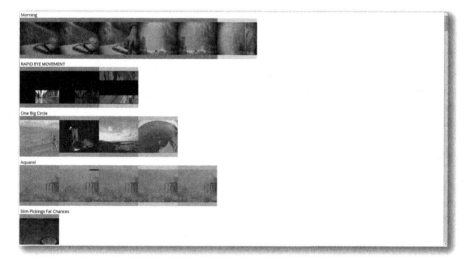

Fig. 2. Screen-shot of the corresponding result page

3.3   Result Limitation and Collaborative Search

When refining a query, it is important to be able to limit the displayed results to a selected few or even one single video. The 2016 edition of IMOTION does not only support such selection based on retrieved results but also provides means to efficiently specify the relevant videos. The latter is important to support collaborative search which is actively supported in this version of the system.

In case one user is sure to have found the correct video but not necessarily the correct sequence, other users can limit their search efforts to this video. This is achieved by representing the video designation to which the results should be limited in an efficient alphanumerical encoding.

3.4   Result Presentation and Browsing

The way retrieval results are displayed has been made more intuitive. Rather than showing isolated shots grouped by similarity measure, we now show the results grouped by video. The shots within a video are ordered chronologically and their score is indicated both with an overlay and by the color of their border. The videos are sorted based on the maximum score of their shots. It is also possible to perform a sequence segmentation of the results, breaking videos into multiple sequences with multiple shots each. This is particularly useful in cases where a query matches more than one sequence of the same video.

Additionally, improvements have been made to the way results are transferred from the back-end to the front-end. Results can now be streamed as they are generated which reduces the time required for the first results to appear. The front-end can display retrieved sequences as they come and re-order them to reflect their appropriate position in the growing set of results.

3.5   Additional and Improved Video Features

To improve the flexibility in query specification, the previously used features have been extended to be able to deal with transparency in query images. This enables the user to have incomplete sketches which focus only on a part of the frame while ignoring the rest. The previous version of the system would not differentiate between an empty and a white area which could lead to unwanted results.

As in our previous system, we use two different ConvNets types for feature extraction. The first one for spatial information and the second one for temporal information. We use the output of neurons in a selected hidden layer as features. This time, however, and in order to speed up similarity search using those features, we reduced the dimensionality of the vectors by adding a bottleneck layer before the final fully connected classification layer. This solution ensures getting a shorter vector of features without degrading accuracy. This has been applied for the spatial ConvNets. No changes were needed for the temporal ConvNets as the classification task it is trained on only covers a much limited number of classes (about 150 human actions) compared to the spatial on (about 1000 concepts).

For the spatial information, we actually use three different ConvNets enabling to highlight different facets of the content. We still use a ConvNet trained on the ImageNet dataset [5]. The large number of categories helps building good feature extractors. But unfortunately, most categories present in ImageNet are not of great interest for search in generic video databases. To improve on our previous system, we hence used an additional ConvNet trained on images downloaded from the internet and which correspond to the 1000 most frequent synsets of WordNet. In addition to training for object recognition, we use one more ConvNet trained to recognize the context/scene within an image. This ConvNet is trained on the Places dataset [9], containing examples of 205 scene categories and a total of 2.5 million images.

For the temporal feature extractor, we increased the number of recognized categories by merging two action recognition databases, namely HMDB-51 [2] and UCF101 [8]. As before, optical flows are extracted from video shots and used as input to our temporal ConvNet.

Finally, our new system also enables to use the audio channel of the video. The system extracts audio features: MFCC, Chroma and temporal modulation. This enables audio-based similarity search. The formulation of audio queries is also possible as the interface enables the user to record audio (vocal imitations of the sound of the video to be retrieved) using a microphone. We see this as a form of audio sketching, complementary to the image sketching used for specifying the visual content.

As before, depending on the weight that the user gives to the various feature sets, the system returns videos that have similarities according to different facets of the content.

4 Conclusions

The 2015 edition of the IMOTION system has already proven to be highly suitable for the VBS competition, especially for the visual part. With the 2016 edition, several improvements have been added to the functionality of the system, in particular to give a user more flexibility when specifying queries for heterogeneous video sequences, and we have improved the usability of the system.

Acknowlegements. This work was partly supported by the Chist-Era project IMOTION with contributions from the Belgian Fonds de la Recherche Scientifique (FNRS, contract no. R.50.02.14.F), the Scientific and Technological Research Council of Turkey (Tübitak, grant no. 113E325), and the Swiss National Science Foundation (SNSF, contract no. 20CH21_151571).

References

1. Collobert, R., Kavukcuoglu, K., Farabet, C.: Torch7: a matlab-like environment for machine learning. In: BigLearn, NIPS Workshop (2011)
2. Kuehne, H., Jhuang, H., Garrote, E., Poggio, T., Serre, T.: HMDB: a large video database for human motion recognition. In: Proceedings of the International Conference on Computer Vision (ICCV), pp. 2556–2563 (2011)
3. Rossetto, L., Giangreco, I., Schuldt, H.: Cineast: a multi-feature sketch-based video retrieval engine. In: Proceedings of the IEEE International Symposium on Multimedia (ISM 2014), pp. 18–23. IEEE (2014)

4. Rossetto, L., Giangreco, I., Schuldt, H., Dupont, S., Seddati, O., Sezgin, M., Sahillioğlu, Y.: IMOTION - a content-based video retrieval engine. In: He, X., Luo, S., Tao, D., Xu, C., Yang, J., Abul Hasan, M. (eds.) MultiMedia Modeling. LNCS, vol. 8936, pp. 255–260. Springer, Heidelberg (2015)
5. Russakovsky, O., Deng, J., Su, H., Krause, J., Satheesh, S., Ma, S., Huang, Z., Karpathy, A., Khosla, A., Bernstein, M., Berg, A.C., Fei-Fei, L.: ImageNet large scale visual recognition challenge. Int. J. Comput. Vis. (IJCV), 1–42 (2015)
6. Schoeffmann, K., Ahlström, D., Bailer, W., Cobârzan, C., Hopfgartner, F., McGuinness, K., Gurrin, C., Frisson, C., Le, D.-D., Del Fabro, M., et al.: The video browser showdown: a live evaluation of interactive video search tools. Int. J. Multimed. Inf. Retr. **3**(2), 113–127 (2014)
7. Seddati, O., Dupont, S., Mahmoudi, S.: Deepsketch: deep convolutional neural networks for sketch recognition and similarity search. In: Proceedings of the 13th International Workshop on Content-Based Multimedia Indexing (CBMI 2015), pp. 1–6. IEEE (2015)
8. Soomro, K., Zamir, A.R., Shah, M.: UCF101: a dataset of 101 human actions classes from videos in the wild. CoRR, abs/1212.0402 (2012)
9. Zhou, B., Lapedriza, A., Xiao, J., Torralba, A., Oliva, A.: Learning deep features for scene recognition using places database. In: Ghahramani, Z., Welling, M., Cortes, C., Lawrence, N.D., Weinberger, K.Q. (eds.) Advances in Neural Information Processing Systems, vol. 27, pp. 487–495. Curran Associates Inc. (2014)

iAutoMotion – an Autonomous Content-Based Video Retrieval Engine

Luca Rossetto[1]([✉]), Ivan Giangreco[1], Claudiu Tănase[1], Heiko Schuldt[1],
Stéphane Dupont[2], Omar Seddati[2], Metin Sezgin[3], and Yusuf Sahillioğlu[3]

[1] Databases and Information Systems Research Group, Department of Mathematics
and Computer Science, University of Basel, Basel, Switzerland
{luca.rossetto,ivan.giangreco,c.tanase,heiko.schuldt}@unibas.ch
[2] Research Center in Information Technologies, Université de Mons, Mons, Belgium
{stephane.dupont,omar.seddati}@umons.ac.be
[3] Intelligent User Interfaces Lab, Koç University, Istanbul, Turkey
{mtsezgin,ysahillioglu}@ku.edu.tr

Abstract. This paper introduces iAutoMotion, an autonomous video
retrieval system that requires only minimal user input. It is based on
the video retrieval engine IMOTION. iAutoMotion uses a camera to
capture the input for both visual and textual queries and performs
query composition, retrieval, and result submission autonomously. For
the visual tasks, it uses various visual features applied to the captured
query images; for the textual tasks, it applies OCR and some basic natural language processing, combined with object recognition. As the iAutoMotion system does not conform to the VBS 2016 rules, it will participate as unofficial competitor and serve as a benchmark for the manually
operated systems.

1 Introduction

The stated goal of the Video Browser Showdown (VBS) [4] is to evaluate *interactive* video search systems. This is based on the implicit assumption that an
interactive video search system which involves its users directly and intensely
provides benefits in terms of flexibility, accuracy, and/or speed over more traditional, yet less interactive approaches. According to the general rules of the VBS
2016 competition, only interactive video search systems are entitled to participate. However, in order to evaluate the implicit assumptions according to which
interactivitiy actually also implies flexibility, accuracy, and/or performance, we
have developed the iAutoMotion system, an autonomous video retrieval engine.

In this paper, we present the iAutoMotion system which is designed to solve
all types of tasks of the VBS with as little user interaction as possible. Intention-
ally, this system violates some of the rules of VBS. It will thus not participate as
a regular competitor but instead, it is intended to serve as a performance benchmark so as to be able to evaluate the interactive systems against the autonomous
one which is operated without a user in the loop.

© Springer International Publishing Switzerland 2016
Q. Tian et al. (Eds.): MMM 2016, Part II, LNCS 9517, pp. 383–387, 2016.
DOI: 10.1007/978-3-319-27674-8_37

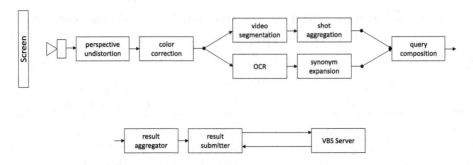

Fig. 1. Architectural overview of iAutoMotion

The iAutoMotion system uses the same back-end as the iMotion system [3] but acquires all necessary information for query formation with a webcam. We detail the system architecture and in particular the processing pipeline of iAuto-Motion in Sect. 2 and discuss the remaining, rather rudimentary user interaction, mainly for configuration, in Sect. 3. Section 4 briefly describes the implementation of iAutoMotion and Sect. 5 concludes.

2   System Architecture

This section describes the overall architecture and processing pipeline of iAuto-Motion as depicted in Fig. 1. The input video which is recorded using a webcam goes through multiple stages of processing which are explained below.

2.1   Perspective Undistortion and Color Correction

The first part of the pipeline aims at restoring the image data originating from the webcam to its original form. It therefore has to correct for the perspective distortion created by the distance of the webcam from the center of the screen. The subsequent step in the pipeline changes color properties to compensate for factors such as illumination to optimize saturation and contrast of the acquired images as well as ensuring their accuracy in hue. This part of the pipeline requires a one-time manual calibration when iAutoMotion is set up.

2.2   Visual Search and Text Search

The processing pipeline comes in two variants: one is tailored to the visual search task where a sequence that needs to be found is shown. The second variant is optimized for the textual search in which a textual description of the search task is given. In what follows, we describe both alternatives of the query pipeline. The switch from one mode (visual or textual) to the other is again one of the few manual tasks in iAutoMotion.

Video Segmentation and Shot Aggregation (Visual Search): The part of the pipeline that is specialized on the visual tasks of the competition consists of a video segmentation and a shot aggregation module. The task of the video segmentation module is to segment the input video into visually continuous sequences which are then aggregated by the shot aggregator. This process reduces redundant queries by ensuring that only one query is performed for every shot in the input sequence while simultaneously reducing the chances of missing visual information contained in a short shot. The resulting shot aggregates are handed off to the query composition module which then queries the retrieval back-end.

OCR and Semantic Text Expansion (Textual Search): To be able to perform queries for the textual tasks of the competition, the query text has to be acquired using optical character recognition. Once the text has been obtained, all nouns are extracted from it. Because the retrieval back-end is only able to recognize a limited number of objects and concepts, synonyms and hypernyms have to be considered as well. The resulting list of detectable objects is passed to the query composition module.

2.3 Query Composition

The query composition module receives data from either of the two sub-chains and builds one or multiple query objects which are then sent to the retrieval back-end. The used retrieval back-end is identical to the one used in the IMOTION system [3].

2.4 Result Aggregator and Submitter

The result aggregator takes all results returned by the retrieval back-end and combines them into a list of sequences which match the original input sequence. Every sequence is processed and a sub-sequence is selected which is guaranteed not to exceed the restrictions in duration imposed on all query sequences. All these results are sorted by score in decreasing order. The result submitter takes the top-n sequences and submits them one after the other to the evaluation server until it either receives feedback that the submitted sequence was indeed the correct one or the number of submitted sequences exceeds the threshold n. Because of the scoring scheme used by the evaluation server, submitting further sequences in the hope of hitting the correct one becomes quickly pointless. Therefore, for VBS n can be set to a value around or below 10.

3   User Interaction

Even tough the system is designed to solve the posed challenges autonomously, it still requires a minimal amount of user interactions.

3.1 Setup and Configuration

The main interaction with a human is required during the initial setup. The camera has to be mounted stably and with a clear view on the screen on which the queries of either type are going to be displayed. The bounds of this screen need to be selected to serve as input for the calibration procedure of the perspective undistortion module. Color correction, while not requiring manual input to work, can be configured to improve its results. Finally, the network needs to be set up in order to enable the system to communicate with the competition server for result submission.

3.2 Query Initialization

During the competition itself, the system needs to know when a query starts and ends, and what type of query – visual or textual – should be processed. This can be signalized using a mouse or a keyboard button which needs to be held for the duration of the query. Depending on which button is pressed, the system will either use the processing sub-pipeline for the visual or the one for the textual task.

4 Implementation

The iAutoMotion system is written in Java and uses BoofCV [1] for image processing and RiTa[1] with WordNet [2] for synonym expansion. It communicates directly with the retrieval back-end bypassing the IMOTION web server. This can be done because the autonomous front-end has no need for static content such as preview images or video files.

5 Conclusions

In this paper, we have presented iAutoMotion, a video retrieval system that is fully automated at retrieval time and that just needs some manual configuration at set-up time. It will participate as unofficial competitor to VBS 2016 and serves as a benchmark for the manually operated systems.

Acknowlegements. This work was partly supported by the Chist-Era project IMO-TION with contributions from the Belgian Fonds de la Recherche Scientifique (FNRS, contract no. R.50.02.14.F), the Scientific and Technological Research Council of Turkey (Tübitak, grant no. 113E325), and the Swiss National Science Foundation (SNSF, contract no. 20CH21_151571).

---

[1] https://rednoise.org/rita/.

References

1. Abeles, P.: Boofcv (2012). http://boofcv.org/
2. Miller, G.A.: Wordnet: a lexical database for English. Commun. ACM **38**(11), 39–41 (1995)
3. Rossetto, L., Giangreco, I., Heller, S., Tănase, C., Schuldt, H., Dupont, S., Omar Seddati, O., Sezgin, M., Altiok, O.C., Sahillioğlu, Y.: IMOTION - searching for video sequences using multi-shot sketch queries. In: MultiMedia Modeling. Springer (2016)
4. Schoeffmann, K., Ahlström, D., Bailer, W., Cobârzan, C., Hopfgartner, F., McGuinness, K., Gurrin, C., Frisson, C., Le, D.-D., Del Fabro, M., et al.: The video browser showdown: a live evaluation of interactive video search tools. Int. J. Multimed. Inf. Retr. **3**(2), 113–127 (2014)

Selecting User Generated Content for Use in Media Productions

Werner Bailer[✉], Wolfgang Weiss, and Stefanie Wechtitsch

DIGITAL – Institute for Information and Communication Technologies,
JOANNEUM RESEARCH Forschungsgesellschaft mbH,
Steyrergasse 17, 8010 Graz, Austria
{werner.bailer,wolfgang.weiss,stefanie.wechtitsch}@joanneum.at

Abstract. This paper describes a video browsing tool for visual media production, enabling users to efficiently find relevant user generated content, while considering its relation to the professional content captured for the production. Users can iteratively cluster the content set by different features, and restrict the content set by selecting a subset of clusters, and perform similarity search. The tool supports metadata captured with the content (e.g., location, device) and provides support for extracting global and local visual features. Content with multiple views is supported.

1 Introduction

Media capture of live events such as concerts can be improved by including user generated content (UGC), adding more perspectives and possibly covering scenes outside the scope of professional coverage. However, UGC only adds value to a production if it does not just repeat what may be covered with high-end professional equipment, but adds another angle or complements the professional content. Thus selecting appropriate UGC video segments poses an interactive search problem. Criteria of interest are visual properties of the content (e.g., camera motion, dominant colors), technical and quality metadata (e.g., location, presence of coding artifacts) and the visual similarity among the UGC clips as well as the overlap with professional content.

The proposed video browsing tool provides a web or desktop user interface to explore such content collections. An earlier version of the tool was used in previous editions of the Video Browser Showdown (VBS) [3]. The functionality of the tool has been extended to plug in other types of metadata (e.g., obtained from mobile devices used for capturing the videos) as well as more efficient ways to determine visual similarity.

The rest of this paper is organized as follows. Section 2 reviews related work. In Sect. 3 the content preparation and feature extraction steps supported by the proposed browsing tool are discussed, and Sect. 4 presents the user interface of the browsing tool. Section 5 concludes the paper.

© Springer International Publishing Switzerland 2016
Q. Tian et al. (Eds.): MMM 2016, Part II, LNCS 9517, pp. 388–393, 2016.
DOI: 10.1007/978-3-319-27674-8_38

2   Related Work

The proposed video browsing tool is an extension of our earlier work [2,3]. Previous editions of the Video Browser Showdown (VBS) and Video Search Showcase (VSS) have fostered research around more efficient interfaces for addressing known-item search tasks. In recent years, mobile clients have received increased attention, e.g., addressing the specific requirements of browsing on small screens [8,9] or using multiple mobile devices for collaborative browsing [6]. However, in our application scenario we are more interested in making use of video contributed from multiple mobile devices, while browsing still takes place on a large screen. User generated video comes with some useful metadata (e.g., location) but is likely to contain redundant segments, if users are in similar areas. Some design choices in our tool such as the key frame based presentation of shots are confirmed by findings of other researchers (e.g. [5]), and are thus kept unchanged. Like other researchers, we observed that the video player is hardly used. However, as it does not consume screen space but opens in a separate window we decided to keep it for the few cases where it may be useful. A graph-based browsing approach has been described in [4], which has some similarities to the proposed approach in terms of the underlying data organization, but could be seen as an alternative visualization, showing multiple steps of the browsing history at once. Unlike tools specifically designed for a known-item scenario, the proposed tool does not include sketch-based query functionalities, as content needs in a non-scripted production scenario are more characterized by the visible objects and the impression of the scene rather than a clear idea of the visual composition.

3   Content Preparation

Automatic content analysis is performed during the ingest of content. Camera motion estimation, visual activity estimation, extraction of global color features and estimation of object trajectories are performed. The extracted features are represented using the MPEG-7 Audiovisual Description Profile [1] and are indexed in an SQLite database. In addition, sensor metadata (e.g., location) and quality metadata from mobile devices such as location can be indexed. If quality metadata is already extracted on the mobile device (e.g., as in the approach described in [7] it can be directly indexed, otherwise quality analysis can be performed during ingest. Despite this range of supported metadata, none of them is required for the operation of the tool. If metadata are not available for a (part of a) content set, some filtering and clustering options will be inactive, but the content can still be found using other types of metadata.

In order to efficiently determine similarity between video segments, we use the VLAD descriptor [10] and apply it to matching key frames (typically every 5th frame) of the video segments. After scaling the input images to SD resolution, up to 500 SIFT [11] key points and descriptors are extracted from each image. In order to compute the VLAD signature of each image we used the VLFeat[1] open

---

[1] http://www.vlfeat.org.

source library. For the VLAD calculation we use a vocabulary generated from over 300 K descriptors by k-means clustering with $k = 256$. The descriptors for building the vocabulary have been extracted from a news data set of the TOSCA-MP project[2].

Based on sum of squared errors the similarities between the VLAD signatures and thus the image similarities between key frames are calculated. For a pair of key frames (I_i, I_j), VLAD yields distances d_{ij}^V, which are transformed into similarities

$$ s_{ij}^V = \begin{cases} \theta_V - d_{ij}^V, \text{if } d_{ij}^V < \theta_V \\ 0, otherwise, \end{cases} \tag{1} $$

where θ_V is a threshold for the maximum distance. The similarity of two video segments is determined as their temporal alignment which the maximizes summed similarities of the aligned key frames.

4   Browsing User Interface

The browsing user interface follows an interaction paradigm of iterative clustering and selecting steps [2], thus narrowing down the subset of content considered relevant for the current content need. In addition, similarity search is supported as a "shortcut", if an item similar to the needed content item has been spotted in the current content set. All clustering, selection and similarity search actions are logged and visualized as a tree, so that a user can always trace back the browsing history and branch off in another direction.

The browsing tool is designed as a framework, backed by an SQLite database. Different types of metadata or features can be added by registering them in the respective configuration files, adding the respective table in the database and (if needed) providing a custom implementation of the desired similarity measure and clustering implementation.

The user interface is available both as a desktop and as a web-based version, offering the same functionality. Both use the same back-end implementation, which is accessible as a SOAP web service for the web-based client. Figure 1 shows a screenshot of the web-based interface. The central component of the video browsing tool's user interface is a light table, showing key frames of the current content set. The colored boxes around sets of key frames indicate the clusters. If the type of metadata provides sufficient textual or numerical information, the clusters will also be labeled. A preview of the timeline around a key frame is available, visible in the screenshot as the two lines of key frames with gray background. This feature helps a user to quickly judge the relevance of the video segment without the need to open a video player. If multiple views of the content are available (e.g., multiple clips recorded with mobile phones with nearby location metadata), they are shown in parallel in the preview.

The left panel of the application contains the browsing history at the top. The tree is automatically populated by the user's actions, and clicking on a node

---

[2] http://www.tosca-mp.eu.

Fig. 1. Screenshot of the web-based video browsing tool

Fig. 2. Example of a similarity search (left) and clustering by motion trajectories (right)

brings up the content set that is the result of the clustering and selection actions from the root to this node. If new user actions are performed when a non-leaf node is active, a new branch is created. The bottom of the left panel contains an area to collect video segments that are relevant for the content need. A user can simply drag & drop key frames from the light table view into the result area in order to add the related video segment. The video segments in the result list can be exported as edit decision list (EDL), in order to work with them in a non-linear editing software.

The middle panel on the left side is used for similarity search. Key frames can be dragged into this area and are then used as queries (see Fig. 2). The type of feature and a minimum similarity threshold can be selected. The results are again displayed as a content set in the light table view, and can be sorted by similarity or other available metadata. Features yielding a similarity measure, such as the newly added visual similarity using VLAD, can also be used for clustering by plugging the similarity scores into a clustering algorithm such as single-linkage clustering. Figure 2 shows an example of clustering by similarity of object trajectories in video.

5   Conclusion

In this paper we have presented an updated version of a video browsing tool following a clustering and content selection paradigm. Support for additional metadata has been added in order to enable handling of user generated content (UGC). This supports use cases where UGC is used to complement professional content, and content selection needs to be performed as part of the production process. In addition, a more efficient visual similarity measure based on VLAD has been added.

Acknowledgments. The research leading to these results has received funding from the European Union's Seventh Framework Programme (FP7/2007-2013) under grant agreements n° 215475, "2020 3D Media – Spatial Sound and Vision" (http://www.20203dmedia.eu/), and n° 287532, "TOSCA-MP - Task-oriented search and content annotation for media production" (http://www.tosca-mp.eu) and n° 610370, ICoSOLE ("Immersive Coverage of Spatially Outspread Live Events", http://www.icosole.eu).

References

1. Information technology - multimedia content description interface - part 9: Profiles and levels, amendment 1: Extensions to profiles and levels. ISO/IEC 15938–9:2005/Amd1:2012 (2012)
2. Bailer, W., Weiss, W., Kienast, G., Thallinger, G., Haas, W.: A video browsing tool for content management in postproduction. Int. J. Digital Multimed. Broadcast. **2010**, 17 (2010). Article ID 856761. doi:10.1155/2010/856761
3. Bailer, W., Weiss, W., Schober, C., Thallinger, G.: Browsing linked video collections for media production. In: Gurrin, C., Hopfgartner, F., Hurst, W., Johansen, H., Lee, H., O'Connor, N. (eds.) MMM 2014, Part II. LNCS, vol. 8326, pp. 407–410. Springer, Heidelberg (2014)
4. Barthel, K.U., Hezel, N., Mackowiak, R.: Graph-based browsing for large video collections. In: He, X., Luo, S., Tao, D., Xu, C., Yang, J., Hasan, M.A. (eds.) MMM 2015, Part II. LNCS, vol. 8936, pp. 237–242. Springer, Heidelberg (2015)
5. Blažek, A., Lokoč, J., Matzner, F., Skopal, T.: Enhanced signature-based video browser. In: He, X., Luo, S., Tao, D., Xu, C., Yang, J., Hasan, M.A. (eds.) MMM 2015, Part II. LNCS, vol. 8936, pp. 243–248. Springer, Heidelberg (2015)
6. Cobârzan, C., Del Fabro, M., Schoeffmann, K.: Collaborative browsing and search in video archives with mobile clients. In: He, X., Luo, S., Tao, D., Xu, C., Yang, J., Hasan, M.A. (eds.) MMM 2015, Part II. LNCS, vol. 8936, pp. 266–271. Springer, Heidelberg (2015)
7. Fassold, H., Wechtitsch, S., Thaler, M., Kozłowski, K., Bailer, W.: Real-time video quality analysis on mobile devices. In: Proceedings of the 7th ACM International Workshop on Mobile Video, pp. 23–24, Portland, March 2015

8. Hudelist, M.A., Xu, Q.: The multi-stripe video browser for tablets. In: He, X., Luo, S., Tao, D., Xu, C., Yang, J., Hasan, M.A. (eds.) MMM 2015, Part II. LNCS, vol. 8936, pp. 272–277. Springer, Heidelberg (2015)

9. Hürst, W., van de Werken, R., Hoet, M.: A storyboard-based interface for mobile video browsing. In: He, X., Luo, S., Tao, D., Xu, C., Yang, J., Hasan, M.A. (eds.) MMM 2015, Part II. LNCS, vol. 8936, pp. 261–265. Springer, Heidelberg (2015)

10. Jegou, H., Perronnin, F., Douze, M., Sanchez, J., Perez, P., Schmid, C.: Aggregating local image descriptors into compact codes. IEEE Trans. Pattern Anal. Mach. Intell. **34**(9), 1704–1716 (2012)

11. Lowe, D.: Distinctive image features from scale-invariant keypoints. Int. J. Comput. Vis. **60**(2), 91–110 (2004)

VERGE: A Multimodal Interactive Search Engine for Video Browsing and Retrieval

Anastasia Moumtzidou[1(✉)], Theodoros Mironidis[1],
Evlampios Apostolidis[1], Foteini Markatopoulou[1,2],
Anastasia Ioannidou[1], Ilias Gialampoukidis[1],
Konstantinos Avgerinakis[1], Stefanos Vrochidis[1], Vasileios Mezaris[1],
Ioannis Kompatsiaris[1], and Ioannis Patras[2]

[1] Information Technologies Institute, Centre for Research
and Technology Hellas, 6th Km. Charilaou - Thermi Road, Thermi,
57001 Thessaloniki, Greece
{moumtzid,mironidis,apostolid,markatopoulou,ioananas,
heliasgj,koafgeri,stefanos,bmezaris,ikom}@iti.gr
[2] School of Electronic Engineering and Computer Science, QMUL, London, UK
i.patras@qmul.ac.uk

Abstract. This paper presents VERGE interactive search engine, which is capable of browsing and searching into video content. The system integrates content-based analysis and retrieval modules such as video shot segmentation, concept detection, clustering, as well as visual similarity and object-based search.

1 Introduction

This paper describes VERGE interactive video search engine[1], which is capable of browsing and retrieving video collections by integrating multimodal indexing and retrieval modules. VERGE supports Known Item Search task, which requires the incorporation of browsing, exploration and retrieval capabilities in a video collection.

Evaluation of earlier versions of VERGE search engine was performed with participation in video retrieval related conferences and showcases such as TRECVID, VideOlympics and Video Browser Showdown (VBS). Specifically, ITI-CERTH participated in several TRECVID tasks such as the Known Item Search (KIS) task, the Instance Search (INS) task, the Surveillance Event Detection (SED) task, in the VideOlympics event, and finally in VBS 2014 and VSS 2015. The proposed version of VERGE aims at participating to the KIS task of the Video Browser Showdown [1].

2 Video Retrieval System

VERGE[2] is an interactive retrieval system that combines advanced browsing and retrieval functionalities with a user-friendly interface, and supports the submission of queries and the accumulation of relevant retrieval results. The following indexing and

---

[1] More information and demos of VERGE are available at: http://mklab.iti.gr/verge/
[2] Latest VERGE system is available at: http://mklab-services.iti.gr/trec2015_v1/

© Springer International Publishing Switzerland 2016
Q. Tian et al. (Eds.): MMM 2016, Part II, LNCS 9517, pp. 394–399, 2016.
DOI: 10.1007/978-3-319-27674-8_39

retrieval modules are integrated in the developed search application: (a) Visual Similarity Search Module; (b) Object-based Visual Search, (c) High Level Concept Detection; and (d) Hierarchical Clustering.

The aforementioned modules allow the user to search through a collection of images and/or video keyframes. However, in the case of a video collection, it is essential that the videos are pre-processed in order to be indexed in smaller segments and semantic information should be extracted. The module that is applied for segmenting videos is shot segmentation.

Figure 1 depicts the general framework realized by VERGE in case of video collection, which contains all the aforementioned segmenting and indexing modules.

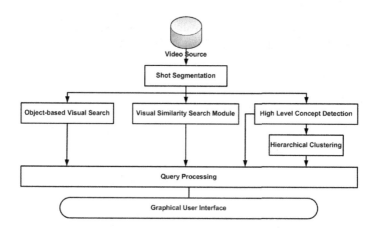

Fig. 1. Screenshot of VERGE video retrieval engine.

2.1   Video Temporal Segmentation

This module is applied for decomposing a video into elementary temporal segments by applying shot segmentation. The shots of the video, i.e., groups of consecutive frames captured without interruption from a single camera, are defined based on a variation of the algorithm described in [2]. According to the utilized method, the visual content of each video frame is represented by extracting an HSV histogram and a set of ORB (Oriented FAST and Rotated BRIEF) descriptors (introduced in [3]), allowing the algorithm to detect differences between a pair of frames, both in color distribution and at a more fine-grained structure level. An image matching strategy is then applied using the extracted descriptors, for assessing the visual similarity between successive or neighboring frames of the video. The computed similarity scores and their pattern over short sequences of frames are then compared against experimentally pre-specified thresholds and models that indicate the existence of abrupt and gradual shot transitions. The defined transitions are re-evaluated with the help of a flash detector which removes erroneously detected abrupt transitions due to camera flashes, and a pair of dissolve and wipe detectors (based on the methods from [4] and [5] respectively) that filter out wrongly identified gradual transitions due to camera and/or object movement. Finally, the union of the resulting sets of detected abrupt and gradual transitions forms the output of the applied technique.

2.2   Visual Similarity Search

The visual similarity search module performs content-based retrieval based on local information. Specifically, simple SURF and color extension of SURF features are extracted. Then, we apply K-Means clustering on the database vectors in order to acquire the visual vocabulary and VLAD encoding for representing the images [6].

For Nearest Neighbour search, the best performing approach between the query and database vectors described in [6, 7] is applied. This approach involves the construction of an IVFADC index for database vectors and then the computation of K-Nearest Neighbours from the query file. Search is realized by combining an inverted file system with the Asymmetric Distance Computation (ADC). Finally, a web service is implemented in order to accelerate the querying process.

2.3   Object-Based Visual Search

This module performs instance-based object retrieval and is based on the Bag-Of-Words (BoW) model. Initially, the fast Hessian detector and SIFT descriptor are applied for local feature extraction from the keyframes. Then, the detected features are randomly sampled and afterwards clustered using Repeated Bisecting K-Means [8] and a 2-layer visual vocabulary is constructed. Vocabularies of various sizes up to 150 K are explored and the final representation of each frame is the result of a hard assignment. An inverted index is built using the open-source Apache Lucene software for fast online search of the image database BoW vectors. For querying the system, the whole keyframe (containing the object of interest) or any object/cropped part of the image can be selected. The similarity score between a query image and a video frame is obtained based on Lucene's scoring function (which exploits the tf-idf weighting scheme) and the ranking position of the frame in the retrieved list, i.e. Borda Counts is computed for all frames in the list in order to form the final ranking.

2.4   High Level Concept Retrieval Module

This module indexes the video shots based on 346 high level concepts (e.g. water, aircraft). To build concept detectors a two-layer concept detection system is employed [9]. The first layer builds multiple independent concept detectors. The video stream is initially sampled, generating for instance one keyframe per shot by shot segmentation. Subsequently, each sample is represented using one or more types of features based on three different pre-trained convolutional neural networks (CNN): (i) The 16-layer deep ConvNet network [10], (ii) the 22-layer GoogLeNet network [11] and (iii) the 8-layer CaffeNet network [12]. Each of these networks is applied on the keyframes and the output of one or more layers is used as a feature. The CNN-based feature vectors are served as input to Support Vector Machine (SVN) classifiers. Specifically, one SVM is trained per concept and per feature type. The output of the classifiers trained for the same concept is combined using a cascade of classifiers [13]. In the second layer of the stacking architecture, the fused scores from the first layer are aggregated in model vectors and refined by two different approaches. The first approach uses a multi-label learning algorithm that incorporates concept correlations [9]. The second approach is a temporal re-ranking method that re-evaluates the detection scores based on video segments as proposed in [14].

2.5   Hierarchical Clustering

This module clusters the video keyframes in a hierarchical agglomerative manner, so as to provide a structured hierarchical view of the video keyframes. We employ hierarchical agglomerative clustering [15], in which each leaf of the generated dendrogram represents a group of keyframes of similar content. We further elaborate the classic hierarchical agglomerative clustering method, in terms of speed and scalability, using skewed-split k-d trees [16]. In the clustering process, we also incorporate the responses of the concept detectors for each video shot, in order to increase the efficiency of the hierarchical representation.

3   VERGE Interface and Interaction Modes

The modules described in Sect. 2 are incorporated into a friendly user interface (Fig. 2) in order to aid the user to interact with the system. The existence of a friendly and smartly designed graphical interface (GUI) plays a vital role in the procedure. The interface features a fast and effective search functionality, partially thanks to the transition from MySQL to MongoDB. Comparing to the previous system, the technique of endless scrolling has been enabled for faster browsing. A RESTful API has also been developed in order to achieve asynchronous data calls. The main results area extends to the 90 % of the screen and the video search toolbar is fixed to the top and covers about 10 % of screen height in order to give the user the opportunity to have instant access to the core search modules. All the other components are collapsible. The interface comprises of three main components: (a) the central component, (b) the left side and (c) the lower panel. We have incorporated the aforementioned modules inside these components. Below we describe briefly the three main components of the VERGE system and then present a simple usage scenario.

The central component of the interface includes a shot-based representation of the video in a grid-like interface. When the user hovers over a shot keyframe, three options appear on its corners (Fig. 3). A selection tool that lets the user select the current keyframe, a cross tool that expands a view of the temporarily related shots, and a magnifier tool that opens up a frame that contains a larger preview of the image and gives the user the opportunity to crop an object from the image in order to make an object based search. All the shots from the main results area are also draggable. When a user starts to drag an image, the Visual Similarity Module area slides up and he can drop the image on this area in order to make a Visual Similarity Search with one or

Fig. 2. Screenshot of VERGE video retrieval engine(Color figure online).

Fig. 3. Shot options.

more images. On the left side of the interface resides the search history, as well as additional search and browsing options (that include a high level visual concepts list and the hierarchical clustering controls). Lastly, the lower panel is a storage structure that holds the shots selected by the user.

Regarding the usage scenario for the KIS task, we suppose that a user is interested in finding a clip containing 'a yellow VW beetle with roofrack' (Fig. 2). Given that there is a high level concept called "car", the user can initiate a search from it. Then, the visual similarity module can be used, if a relative image is retrieved during the first step or if an image that possibly matches the query is found; the temporally adjacent shots can be browsed and retrieved from the desired clip. Finally, the user can select the desirable shots, which will also be available in a basket structure.

4   Future Work

Future work includes the capability of querying the video collection with either one or more colors found in specific place of the shot or with a rough sketch of objects found in the query image. However, it should be noted that both require knowledge of the query image, i.e. the location of the color(s) in the image for the first case and the objects found inside an image in the second case.

Acknowledgements. This work was supported by the European Commission under contracts FP7-600826 ForgetIT, FP7-610411 MULTISENSOR and FP7-312388 HOMER.

References

1. Schoeffmann, K.: A user-centric media retrieval competition: the video browser showdown 2012-2014. IEEE Multimedia **21**(4), 8–13 (2014)
2. Apostolidis, E., Mezaris, V.: Fast shot segmentation combining global and local visual descriptors. In: 2014 IEEE International Conference on Acoustics, Speech and Signal Processing (ICASSP), pp. 6583–6587 (2014)
3. Rublee, E., Rabaud, V., Konolige, K., Bradski, G.: ORB: an efficient alternative to SIFT or SURF. In: 2011 IEEE International Conference on Computer Vision (ICCV), pp. 2564–2571 (2011)
4. Su, C.-W., Liao, H.-Y.M., Tyan, H.-R., Fan, K.-C., Chen, L.-H.: A motion-tolerant dissolve detection algorithm. IEEE Trans. Multimed. **7**, 1106–1113 (2005)
5. Seo, K.-D., Park, S., Jung, S.-H.: Wipe scene-change detector based on visual rhythm spectrum. IEEE Trans. Consum. Electron. **55**(2), 831–838 (2009)
6. Jegou, H., Douze, M., Schmid, C., Perez, P.: Aggregating local descriptors into a compact image representation. In: Proceedings of CVPR (2010)
7. Jegou, H., Douze, M., Schmid, C.: Product quantization for nearest neighbor search. IEEE Trans. Pattern Anal. Mach. Intell. **33**, 117–128 (2011)
8. Steinbach, M., Karypis, G., Kumar, V.: A comparison of document clustering techniques. In: KDD Workshop on Text Mining (2000)
9. Markatopoulou, F., Mezaris, V., Pittaras, N., Patras, I.: Local features and a two-layer stacking architecture for semantic concept detection in video. IEEE Trans. Emerg. Topics Comput. **3**(2), 193–204 (2015)
10. Simonyan, K., Zisserman, A.: Very deep convolutional networks for large-scale image recognition. arXiv technical report (2014)
11. Szegedy, C., et al.: Going deeper with convolutions. In: CVPR 2015 (2015). http://arxiv.org/abs/1409.4842
12. Krizhevsky, A., Ilya, S., Hinton, G.: Imagenet classification with deep convolutional neural networks. In: Advances in Neural Information Processing System, vol. 25, pp. 1097–1105. Curran Associates, Inc., (2012)
13. Markatopoulou, F., Mezaris, V., Patras, I.: Cascade of classifiers based on binary, non-binary and deep convolutional network descriptors for video concept detection. In: IEEE International Conference on Image Processing (ICIP 2015). IEEE, Canada (2015)
14. Safadi B., Quénot, G.: Re-ranking by local re-scoring for video indexing and retrieval. In: 20th ACM International Conference on Information and Knowledge Management, pp. 2081–2084 (2011)
15. Murtagh, F., Legendre, P.: Ward's hierarchical agglomerative clustering method: which algorithms implement ward's criterion? J. Classif. **31**(3), 274–295 (2014)
16. Gialampoukidis, I., Vrochidis, S., Kompatsiaris, I.: Fast visual vocabulary construction for image retrieval using skewed-split k-d trees. In: Proceedings of the 22nd International Conference on MultiMedia Modeling (MMM 2016), Miami (2016)

Collaborative Video Search Combining Video Retrieval with Human-Based Visual Inspection

Marco A. Hudelist[1]([✉]), Claudiu Cobârzan[1],
Christian Beecks[3], Rob van de Werken[2], Sabrina Kletz[1], Wolfgang Hürst[2],
and Klaus Schoeffmann[1]

[1] Klagenfurt University, Universitätsstrasse 65-67, 9020 Klagenfurt, Austria
{marco,claudiu,ks}@itec.aau.at, skletz@edu.aau.at
[2] Utrecht University, PO Box 80.089, 3508TB Utrecht, The Netherlands
{vandewerken,huerst}@cs.uu.nl
[3] RWTH Aachen University, Aachen, Germany
beecks@cs.rwth-aachen.de

Abstract. We propose a novel video browsing approach that aims at optimally integrating traditional, machine-based retrieval methods with an interface design optimized for human browsing performance. Advanced video retrieval and filtering (e.g., via color and motion signatures, and visual concepts) on a desktop is combined with a storyboard-based interface design on a tablet optimized for quick, brute-force visual inspection. Both modules run independently but exchange information to significantly minimize the data for visual inspection and compensate mistakes made by the search algorithms.

Keywords: Video retrieval · Interactive search · Interaction design · Feature signatures

1 Introduction and Related Work

We introduce an approach that is inspired by two successful systems from last year's edition of the Video Search Showcase (VSS) [16,17] as well as our earlier work on mobile video browsers [8,9]. Blažek et al. showed a system based on signature-based similarity models that won the contest. Huerst et al. [13] provided an interface for efficient visual inspection of the whole database. Although the latter has demonstrated that such a pure human-based approach can compete quite well with traditional retrieval systems for surprisingly large databases [12], there is a natural limit to how many files can be inspected in a given time. Our new approach therefore aims at combining the best of both worlds. A desktop-based video retrieval tool searches for relevant files using traditional querying and filtering. The results are used to change the order in which videos are inspected on a tablet-based visualization of the data. Moreover, the results of the human inspection are in turn considered in future filtering iterations.

The collaborative aspects of our work are inspired by [6,7], both using tablets for collaborative search in single videos and archives, respectively. A server

© Springer International Publishing Switzerland 2016
Q. Tian et al. (Eds.): MMM 2016, Part II, LNCS 9517, pp. 400–405, 2016.
DOI: 10.1007/978-3-319-27674-8_40

storing preprocessed information about archived videos is queried by multiple clients on tablets using content-related criteria. The clients share information like already inspected frames, and submitted queries. Furthermore, our work incorporates signature-based similarity models, which adapt to individual multimedia contents by using an adaptive feature quantization. They have been utilized in many different domains ranging from multimedia data [4,5,19] to scientific data [2].

2   Proposed Approach

The main idea of our approach is to combine human-based visual inspection with content-based filtering in order to facilitate quick navigational access into large video archives. The overall system architecture consists of a desktop-based retrieval tool and a tablet app for visual browsing that exchange information continuously.

2.1   Content-Based Video Retrieval (CBVR) Tool

The main objective of the desktop-based CBVR tool is to retrieve video segments based on content characteristics, such as visual features and semantic visual concepts. For the visual features we utilize a signature-based similarity model relying on the Signature Matching Distance [3], while for the visual concepts we use convolutional neural networks (CNNs) [15].

Signature-Based Similarity Model. Given a video segment, we first extract the characteristic key frames and model the content-based properties of each single key frame by means of features $f_1, \ldots, f_n \in \mathbb{F}$ in a feature space $\mathbb{F}$. In order to reflect the perceived visual properties of the frames, we utilize a 7-dimensional feature space $\mathbb{F} = \mathbb{R}^7$ comprising spatial information, CIELAB color information, coarseness, and contrast information. By clustering the extracted local feature descriptors with the k-means algorithm, we obtain a feature signature $S : \mathbb{F} \to \mathbb{R}$ subject to $|\{f \in \mathbb{F}|S(f) \neq 0\}| < \infty$ for each single key frame, where the representatives $R_S = \{f \in \mathbb{F}|S(f) \neq 0\} \subseteq \mathbb{F}$ are determined by the cluster centroids and their weights $S(f)$ by the relative frequencies of the cluster centroids (for further details see Beecks [1]). Based on this adaptive-binning feature representation model, we propose to utilize the Signature Matching Distance as distance-based similarity measure due to its superior retrieval performance [3]. The Signature Matching Distance defines a distance value between two feature signatures $X, Y \in \mathbb{R}^{\mathbb{F}}$ by making use of a matching $m \subseteq \mathbb{F} \times \mathbb{F}$ that relates similar features to each other, a cost function $c : 2^{\mathbb{F} \times \mathbb{F}} \to \mathbb{R}$ that determines the dissimilarity of a matching, and a ground distance $\delta : \mathbb{F} \times \mathbb{F} \to \mathbb{R}$ that models the dissimilarity between two features as $\mathrm{SMD}_\delta(X, Y) = c(m_{X \to Y}) + c(m_{Y \to X}) - 2\lambda \cdot c(m_{X \leftrightarrow Y})$ for $0 \leq \lambda \leq 1$. The parameter λ models the exclusion of bidirectional matches and is used to parameterize the similarity model accordingly.

Concept Search Based on Deep Learning. Our video retrieval application also supports concept-based search according to visual classes trained on ImageNet with convolutional neural networks (CNNs) [18]. For that purpose, we use a pre-trained model on the ImageNet "ILSVRC-2012" dataset [15], available in the Caffe framework [14]. Each key frame is classified according to this approach and detected ImageNet classes are stored for the corresponding segment if their confidence value is above a given threshold (0.5 in our case). From the entire set of visual concepts (up to 1,000 different ones) only the detected ones are made available for search in the interface of the desktop-based CVBR tool.

Fig. 1. Screenshot of the CBVR tool with display of color signature selector, concept textbox and preview options (Color figure online).

Interface. To define feature signatures users can utilize a blank canvas on the right hand side of the interface (see Fig. 1). The canvas represents a typical video frame. Location and color of a feature can be set by clicking on the canvas. This opens a color and size picker. Moreover, it is possible to define motion of the color feature by pressing and then dragging the mouse cursor in any direction. Motion is visualized with a red arrow. To define concepts users just have to start typing in the concept textbox at the top of the interface. Appropriate concepts start to appear automatically. Already selected concepts are displayed directly in the textbox as labels. Changes made in the filtering controls immediately update the result list. To further investigate a video segment in the result list users can use overlayed playback controls. Furthermore, a segments' screenshot is updated automatically when users hover over it, using the cursors x-coordinate for a temporal mapping. Moreover, a submit button is displayed as soon as users hover over a segment.

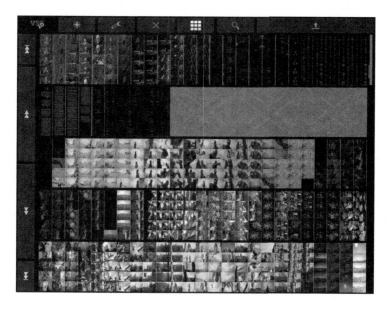

Fig. 2. Interface of the tablet app for human-based visual inspection.

2.2 Tablet App

The tablet app aims at enabling fast sequential search within the video archive by using an approach that proved successful in the VSS 2015. It employs a storyboard layout to present the whole video archive as a series of temporal arranged thumbnail images. The thumbnails are uniformly sampled every second from all the files within the archive. The interface (see Fig. 2) displays 625 images on one screen in accordance with previous research on optimal image sizes on mobiles [10,11]. The thumbnails are arranged in a mix of up/down-left/right directions in order to better identify scenes. The interaction options are kept to a minimum by employing only *Up* and *Down* actions; either across single screens or across files.

2.3 Collaboration Mechanism

All of the interaction between the two modules is performed automatically in the background. On one hand, the CVBR tool informs the tablet app about potentially promising videos, on the other hand, the tablet app informs the desktop-based CVBR tool about already inspected videos (Fig. 3). When the retrieval application generates a new segment result list it also generates a ranked list of videos for the tablet app. This list is created as follows: first, the original result list of segments is restricted to the top 250 results. Outgoing from this smaller list it is then counted how many segments are included in each video. The videos are then sorted accordingly to create a new ranked list for the tablet app. Videos with no matching segments are excluded from this list. The list

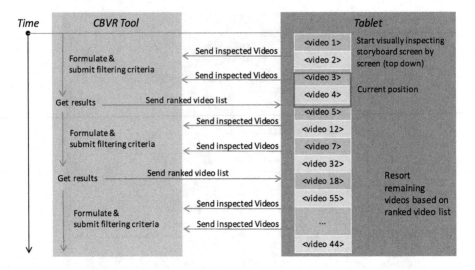

Fig. 3. Diagram explaining the collaboration between CBVR tool and tablet app.

is transmitted to the tablet app, which rearranges the remaining and not yet inspected videos accordingly. Furthermore, the tablet app gives continuous feedback to the desktop-based CVBR tool, about which videos have already been inspected by the tablet user. The CVBR tool then rearranges all corresponding video segments to the bottom of its result lists.

Acknowledgments. The work was funded by the Federal Ministry for Transport, Innovation and Technology (bmvit) and Austrian Science Fund (FWF): TRP 273-N15, supported by Lakeside Labs GmbH, Klagenfurt, Austria and funded by the European Regional Development Fund and the Carinthian Economic Promotion Fund (KWF) under grant 20214/26336/38165.

References

1. Beecks, C.: Distance-based similarity models for content-based multimedia retrieval. Ph.D. thesis, RWTH Aachen University (2013)
2. Beecks, C., Hassani, M., Hinnell, J., Schüller, D., Brenger, B., Mittelberg, I., Seidl, T.: Spatiotemporal similarity search in 3D motion capture gesture streams. In: Claramunt, C., Schneider, M., Wong, R.C.-W., Xiong, L., Loh, W.-K., Shahabi, C., Li, K.-J. (eds.) SSTD 2015. LNCS, vol. 9239, pp. 355–372. Springer, Heidelberg (2015)
3. Beecks, C., Kirchhoff, S., Seidl, T.: Signature matching distance for content-based image retrieval. In: ICMR, pp. 41–48 (2013)
4. Beecks, C., Kirchhoff, S., Seidl, T.: On stability of signature-based similarity measures for content-based image retrieval. MTAP **71**(1), 349–362 (2014)
5. Blažek, A., Lokoč, J., Matzner, F., Skopal, T.: Enhanced signature-based video browser. In: He, X., Luo, S., Tao, D., Xu, C., Yang, J., Hasan, M.A. (eds.) MMM 2015, Part II. LNCS, vol. 8936, pp. 243–248. Springer, Heidelberg (2015)

6. Cobârzan, C., Del Fabro, M., Schoeffmann, K.: Collaborative browsing and search in video archives with mobile clients. In: He, X., Luo, S., Tao, D., Xu, C., Yang, J., Hasan, M.A. (eds.) MMM 2015, Part II. LNCS, vol. 8936, pp. 266–271. Springer, Heidelberg (2015)

7. Cobârzan, C., Hudelist, M.A., Del Fabro, M.: Content-based video browsing with collaborating mobile clients. In: Gurrin, C., Hopfgartner, F., Hurst, W., Johansen, H., Lee, H., O'Connor, N. (eds.) MMM 2014, Part II. LNCS, vol. 8326, pp. 402–406. Springer, Heidelberg (2014)

8. Hudelist, M.A., Schoeffmann, K., Boeszoermenyi, L.: Mobile video browsing with the thumbbrowser. In: Proceedings of the 21st ACM International Conference on Multimedia, MM 2013, pp. 405–406. ACM, New York (2013)

9. Hudelist, M.A., Schoeffmann, K., Xu, Q.: Improving interactive known-item search in video with the keyframe navigation tree. In: He, X., Luo, S., Tao, D., Xu, C., Yang, J., Hasan, M.A. (eds.) MMM 2015, Part I. LNCS, vol. 8935, pp. 306–317. Springer, Heidelberg (2015)

10. Hürst, W., Snoek, C.G.M., Spoel, W.-J., Tomin, M.: Size matters! how thumbnail number, size, and motion influence mobile video retrieval. In: Lee, K.-T., Tsai, W.-H., Liao, H.-Y.M., Chen, T., Hsieh, J.-W., Tseng, C.-C. (eds.) MMM 2011 Part II. LNCS, vol. 6524, pp. 230–240. Springer, Heidelberg (2011)

11. Hürst, W., Snoek, C.G., Spoel, W.-J., Tomin, M.: Keep moving!: revisiting thumbnails for mobile video retrieval. In: Proceedings of the International Conference on Multimedia, MM 2010, pp. 963–966. ACM, New York (2010)

12. Hürst, W., van de Werken, R.: Human-based video browsing - invesitgating interface design for fast video browsing. In: IEEE ISM 2015 (2015, to appear)

13. Hürst, W., van de Werken, R., Hoet, M.: A storyboard-based interface for mobile video browsing. In: He, X., Luo, S., Tao, D., Xu, C., Yang, J., Hasan, M.A. (eds.) MMM 2015, Part II. LNCS, vol. 8936, pp. 261–265. Springer, Heidelberg (2015)

14. Jia, Y., Shelhamer, E., Donahue, J., Karayev, S., Long, J., Girshick, R., Guadarrama, S., Darrell, T.: Caffe: Convolutional architecture for fast feature embedding. In: Proceedings of the ACM International Conference on Multimedia, MM 2014, pp. 675–678. ACM, New York (2014)

15. Krizhevsky, A., Sutskever, I., Hinton, G.E.: Imagenet classification with deep convolutional neural networks. In: Pereira, F., Burges, C., Bottou, L., Weinberger, K. (eds.) Advances in Neural Information Processing Systems 25, pp. 1097–1105. Curran Associates Inc. (2012)

16. Schoeffmann, K.: A user-centric media retrieval competition: the video browser showdown 2012–2014. IEEE MultiMedia 21(4), 8–13 (2014)

17. Schoeffmann, K., Ahlström, D., Bailer, W., Cobarzan, C., Hopfgartner, F., McGuinness, K., Gurrin, C., Frisson, C., Le, D.-D., Fabro, M., Bai, H., Weiss, W.: The video browser showdown: a live evaluation of interactive video search tools. Int. J. Multimedia Inf. Retrieval 3, 113–127 (2014)

18. Simonyan, K., Zisserman, A.: Very deep convolutional networks for large-scale image recognition. CoRR, abs/1409.1556 (2014)

19. Uysal, M.S., Beecks, C., Seidl, T.: On efficient content-based near-duplicate video detection. In: CBMI, pp. 1–6 (2015)

Multi-sketch Semantic Video Browser

David Kuboň[(✉)], Adam Blažek, Jakub Lokoč, and Tomáš Skopal

SIRET Research Group, Department of Software Engineering,
Faculthy of Mathematics and Physics, Charles University in Prague,
Prague, Czech Republic
kubondavid@seznam.cz, blazekada@gmail.com,
{lokoc,skopal}@ksi.mff.cuni.cz

Abstract. This paper presents a tool for interactive filtering and browsing of up to hundreds of hours of video content. In particular, we address the known-item search, i.e., searching for a short video clip known visually or by textual description. Video content is filtered with simple user-defined sketches of the searched scenes consisting of its distinct color regions and significant edges. Furthermore, the filtered content might be browsed with the query-by-example paradigm utilizing either visual or semantic similarity.

1 Introduction

The amount of video data all around us is increasing every day and that's why the ability to effectively browse this content is becoming more and more important. The features required from video browsing tools and their usage vary greatly, from state security searching for a known criminal to a friend telling you about a particular great scene in a movie.

The Video Browser Showdown (VBS) [13] is an annual competition where state-of-the-art tools in video search and browsing are compared. It focuses on interaction with the user and how the tools support interactive navigation within video content. For this reason, textual and photo queries are forbidden; thus, innovative interfaces are being developed by the competitors. More concretely, the task is to find known-items (i.e., a short video clips) described either textually or visually in an archive of hundreds of hours of various video content. The teams receive points according to how quickly and accurately they find the searched clip.

The recent results from VBS have shown that there is none approach that clearly outperforms all others. The participants employed wide variety of techniques, features and user interfaces and, surprisingly, even the simplest approaches such as sequential search in a storyboard [3] preserve performance comparable with the most complex alternatives. Several teams, including us, built their system on the query-by-sketch paradigm, where users may specify the query intent in a form of a color/edge sketch [2,8,11]. Others took advantage of intuitive way of hierarchical exploration of the whole dataset [1] or enabled users to filter the video by numerous features such as presence of a human face, etc. [15].

© Springer International Publishing Switzerland 2016
Q. Tian et al. (Eds.): MMM 2016, Part II, LNCS 9517, pp. 406–411, 2016.
DOI: 10.1007/978-3-319-27674-8_41

In the previous years, our tool was based solely on feature signatures [6] – a flexible descriptor capturing the color distribution in the selected key-frames. To obtain result candidates users specified simple sketches of the searched scene by positioning colored circles. The tool provided immediate responses even to the slightest adjustments of the query, which makes the search indeed interactive and effective. In fact, our tool was the best performing in the last two VBSs although many of the participants from last year achieved noteworthy results. In order to further improve our tool, we have decided to incorporate the following:

Edge sketches

Although color distribution seems to be a robust feature, in some cases it may not be very discriminative. For example, when searching for a scene with no distinct color region or a scene from a black and white movie, we may find the feature signature sketches not powerful enough. For this reason, we newly allow users to specify dominant edges as a part of the color-position sketches (inspired by [11]). We believe that another low-level robust feature especially in the combination with already well-performing feature signatures will improve the filtering power of our tool.

Semantic similarity search

We considered automated semantic concept detection pipe-lines complicated and not performing well in general; thus, we did not utilize any high-level concepts so far. Furthermore, as textual queries are not allowed at VBS, it might be cumbersome to specify a query with these concepts (as there are typically hundreds or thousands to choose from). On the other hand, it is highly natural and intuitive to specify a query such as "Display all scenes with dogs" etc. With deep convolutional neural networks (CNN) emerging in the last years as the state-of-the-art in image recognition and concept detection, it is perhaps time to reconsider the utilization of semantic similarity search in our tool.

Indeed, we have decided to utilize a descriptor from such CNN [9], which can be understood as a semantic characterization of the key-frame, in query-by-example similarity search. More precisely, users may select any key-frame or its subregion as a query in order to retrieve scenes with similar semantic meaning. For example, after selecting a subregion of a key-frame containing a dog we expect to retrieve more scenes with dogs.

The three main filtering methods (semantic similarity, feature signatures and edge sketches) might be combined together or used exclusively. For each task we may find effective particular method and prefer it for the filtering.

We begin by describing the filtering part of the tool, which is based on low-level visual features, in Sect. 2. The paper continues in Sect. 3, where the browsing features including semantic similarity search are discussed. Next, we briefly enumerate remaining features easing the exploration and navigation through the results. Section 4 concludes the paper.

2   Multi-sketch Filtering

Nowadays, wide variety of visual features is available and each excels in some properties such as expressiveness, efficient extraction and/or matching, while it fails in some others. It is clear, though, that the feature selection plays a crucial role in the design of every video retrieval system. We have decided to prefer features that are robust, that can be easily specified by users and that allow efficient extraction and matching. These conditions were satisfied by feature signatures [7] and in the latest version we further incorporated edge histograms [10]. These two features are able to capture color distribution and overall composition of every video key-frame. Additionally, users are able to specify these features directly and independently on each other through a sketch of the searched scene. As these sketches may comprise both edges and color regions, we call them multi-sketches.

2.1   Feature Signatures

Feature signatures represent a key-frame as a set of its distinct color regions. The number of regions held for a particular key-frame varies with its complexity. Users may specify these regions by positioning a colored circle in the sketch (Fig. 1 – right). We assume the query sketches will consist only of a few memorized color regions. These are matched with the regions extracted from the videos and the scenes with the best matches are favored among the results.

Although this feature seems to be fairly robust and has brought us satisfactory results in the previous years, it is dependent on good placement of colored circles and good estimation of their color. Furthermore, scenes with no distinct color region might be hard to find using solely this method.

 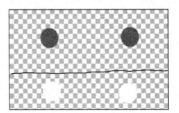

Fig. 1. An example of a key-frame and one of the possible multi-sketches. Users may specify arbitrary amount of distinct color regions and/or edges (Color figure online).

2.2   Edge Histograms

To overcome the problem when the searched scene is hard to be characterized by its color distribution (e.g., a scene from black and white movie) we have searched for some additional low-level feature. Actually, another feature complying to the same demands as feature signatures (robustness, efficient extraction etc.) are edges or – to be more precise – edge histograms.

Being inspired by the IMOTION system [11], we decided to incorporate edge detection and comparison using edge histograms that are extracted from key-frames during preprocessing. A histogram of edges is one of the most used characteristics representing a global composition of an image. As it is invariant to both rotation and translation, it is considered useful for retrieving images.

The usage is simple. The user may draw a line in the sketch (Fig. 1 – right) along the colored circles, at the place where he recalls an edge in the searched scene. Each time the edge part of the sketch is updated, the edge histogram is recalculated causing immediate refining of the results. The similarity measure of a key-frame and the sketch is evaluated by comparing the global, semi-global, and local histograms. As both the global composition and the absolute locations of edges are explored, this technique is less prone to position and direction inaccuracies that are likely to be found in the sketch.

3   Browsing

The filtering produces an ordered list of candidate results. To enable rapid identification of the searched scene, the tool must offer an effective way to explore these results and perhaps refine the initial query. This section describes how we address this problem in our tool.

3.1   Semantic Similarity Search

A relatively frequent issue in our previous versions was that while looking for a particular object, say brown dog, the high ranked results contained many other brown animals and objects. We realized it would be useful to filter the results based on semantic similarity. Therefore, we newly allow users to select a semantically similar key-frame or its subregion in order to form a semantic similarity search query.

Extraction of semantic information from pictures is a complicated task, however, several frameworks utilizing deep CNNs with high accuracy in image recognition are available. One of them is Caffe [4], an already trained NN which classifies images into thousands of categories with a good accuracy on common images. Another example is GoogLeNet [14].

The CNN also produces features (CNN descriptors) which are similar for semantically similar images [5] and even standard Euclidean distance works well with these descriptors [9]; thus, it is possible to implement an efficient and effective semantic similarity search system.

At any point, the user may select one of the retrieved key-frames or its subregion to start a semantic similarity search. The feature vector is extracted on-the-fly and its similarity to the dataset feature vectors is used to refine the search results. As mentioned earlier, users may define the effect of the semantic similarity filtering with respect to the other features (feature signatures, edges). In particular, it is possible to explore the dataset with semantic queries only.

3.2   Browsing Features

Apparently, users spend significant amount of time browsing the results and identifying the searched scene. It is therefore crucial, to design this part of the tool properly. To ease the identification of the searched scene, each match is displayed together with the following and preceding key-frames. Key-frames are compacted into shots, each having only one key-frame fully visible, while the others are cropped and arranged into pseudo key-frames (see Fig. 2). Furthermore, one result row is accompanied with Interactive navigation summary [12], which provides an overview of the whole video. Lastly, the user can exclude any retrieved scene from further displaying or exclude a whole video from the search. Also, the search can be restricted to a single most promising video.

Fig. 2. A video shot compacted (left) and non-compacted (right). Compactation significantly reduces the area yet preserves both temporal and visual information.

Another addition are the sliders allowing the user to change the weight of different search methods – the semantic features, feature signatures and edge histograms. Any of these methods can be turned off when not needed.

4   Conclusion

The paper presents the third generation of our video retrieval tool – Multi-sketch Semantic Video Browser (formerly Signature-based Video Browser). The tool allows effective and interactive filtering and browsing of up to hundreds hours of arbitrary video content. The additions include the possibility of specifying significant edges in the sketches and semantic similarity queries. The three ways of retrieving results – semantic similarity, feature signatures and edge sketches may be combined and balanced as needed. The tool also offers numerous ways for interactive exploration and browsing of results.

Acknowledgments. This research has been supported in part by Czech Science Foundation project 15-08916S and by project SVV-2015-260222.

References

1. Barthel, K.U., Hezel, N., Mackowiak, R.: Graph-based browsing for large video collections. In: He, X., Luo, S., Tao, D., Xu, C., Yang, J., Hasan, M.A. (eds.) MMM 2015, Part II. LNCS, vol. 8936, pp. 237–242. Springer, Heidelberg (2015)

2. Blažek, A., Lokoč, J., Matzner, F., Skopal, T.: Enhanced signature-based video browser. In: He, X., Luo, S., Tao, D., Xu, C., Yang, J., Hasan, M.A. (eds.) MMM 2015, Part II. LNCS, vol. 8936, pp. 243–248. Springer, Heidelberg (2015)

3. Hürst, W., van de Werken, R., Hoet, M.: A storyboard-based interface for mobile video browsing. In: He, X., Luo, S., Tao, D., Xu, C., Yang, J., Hasan, M.A. (eds.) MMM 2015, Part II. LNCS, vol. 8936, pp. 261–265. Springer, Heidelberg (2015)

4. Jia, Y., Shelhamer, E., Donahue, J., et al.: Caffe: convolutional architecture for fast feature embedding. arXiv preprint arXiv:1408.5093 (2014)

5. Karayev, S., Hertzmann, A., Winnemoeller, H., et al.: Recognizing image style. CoRR, abs/1311.3715 (2013)

6. Kruliš, M., Lokoč, J., Skopal, T.: Efficient extraction of clustering-based feature signatures using gpu architectures. Multimedia Tools Appl. **74**, 1–33 (2015)

7. Lokoč, J., Blažek, A., Skopal, T.: On effective known item video search using feature signatures. In: ICMR, p. 524 (2014)

8. Ngo, T.D., Nguyen, V.-T., Nguyen, V.H., Le, D.-D., Duong, D.A., Satoh, S.: NII-UIT browser: a multimodal video search system. In: He, X., Luo, S., Tao, D., Xu, C., Yang, J., Hasan, M.A. (eds.) MMM 2015, Part II. LNCS, vol. 8936, pp. 278–281. Springer, Heidelberg (2015)

9. Novak, D., Batko, M., Zezula, P.: Large-scale image retrieval using neural net descriptors. In: Proceedings of the 38th International ACM SIGIR Conference on Research and Development in Information Retrieval, SIGIR 2015, pp. 1039–1040. ACM, New York (2015)

10. Park, D.K., Jeon, Y.S., Won, C.S.: Efficient use of local edge histogram descriptor. In: Proceedings of the 2000 ACM Workshops on Multimedia, MULTIMEDIA 2000, pp. 51–54. ACM, New York (2000)

11. Rossetto, L., Giangreco, I., Schuldt, H., Dupont, S., Seddati, O., Sezgin, M., Sahillioğlu, Y.: IMOTION — a content-based video retrieval engine. In: He, X., Luo, S., Tao, D., Xu, C., Yang, J., Hasan, M.A. (eds.) MMM 2015, Part II. LNCS, vol. 8936, pp. 255–260. Springer, Heidelberg (2015)

12. Schoeffmann, K., Boeszoermenyi, L.: Video browsing using interactive navigation summaries. In: Seventh International Workshop on Content-Based Multimedia Indexing. CBMI 2009, pp. 243–248, June 2009

13. Schoeffmann, K., Ahlström, D., Bailer, W., et al.: The video browser showdown: a live evaluation of interactive video search tools. Int. J. Multimedia Inf. Retrieval **3**(2), 113–127 (2014)

14. Szegedy, C., Liu, W., Jia, Y., et al.: Going deeper with convolutions. CoRR, abs/1409.4842 (2014)

15. Zhang, Z., Albatal, R., Gurrin, C., Smeaton, A.F.: Interactive known-item search using semantic textual and colour modalities. In: He, X., Luo, S., Tao, D., Xu, C., Yang, J., Hasan, M.A. (eds.) MMM 2015, Part II. LNCS, vol. 8936, pp. 282–286. Springer, Heidelberg (2015)

Faceted Navigation for Browsing Large Video Collection

Zhenxing Zhang[✉], Wei Li, Cathal Gurrin, and Alan F. Smeaton

School of Computing, Insight Centre for Data Analytics, Dublin City University,
Glasnevin, Co. Dublin, Ireland
{zzhang,wli,cgurrin,asmeaton}@computing.dcu.ie
https://www.insight-centre.org

Abstract. This paper presents a content-based interactive video browsing system, developed for the Video Browser Showdown 2016, with the aim of supporting a user to find specific video clips from a large video collection under time constraints. Since the target of this evaluation forum is to evaluate and demonstrate the development of interactive video search tools, we focus on known-item search tasks, rather than query-by-example or query-by-text approaches for large-scale image/video retrieval. In this paper, we describe an interactive video retrieval system which employs the concept filters and faceted navigation to aid users quickly and intuitively locate the interested content when browsing in a large video collections based on automatically extracted semantic concepts, object labels and attributes from video content.

Keywords: Multimedia information retrieval · Faceted navigation

1 Introduction

The Video Browser Showdown (VBS) [1] is a collaborative evaluation task for video browsing systems. In 2016 the VBS task is to design and develop an interactive video retrieval systems which could assist users to locate required video clips quickly and effectively from a large content archives. The evaluation dataset contains around 100–200 h of video with the content comprising of various BBC TV programmes.

The search tasks at VBS 2016 simulate Known-Item Search (KIS) tasks which are common in the real-world. There are two categories of search task explored. The Visual KIS (VKIS) task which selects a random video clip (of about 30 s) from the archive as a query. Participants must generate a query manually that can help their software to locate the desired video clip. The Textual KIS (TKIS) task, on the other hand, provides participants with a textual search topic. The participant must generate a search query from the textual description and use their search system to locate the video clip described by the textual query. The VBS search tasks are undertaken in a competitive manner by all participating groups during a dedicated session of the MMM conference. During the competition, each team

© Springer International Publishing Switzerland 2016
Q. Tian et al. (Eds.): MMM 2016, Part II, LNCS 9517, pp. 412–417, 2016.
DOI: 10.1007/978-3-319-27674-8_42

has 3 min to locate the known-item video clip and the performance score will be calculated based on a combination of search accuracy and search efficiency.

In order to motivate participants to develop more creative content exploring and browsing tools, state-of-the-art search tools, such as query-by-example or query-by-text, are not allowed in this competition.

In previous years, VBS participants have developed various different approaches to support interactive video retrieval. An overall analysis on these approaches reveals that most of them focus on the idea of searching/filtering and subsequent browsing the results, either though high-level semantic concepts or low-level visual descriptors. Based on the wide range of approaches used by participants in previous years [2], it is our observation that subjective and imprecise user expressions pose challenges for teams who wish to provide appropriate tools to assist in the formation of suitable queries that can make use of the advanced visual retrieval capabilities of these interactive systems. Only the SIRET team [3] gained success for visual retrieval tasks by making good use of position-color feature signatures and designed a special color picker interface to allow users to draw coloured regions when searching for the interested video clip.

Based on these observations [2], in this competition, we'll focus on visual content exploration and navigation technologies, using an interactive user interface with the ability to browse large sets of video content quickly and effectively. To achieve this goal, we developed an interactive exploratory search system with faceted navigation [4] based on the application of visual content analysis, such as automatically extracted semantic concepts, object labels, and object attributes from image/video content. To fulfill the requirement of finding the specific video segments, our system allows users to locate a desired video clip using multiple facets and different combinations of semantic concepts in various orders.

2 Video Segmentation and Representation

To reduce computing complexity and remove near-duplicate frames, a common method used in content-based video analysis is to segment a video into constituent shots, each of which is a temporally sequential set of video frames, with an appropriate keyframe to represent the content of each shot. In this work, we segmented the videos into shots based on the shot boundaries detection technologies outlined by Pickering et al. [5]. Following this, the middle frame was extracted to provide a quick overview storyboard of the video content; this strategy has proved successful in the TRECVid instance search task [6]. The average duration of a shot in our system is around 10 s which means that a 100 h video collection would abstract to about 36,000 shots. In addition, we also extracted dense frames (one frame per second) to allow users to inspect the detailed context of each video shot whenever necessary.

3 Visual Content Analysis

Image classification and object recognition using Deep Convolutional Networks (DCNs) has been shown to perform well on various evaluation tasks in recent years. The learning framework, Caffe [11], along with learned models is open source, which encourages researchers to use and contribute to the framework. In our work, we employ DCNs to extract meaningful information, such as semantic concepts, object labels and object attributes, to describe the visual content of video shots. More specifically, we chose two pre-trained models to cover the wide range of possible topics in the BBC programming videos, these two models are:

R-CNN ILSVRC-2013 [7] this model includes 200 object categories, such as person, dog, chair and so on. We chose this model to describe desired object information that could be easily captured by users while interpreting the search topics.

Places-CNNs [9]. The second model introduced 205 scene categories to describe the overall background context for any given images, such as office, restaurant, valley, desert and so on. We chose this model to allow users to navigate the video content with straightforward environments information.

We used the Caffe Library[1] running on a server machine equipped with a GeForce GTX 970 Graphic card and 16 GB RAM to do the visual processing tasks. Thanks to the efficient algorithm implementation and GPU computational power, it took about two seconds to extract the visual information and label a frame into a set of textual terms.

Places-CNNs Model results:
- *Indoor*
- *Kitchen, restaurant_kitchen, Kitchenette*

R-CNNs Model results:
- *People*   *Table*
- *Flower Pot*   *Refrigerator*

Fig. 1. An example results from visual content analysis using DCNs models. The Place-CNNs model labeled the image with overall background information, and the R-CNN model focused on label image with object categories.

Figure 1 illustrates an example result after applying the semantic visual analysis process for one random frame from the video collection. When using the R-CNN model, the bottom up region proposals approach [7] was employed to provide the localization of labeled objects from frames with complex background. For background place recognition with Places-CNN model, the frame image was used as one input to label the most likely place categories from various informative regions.

---

[1] http://caffe.berkeleyvision.org/.

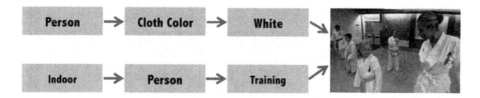

Fig. 2. This example demonstrates two possible routes to find the target video clip with text query: *A group of mostly kids practicing Karate moves indoors (in white clothes), including close-ups of a blond young woman talking to a girl, and shots showing the instructor, a bald man with glasses.*

4   Faceted Navigation

Faceted navigation is implemented in this work to support exploratory search and navigation, while aiding users to find the required video clips quickly and effectively. Faceted navigation reflects the fact that users may seek information in a number of different ways, based on their own understanding of the query topics. In our standard faceted navigation system, the facets are built from different object labels or scene categories which are generated during the video content analysis process. These facets in turn contain attributes by which the list of the same category can be further filtered.

Compared to the previous proposed tools for this task, the main advantages of our faceted navigation system could be summarised into the following three points:

- It does not require users to manually input search query to match the high-level semantic concepts or low-level feature descriptors. This frees users from manually interpret the complex visual or text topics.
- It could take the full advantage of the advanced visual analysis approaches. Since the facets was organized and presented as filters for users to narrow down the range of search results, it could benefits from more precise and detailed content descriptions.
- Finally, it provides multiple navigation routes to help users to identify the same video clip.

Our faceted navigation relies on an underlying infrastructure that enables associations among elements of multiple types and allow users to drill down through categories and attributes naturally. The taxonomy structure of our system is constructed from the visual content of key frames and specifically addressed on three aspects: Similarity clusters, Color, and Semantic labels. To explore the content of video collection, users could choose different facets and allow the system to narrow down the candidate collection for inspection. Figure 2 demonstrates an example of using our faceted navigation to find the interested video clips from two different routes. After the general understanding of example text query topic in the first step, users can very easily to locate the target video clips by following the provided various facets.

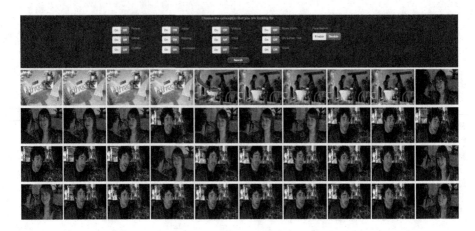

Fig. 3. Screenshot for interactive browser interface. Concept filters are provided on the top panel, dense frames are displayed in the main window for interpret video clip.

5   Browsing Interface

Based on our experience of previous participation in the Video Browser Showdown [2], we followed two main guidelines to design the interactive user interface for this system. Firstly, the interface should be simple and straightforward to allow users efficiently browse when dealing with a large amount of scenes. Secondly, the interface should allow users to quickly inspect of the video clips by providing an overview of the context, such as the previous and following shots.

Figure 3 illustrates the initial design of our interactive interface for faceted navigation system. The user interface allows the users to trigger the search process by selecting the interested concept filters, and navigate them to explore the content by continuously ask them to contribute input to further narrow down the search range by suggesting useful facets.

Acknowledgments. This publication has emanated from research conducted with the financial support of Science Foundation Ireland (SFI) under grant number SFI/12/RC/2289.

References

1. Schoeffmann, K.: A user-centric media retrieval competition: the video browser showdown 2012–2014. IEEE Multimedia **21**(4), 8–13 (2014)
2. Zhang, Z., Albatal, R., Gurrin, C., Smeaton, A.F.: Interactive known-item search using semantic textual and colour modalities. In: He, X., Luo, S., Tao, D., Xu, C., Yang, J., Hasan, M.A. (eds.) MMM 2015, Part II. LNCS, vol. 8936, pp. 282–286. Springer, Heidelberg (2015)
3. Blažek, A., Lokoč, J., Matzner, F., Skopal, T.: Enhanced signature-based video browser. In: He, X., Luo, S., Tao, D., Xu, C., Yang, J., Hasan, M.A. (eds.) MMM 2015, Part II. LNCS, vol. 8936, pp. 243–248. Springer, Heidelberg (2015)

4. Hearst, M.A. : UIs for faceted navigation: recent advances and remaining open problems. In: The Workshop on Computer Interaction and Information Retrieval, HCIR 2008 (2008)
5. Pickering, M.J., Ruger, S.M.: Evaluation of key frame-based retrieval techniques for video. Comput. Vis. Image Underst. **92**(2–3), 217–235 (2003)
6. Zhang, Z., Albatal, R., Gurrin, C., Smeaton, A.F.: TRECVid 2013 experiments at Dublin city University. In: TREC Video Retrieval Evaluation (2013)
7. Girshick, R., Donahue, J., Darrell, T., Malik, J.: Rich feature hierarchies for accurate object detection and semantic segmentation. In: IEEE Conference on Computer Vision and Pattern Recognition (2014)
8. Guadarrama, S., Rodner, E., Saenko, K., Zhang, N., Farrell, R., Donahue J., Darrell, T.: Open-vocabulary object retrieval. In: Robotics: Science and Systems (2014)
9. Zhou, B., Lapedriza, A., Xiao, J., Torralba, A., Oliva, A.: Learning deep features for scene recognition using places database. In: Advances in Neural Information Processing Systems 27 (2014)
10. Patterson, G., Xu, C., Su, H., Hays, J.: The SUN attribute database: beyond categories for deeper scene understanding. Int. J. Comput. Vis. **108**(1), 59–81 (2014)
11. Jia, Y., Shelhamer, E., Donahue, J., Karayev, S., Long, J., Girshick, R., Guadarrama, S., Darrell, T.: Caffe: convolutional architecture for fast feature embedding. In: ACM MM Open Source Competition (2014)

Navigating a Graph of Scenes
for Exploring Large Video Collections

Kai Uwe Barthel[(⊠)], Nico Hezel, and Radek Mackowiak

Visual Computing Group, HTW Berlin, University of Applied Sciences,
Wilhelminenhofstraße 75a, 12459 Berlin, Germany
barthel@htw-berlin.de

Abstract. We present a novel approach to browse huge sets of video scenes using a hierarchical graph and visually sorted image maps allowing the user to explore the graph similar to navigation services. In a previous paper [1] we proposed a scheme to generate such a graph of video scenes and investigated several browsing and visualization concepts. In this paper we extend our work by adding semantic features learned from a convolutional neural network. In combination with visual features we constructed an improved graph where related images (video scenes) are connected with each other. Different images or areas in the graph may be reached by following the most promising path of edges. For efficient navigation we propose a method which projects images onto a 2D plane preserving their complex inter-image relationships. To start a search process, the user may either choose from a selection of typical videos scenes or use tools such as search by sketch or category. The retrieved video frames are arranged on a canvas and the view of the graph is directed to a location where matching frames can be found.

Keywords: Content-based video retrieval · Exploration · Image browsing · Visualization · Navigation · Convolutional neural networks

1 Introduction

Searching in large unannotated video collections for particular scenes is not trivial and time consuming. In the Video Browser Showdown 2016 its participants have to develop interactive video browsing systems to find approximately 20 s long shots in a video archive containing 100 to 200 h of various BBC programs. The given tasks are either visual or textual KIS (known item search). Meaning the moderator presents a short video clip or textual description about a specific segment respectively. The target segment is randomly selected out of the entire collection and it has to be found within 6 min.

The results of past Video Browser Showdowns [2] have shown an overall better performance in the visual and a lot of space for improvements in the textual tasks. The latter requires a clean interface, displaying many video frames without losing overview and the support for near real-time interactive input to quickly explore large video collections.

© Springer International Publishing Switzerland 2016
Q. Tian et al. (Eds.): MMM 2016, Part II, LNCS 9517, pp. 418–423, 2016.
DOI: 10.1007/978-3-319-27674-8_43

In [3] we have proposed *ImageMap* and demonstrated its abilities in [4] by hierarchically arranging one million stock images from Fotolia according to their visual and semantic similarities. Projecting large amounts of images onto a two dimensional space cannot always preserve the complex relationships between the images. A solution for this problem has been presented in [1] by creating a hierarchical set of high dimensional graphs where edges connect related images. Different hierarchical levels contain abstractions of the graphs, resulting in a level of detail mechanism. Lower layers connect very similar images while higher layers far related images.

Navigation concepts for such structures are very challenging, because they either confront the user with too many dimensions of the graph or suggest a none existing two dimensional world limiting the relations. In this paper we present an approach where a subset of images is constantly extracted from the graph and arranged onto a surface while preserving the previous display of arranged images. This approach creates "an endless map" but allows to reconnect and display images multiple times, while at the same time it reduces the amount of dimensions the user has to deal with.

2   Generating Semantic Features

In [5] Krizhevsky shows that convolutional neural networks are not only able to achieve high accuracy rates regarding image classification tasks, but also that the produced activations of a hidden layer can serve as good feature vectors in the context of image retrieval. ConvNets are typically trained from visual images to semantic categories. For a trained network the different layer activations represent abstraction levels depending on their depth. Activations produced by earlier layers contain mainly visual features while deeper layers represent semantic information [6]. Comparing raw pixels of images using the L1 metric leads to bad results in image retrieval tasks. On the other hand, the output of a neural network trained on a different target set does not have the right categories to perform well. Intermediate layers can be seen as abstract representations of both information and are therefore less dependent to the labels of the pretrained network. In our case those labels are only needed to unravel the complex patterns of the input during the training phase of a network.

Training convolutional neural networks takes a lot of time even when using powerful computer hardware. However, existing networks can be reused for feature extraction. In order to get good feature vectors, we tested all fully connected layers of selected state of the art convolutional neural networks regarding their image retrieval quality. The tested models were chosen by their classification accuracy and their diversity. All of them were trained on the ILSVRC2012 data set [7].

For evaluation we have used the STL10 set [8] which contains other labels than the ILSVRC2012 set. Figure 1b shows that the GoogleNet model performs generally better than the Delving model. Also the best results were obtained by the last layer of this model. We could have used the SoftMax of the last layers as well, but since the test set is a subset of the ImageNet challenge the results would be misleading.

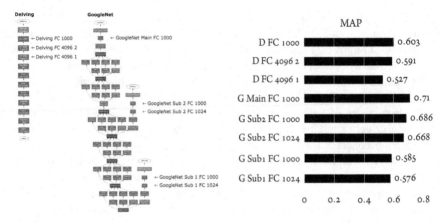

Fig. 1. (**a**) Layers used for the testing. (**b**) Obtained mean average precision (MAP) results for the different layers of the STL10 set.

3   Enriching a Graph of Scenes with Semantic Information

In [1] we extracted keywords for unknown scenes from similar looking annotated images. We fused the cosine similarity of the keywords for each image with their visual similarity. We retrieved better results with this approach compared to using only one of them. Since the annotated images were from a stock agency their visual appearance is very different to movie scenes (Fig. 2).

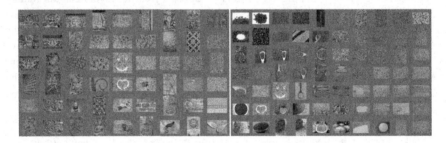

Fig. 2. Left: images are visually sorted, right: sorting using semantic and visual features

To overcome this issue fusing the similarities of the visual features with the similarities of the semantic information, created by a hidden layer of the GoogleNet model, helps. In order to proof our thesis, we conducted an experiment were two graphs are compared by their distortion. Since the weights of the edges in both graphs are calculated with different algorithms a direct comparison is not possible. Therefore, we used a labeled set and the cosine similarity of the annotations. Each node in the graph is represented by an image and its keywords are compared to the keywords of the neighboring nodes. The graph quality is the sum of all the similarities.

The provided material does not contain any semantic information which could be used for an automatic categorization of the extracted scenes. We have used the provided video archive of the VSS2015 and extracted all scenes from it. In a next step the feature vectors for them were calculated from the Main FC 1000 layer of the GoogleNet model. Using the Linde-Buzo-Gray algorithm [9] on the features cluster centroids were computed. We manually labeled each cluster by analyzing its scenes. A cluster can have multiple labels. In a cleaning step we discarded some of them if they were less frequent or not easily definable. New scenes of the VSS 2016 challenge will calculate their shares to all centroids by using the inverse distance weighted interpolation [10] and weight the corresponding labels with it.

4 Projecting and Navigating a Graph in a 2D Manner

This section explains the mechanism to project parts of a graph onto a 2D canvas. In [3] we have already proposed a method which describes how a visual navigation through millions of images can be achieved by using a pyramid structure containing different levels-of-detail on which the images got sorted according to their similarities. Unfortunately, this resulted in some similarity discontinuities which is why we changed the internal structure. By dynamically querying a previously constructed graph it is possible to preserve the complex inter-image relationships. Reaching different images or areas of the graph can now be achieved by following the most promising path of edges (Fig. 3).

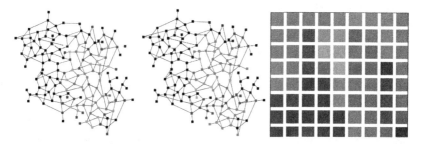

Fig. 3. Displaying a sub graph on a 2D canvas. Left: recursively retrieving the required amount of images starting from the red vertex, Middle: colorized vertices which represents the visual and semantic similarity, Right: arranged vertices on a 2D canvas (Color figure online)

Starting from a location in the graph we recursively retrieve the neighboring vertices until we reached a desired amount. In a next step they get arranged onto a 2D plane like in [3], such that similar images are close to each other. The result may not contain the images the user was looking for. Therefor a user action is necessary to navigate to another part of the graph. This is achieved by dragging the surface into the direction where anticipated images may be located. The movement of the plane moves some images out of the viewport and leaves some empty space on the other side. The inverse dragging direction indicates the important images and instructs the system to retrieve their neighboring vertices. Already displayed images are ignored. The sorting of the newly retrieved images is achieved by searching for the most suited empty space, which is defined by the highest surrounding accumulated image similarities.

The similarities of the graph do not remain the same when using categories to boost certain edges. In the retrieval process edges get boosted if they belong to new images which have those categories assigned. For certain regions of the graph this leads to an increased chance of being displayed (Fig. 4).

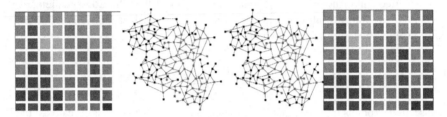

Fig. 4. Navigating the graph. (A) Dragging the canvas to the left. (B) Vertices (blue) representing disappeared images in A. (C) Retrieving neighboring vertices (green) of images in the inverse dragging direction. (D) Appending the new images into the two dimensional projection (Color figure online)

5   Vibro System

We extended our video browsing system (*Vibro*) [1] with the new ideas described in the previous sections. By using the features created by the GoogleNet model we were able to improve the overall graph quality. In order to boost scenes containing certain concepts we enriched the video frames with categories. The graph in combination with the new navigation technique results in an arrangement of images (Fig. 5) without similarity discontinuities. We are also using the timed-sketch approach similar like the one in [11] with our own sketch feature vector. In the search options of the Vibro system are two sketch panels for inputting two timely separated scenes. This way

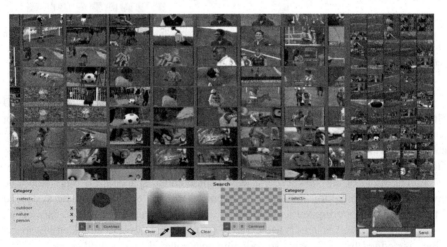

Fig. 5. User interface of the vibro system. Top Left: 2D projection of the graph of scenes. Top right: search results displayed as sequences. Bottom: search options.

visually succeeding scenes can be found more easily. The reason we are using our own sketch vector instead of the semantic feature vectors produced by the GoogleNet model is because they can ignore transparent regions. Since our application tries to respond in real-time we have developed an efficient feature vector and retrieval approach.

In order to search for scenes with a specific visual appearance the user starts to draw onto the left canvas. The interface changes in real-time according to the user's action and updates all views simultaneously. To increase the success of the retrieved results the user can enrich the sketch with categories. After inspecting the projected graph for potential scenes the user checks adjacent images in the sequence panel. If a supposed scene is found a double click plays the corresponding video at the corresponding position.

References

1. Barthel, K.U., Hezel, N., Mackowiak, R.: Graph-based browsing for large video collections. In: He, X., Luo, S., Tao, D., Xu, C., Yang, J., Hasan, M.A. (eds.) MMM 2015, Part II. LNCS, vol. 8936, pp. 237–242. Springer, Heidelberg (2015)
2. Schoeffmann, K., et al.: The video browser showdown: a live evaluation of interactive video search tools. Int. J. Multimed. Inf. Retr. (MMIR) 3(2), 113–127 (2014)
3. Barthel, K.U., Hezel, N., Mackowiak, R.: ImageMap - visually browsing millions of images. In: He, X., Luo, S., Tao, D., Xu, C., Yang, J., Hasan, M.A. (eds.) MMM 2015, Part II. LNCS, vol. 8936, pp. 287–290. Springer, Heidelberg (2015)
4. http://www.picsbuffet.com. Accessed 21 September 2015
5. Krizhevsky, A., Sutskever, I. Hinton, G.E.: ImageNet classification with deep convolutional neural networks. In: NIPS 2012, Neural Information Processing Systems, Lake Tahoe, Nevada (2012)
6. Razavian, A.S., Azizpour, H., Sullivan, J., Carlsson, S.: CNN features off-the-shelf: an astounding baseline for recognition. In: CVPR Workshops 2014, pp. 512–519 (2014)
7. Russakovsky, O., Deng, J., Su, H., Krause, J., Satheesh, S., Ma, S., Huang, Z., Karpathy, A., Khosla, A., Bernstein, M., Berg, A.C., Fei-Fei, L.: ImageNet large scale visual recognition challenge. IJCV 115(3), 211–252 (2015)
8. Coates, A., Lee, H., Ng, A.Y.: An analysis of single layer networks in unsupervised feature learning. In: AISTATS (2011)
9. Linde, Y., Buzo, A., Gray, R.: An algorithm for vector quantizer design. IEEE Trans. Commun. 28, 84 (1980)
10. Donald, S.: A two-dimensional interpolation function for irregularly-spaced data. In: Proceedings of the 1968 ACM National Conference, pp. 517–524 (1968)
11. Lokoč, J., Blažek, A., Skopal, T.: Signature-based video browser. In: Hopfgartner, F., Hurst, W., Johansen, H., Lee, H., O'Connor, N., Gurrin, C. (eds.) MMM 2014, Part II. LNCS, vol. 8326, pp. 415–418. Springer, Heidelberg (2014)

Mental Visual Browsing

Jun He, Xindi Shang$^{(\boxtimes)}$, Hanwang Zhang, and Tat-Seng Chua

School of Computing, National University of Singapore, Singapore, Singapore
{junhe,xindi.shang,hanwang,chuats}@comp.nus.edu.sg

Abstract. We present a surprisingly easy-to-use video browser for helping users to pinpoint a specific video shot in mind, within a long video. At each interactive iteration, the only user effort required is to click 1 shot, which most visually relates to the user's mental target, out of 8 displayed shots. Then, the system updates the browsing model and display another 8 shots for the next iteration. The proposed system is underpinned by a theoretically-sound Bayesian framework that maintains the probabilities of all the video shots segmented from the long video. This framework guarantees that we can find the target shot out of around 1-h video within 3–5 iterations. We believe that our system will perform well in the Video Broswer Showdown game of MMM 2016.

Keywords: Relevance feedback · Bayesian system · Video browsing · Mental search

1 Introduction

Due to the recent advances in visual concept detection [5], we have witnessed a great progress in video search and classification [6]. In these tasks, we generally cast videos into a set of image frames and then consider the whole video as the smallest unit to be operated. However, we neglect that there are also demands for browsing a single long video. This application scenario is typically common in many video editing tasks (*e.g.*, movie or TV post-production). Video Browser Showdown in joint with MMM conference is an interesting venue for researchers to evaluate their systems for video browsing. Video browsing is the interactive process of exploring video content in order to find particular segments [8]. Users are first shown a short video clip (*e.g*, 20 s) and then pinpoint the clip in a long video (*e.g.*, 1–2 h). Most previous browsing tools are designed with sophisticated concept selection [7]. Although these designs are effective for expert users, they are not easy-to-use for novice users and are not flexible for users' nuanced search intentions.

In this paper, we present an easy-to-use video browsing system, called Mental Visual Browsing, which only require user interaction in the way of "click" on the most visually similar shots in 8 displayed shots. This browsing session usually terminates in 3–5 rounds when users find the exact match to their imaginary queries (see Fig. 1). That is to say, users are only required to quickly scan 24–40 images to pinpoint the target shot in a long video containing around 1000–3000

© Springer International Publishing Switzerland 2016
Q. Tian et al. (Eds.): MMM 2016, Part II, LNCS 9517, pp. 424–428, 2016.
DOI: 10.1007/978-3-319-27674-8_44

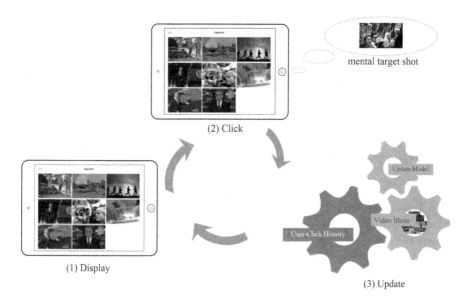

Fig. 1. Our interactive system contains three key parts: (1) Display. 8 shots are displayed to the user; (2) Click. The user click one shot that most relates to the mental target shot; and (3) Update. The underlying algorithm updates the model for the next display. Note that the user effort is considerably limited.

Fig. 2. When the user finds the target shot, the shot can be considered as a starting point for precise localization in the long video.

shots. The interaction is illustrated in Fig. 1. Suppose the long video is segmented into N shots. At the first iteration, the system just randomly select 8 shots to be displayed. After reviewing the 8 shots (by seeing the 8 representative frame image of the shots), the user click the shot that is most visually similar to her target shot in mind. After receiving the click, the system updates the underlying user intention model and then the model selects the most 8 informative shots (exclusive of the former 8 shots) for the display at next iteration. After the user meets the target shots, she can zoom in to fine-tune the exact locations in the video (see Fig. 2). In summary, we want to highlight that the proposed browsing method is very user-friendly to both experts and novices, and the underlying algorithm can effectively capture users' mental intention through very limited clicks.

2   Video Visual Features

We first segment the long video into shots by using any off-the-shelf shot transition detection softwares. For a typical 1-h video, we obtain 1,000–3,000 shots as our basic units for display. Since our system relies on high-performance visual features for visual representations of users' mental intention, we adopt the recent video features proposed by Xu *et al.* [10]. Specifically, we first extract the frame-level CNN descriptors using Caffe toolkit [4]. The CNN features are using the 7-th fully- connected layer output of VGG-16 networks [9]. Then, we utilize Vector 0f Locally Aggregated Descriptors (VLAD) [3] with $k = 5$ to encode them. We also use Latent Concept Descriptors since it is shown to outperform pure CNN features. We use 256-component k-means centers for VLAD encoding and for all CNN features, we apply PCA-whitening to reduce their dimension into 512. In the same time, signed square root and intro normalization, and L2 normalization are used for post-processing [1]. As a result, each shot is represented by a 256×512-d vector.

3   Statistical Feedback Model

Our goal is to assist users to find their mental target video shot via minimum interactions. With users in the loop, the system will learn how to describe users' intention by modeling the conditional probabilities given users' click on shots. The statistical model presented here is inspired by the one proposed by Ferecatu and Geman [2].

Suppose there are N shots segmented from a long video, denoted as $\mathcal{V} = \{1...i...N\}$ for simplicity. The objective is to identify a shot that matches the semantic and visual impressions in the mind of the user. Let $\mathcal{M} \subset \mathcal{V}$ denote the possible subset of the long video shots, *i.e.*, it can be considered as candidates of users' target mental shots. Of course, $\mathcal{M}$ is unknown or even imaginary to the system. We assume that if a member of $\mathcal{M}$ is displayed, the user will be satisfied and terminates the browsing. At that point, this shot can be used as a starting point for precise localization in the long video. At the t-th iteration, the system displays a shot set $\mathcal{D}_t \subset \mathcal{V}$ to the user. Then, the user can make a feedback to the system by clicking on one of the displayed shots. We denote this feedback as $c_t = k$, where $k \in \mathcal{D}_t$. Moreover, we denote the history displays and user clicks as $\mathcal{H}_t = \mathcal{H}_{t-1} \cup \{c_t = k, k \in \mathcal{D}_t\}$. Our system maintains N-independent

Bayesian system $p_t(i)$ for each video shot i. In particular, $p_t(i)$ is the estimated probability of the event that shot i belongs tho the target set $\mathcal{M}$:

$$p_t(i) = P(i \in \mathcal{M}|\mathcal{H}_t). \tag{1}$$

Next, we detail three key components of our system: (1) **Update**: how to update $p_t(i)$; (2) **Click**: how the system responds to user click $c_t = k$; and (3) **Display**: how to choose shots for displaying $\mathcal{D}_t$.

3.1   Update

By applying Bayes rule, we can rewrite the definition in Eq. (1) as:

$$
\begin{aligned}
p_t(i) &= P\left(i \in \mathcal{M}|\mathcal{H}_{t-1}, c_t = k, k \in \mathcal{D}_t\right) \\
&= \frac{P\left(c_t = k|\mathcal{H}_{t-1}, i \in \mathcal{M}, k \in \mathcal{D}_t\right) P\left(i \in \mathcal{M}|\mathcal{H}_{t-1}, k \in \mathcal{D}_t\right)}{P\left(c_t = k|\mathcal{H}_{t-1}, k \in \mathcal{D}_t\right)}.
\end{aligned}
\tag{2}
$$

In order to further simplify the above model for computability, we should note several statistical sufficiency and assumptions. First, $c_t = k$ is a sufficient statistic for $k \in \mathcal{D}_t$ since it is obvious that once the user clicks on k, k must be in $\mathcal{D}_t$. Therefore, $P(c_t = k|\mathcal{H}_{t-1}, k \in \mathcal{D}_t) = P(c_t = k|\mathcal{H}_{t-1})$. Second, the fact $i \in \mathcal{M}$ is unrelated to the display set $\mathcal{D}_t$, so we have $P(i \in \mathcal{M}|\mathcal{H}_{t-1}, k \in \mathcal{D}_t) = P(i \in \mathcal{M}|\mathcal{H}_{t-1})$. Third, once given the display $\mathcal{D}_t$, the history $\mathcal{H}_{t-1}$ is no longer informative to the user click $c_t - k$, so we have $P(c_t = k|\mathcal{H}_{t-1}, i \in \mathcal{M}, k \in \mathcal{D}_t) = P(c_t = k|i \in \mathcal{M}, k \in \mathcal{D}_t)$ and $P(c_t = k|\mathcal{H}_{t-1}, k \in \mathcal{D}_t) = P(c_t = k|k \in \mathcal{D}_t)$. Based on these statistical properties, we have the update model for $p_t(i)$ as:

$$
\begin{aligned}
p_t(i) &= \frac{P\left(c_t = k|i \in \mathcal{M}, k \in \mathcal{D}_t\right) p_{t-1}(i)}{P\left(c_t = k|k \in \mathcal{D}_t\right)} \\
&= \frac{P\left(c_t = k|i \in \mathcal{M}, k \in \mathcal{D}_t\right) p_{t-1}(i)}{P\left(c_t = k|i \in \mathcal{M}, k \in \mathcal{D}_t\right) p_{t-1}(i) + P\left(c_t = k|i \notin \mathcal{M}, k \in \mathcal{D}_t\right) \left(1 - p_{t-1}(k)\right)}.
\end{aligned}
\tag{3}
$$

Therefore, all we need to update $p_t(i)$ is to model the click $P(c_t = k|i \in \mathcal{M}, k \in \mathcal{D}_t)$ and $P(c_t = k|i \notin \mathcal{M}, k \in \mathcal{D}_t)$.

3.2   Click

Let $d(\cdot, \cdot)$ and $s(\cdot, \cdot)$ denote the distance and similarity in the feature space, respectively. Our click probability is modeled as:

$$
\begin{cases}
P\left(c_t = k|i \in \mathcal{M}, k \in \mathcal{D}_t\right) = \dfrac{s(i, k)}{\sum_{j \in \mathcal{D}_t} s(j, k)}, \\[3mm]
P\left(c_t = k|i \notin \mathcal{M}, k \in \mathcal{D}_t\right) = \dfrac{d(i, k)}{\sum_{j \in \mathcal{D}_t} d(j, k)}.
\end{cases}
\tag{4}
$$

The intuition behind the above definitions is that if i is the target shot, the probability is enhanced by the similarity between i and click k; if i is not the target, the probability is depressed by the distance between i and k.

3.3 Display

We select the most informative subset $\mathcal{D}_t \subset \mathcal{V}$ by minimizing $\min_{\mathcal{D} \subset \mathcal{V}} \text{Entropy}$ $(z_t|\mathcal{H}_{t-1})$. Intuitively, the selection attempts to make an optimal Voronoi partition for the video shots. The detailed calculation of the probability z_t is given in [2].

References

1. Arandjelovic, R., Zisserman, A.: All about VLAD. In: CVPR (2013)
2. Ferecatu, M., Geman, D.: A statistical framework for image category search from a mental picture. TPAMI **31**(6), 1087–1101 (2009)
3. Jégou, H., Perronnin, F., Douze, M., Sanchez, J., Perez, P., Schmid, C.: Aggregating local image descriptors into compact codes. TPAMI **34**(9), 1704–1716 (2012)
4. Jia, Y.: Caffe: an open source convolutional architecture for fast feature embedding (2013). http://caffe.berkeleyvision.org
5. Krizhevsky, A., Sutskever, I., Hinton, G.E.: Imagenet classification with deep convolutional neural networks. In: NIPS (2012)
6. Over, P., Awad, G., Michel, M., Fiscus, J., Sanders, G., Shaw, B., Kraaij, W., Smeaton, A.F., Quenot, G.: Trecvid 2012 - an overview of the goals, tasks, data, evaluation mechanisms and metrics. In: TRECVID (2012)
7. Schoeffmann, K.: A user-centric media retrieval competition: the video browser showdown 2012–2014. IEEE MultiMedia **21**, 8–13 (2014)
8. Schoeffmann, K., Ahlström, D., Bailer, W., Cobârzan, C., Hopfgartner, F., McGuinness, K., Gurrin, C., Frisson, C., Le, D.-D., Del Fabro, M., et al.: The video browser showdown: a live evaluation of interactive video search tools. IJMIR **3**(2), 113–127 (2014)
9. Simonyan, K., Zisserman, A.: Very deep convolutional networks for large-scale image recognition (2014). arXiv preprint arXiv:1409.1556
10. Xu, Z., Yang, Y., Hauptmann, A.G.: A discriminative cnn video representation for event detection (2014). arXiv preprint arXiv:1411.4006

Author Index

Printed in the United States
By Bookmasters